ACS SYMPOSIUM SERIES 662

Antinutrients and Phytochemicals in Food

Fereidoon Shahidi, EDITOR
Memorial University of Newfoundland

Developed from a symposium sponsored
by the Division of Agricultural and Food Chemistry

American Chemical Society, Washington, DC

Library of Congress Cataloging-in-Publication Data

Antinutrients and phytochemicals in food / Fereidoon Shahidi.

p. cm.—(ACS symposium series, ISSN 0097–6156; 662)

"Developed from a symposium sponsored by the Division of Agricultural and Food Chemistry at the 210th National Meeting of the American Chemical Society, Chicago, Illinois, August 20–25, 1995."

Includes bibliographical references and indexes.

ISBN 0–8412–3498–1

1. Food—Toxicology—Congresses. 2. Plant toxins—Congresses. 3. Botanical chemistry—Congresses. 4. Food—Composition—Congresses. 5. Cancer—Chemoprevention—Congresses. 6. Nutritionally induced diseases—Congresses.

I. Shahidi, Fereidoon, 1951– . II. American Chemical Society. Division of Agricultural and Food Chemistry. III. American Chemical Society. Meeting (210th: 1995: Chicago, Ill.) IV. Series.

RA1260.A58 1995
615.9′54—dc21

97–6913
CIP

This book is printed on acid-free, recycled paper.

PRINTED IN THE UNITED STATES OF AMERICA

Foreword

THE ACS SYMPOSIUM SERIES was first published in 1974 to provide a mechanism for publishing symposia quickly in book form. The purpose of this series is to publish comprehensive books developed from symposia, which are usually "snapshots in time" of the current research being done on a topic, plus some review material on the topic. For this reason, it is necessary that the papers be published as quickly as possible.

Before a symposium-based book is put under contract, the proposed table of contents is reviewed for appropriateness to the topic and for comprehensiveness of the collection. Some papers are excluded at this point, and others are added to round out the scope of the volume. In addition, a draft of each paper is peer-reviewed prior to final acceptance or rejection. This anonymous review process is supervised by the organizer(s) of the symposium, who become the editor(s) of the book. The authors then revise their papers according to the recommendations of both the reviewers and the editors, prepare camera-ready copy, and submit the final papers to the editors, who check that all necessary revisions have been made.

As a rule, only original research papers and original review papers are included in the volumes. Verbatim reproductions of previously published papers are not accepted.

ACS BOOKS DEPARTMENT

Contents

Preface vii

1. **Beneficial Health Effects and Drawbacks of Antinutrients and Phytochemicals in Foods: An Overview** 1
   Fereidoon Shahidi

2. **Protease and α-Amylase Inhibitors of Higher Plants** 10
   John R. Whitaker

3. **Plant Lectins: Properties, Nutritional Significance, and Function** 31
   Irvin E. Liener

4. **Antinutritional and Allergenic Proteins** 44
   H. Frøkiær, T. M. R Jørgensen, A. Rosendal, M. C. Tonsgaard, and V. Barkholt

5. **Potato Polyphenols: Role in the Plant and in the Diet** 61
   Mendel Friedman

6. **Potato Glycoalkaloids: Chemical, Analytical, and Biochemical Perspectives** 94
   S. J. Jadhav, S. E. Lutz, G. Mazza, and D. K. Salunkhe

7. **Biological Activities of Potato Glycoalkaloids** 115
   Leslie C. Plhak and Peter Sporns

8. **α-Galactosides of Sucrose in Foods: Composition, Flatulence-Causing Effects, and Removal** 127
   Marian Naczk, Ryszard Amarowicz, and Fereidoon Shahidi

9. **Glucosinolates in *Brassica* Oilseeds: Processing Effects and Extraction** 152
   Fereidoon Shahidi, James K. Daun, and Douglas R. DeClercq

10. **Cyanogenic Glycosides of Flaxseeds** 171
    Fereidoon Shahidi and P. K. J. P. D. Wanasundara

11. Nutritional Implications of Canola Condensed Tannins 186
Marian Naczk and Fereidoon Shahidi

12. Methods for Determination of Condensed and Hydrolyzable Tannins 209
Ann E. Hagerman, Yan Zhao, and Sarah Johnson

13. Lawsone: Phenolic of Henna and Its Potential Use in Protein-Rich Foods and Staining 223
Rashda Ali and Syed Asad Sayeed

14. Anticarcinogenic Activities of Polyphenols in Foods and Herbs 245
Ken-ichi Miyamoto, Tsugiya Murayama, Takashi Yoshida, Tsutomu Hatano, and Takuo Okuda

15. Chemiluminescence of Catechins and Soybean Saponins in the Presence of Active Oxygen Species 260
Kazuyoshi Okubo, Yumiko Yoshiki, Kiharu Igarashi, and Kazuhiko Yotsuhashi

16. Phytoestrogens and Lignans: Effects on Reproduction and Chronic Disease 273
Sharon E. Rickard and Lilian U. Thompson

17. Interactions and Biological Effects of Phytic Acid 294
Sharon E. Rickard and Lilian U. Thompson

18. Anticarcinogenic Effects of Saponins and Phytosterols 313
A. V. Rao and R. Koratkar

Author Index 325

Affiliation Index 325

Subject Index 325

Preface

ANTINUTRIENTS IN FOODS are responsible for the deleterious effects that are related to the absorption of nutrients and micronutrients which may interfere with the function of certain organs. Most of these antinutrients are present in foods of plant origin. Thus, the presence of glucosinolates, cyanogenic glycosides, enzyme inhibitors, lectins, lignans, alkaloids, and phenolic compounds in foods may induce undesirable effects in humans if their consumption exceeds an upper limit. Certain harmful effects might also be due to the breakdown products of these compounds.

However, some antinutrients as well as their breakdown products may possess beneficial health effects if present in small amounts. The mechanism through which the antinutritional and beneficial effects of food antinutrients are exerted is the same. Thus, manipulating processing conditions, in addition to removing certain unwanted compounds in foods, may be required to eliminate the deleterious effects of antinutrients and take advantage of their health benefits.

This book is based on a symposium presented at the 210th National Meeting of the American Chemical Society, titled "Antinutrients and Phytochemicals in Foods", sponsored by the ACS Division of Agricultural and Food Chemistry, in Chicago, Illinois, August 20–25, 1995. A group of internationally recognized scientists worked together to prepare the 18 chapters contained herein. This book is of interest to chemists, biochemists, nutritionists, and food scientists in academia, industry, and government laboratories. It may also serve as a reference text for graduate and senior undergraduate students interested in food chemistry and biochemistry.

FEREIDOON SHAHIDI
Department of Biochemistry
Memorial University of Newfoundland
St. John's, Newfoundland A1B 3X9
Canada

December 17, 1996

Chapter 1

Beneficial Health Effects and Drawbacks of Antinutrients and Phytochemicals in Foods

An Overview

Fereidoon Shahidi

Department of Biochemistry, Memorial University of Newfoundland, St. John's, Newfoundland A1B 3X9, Canada

Antinutrients in foods are responsible for deleterious effects related to the absorption of nutrients and micronutrients. However, some antinutrients may exert beneficial health effects at low concentrations. The mechanisms by which adverse and beneficial effects of food antinutrients operate are the same. Thus, manipulation of processing conditions and/or removal of certain unwanted components of foods may be required.

Antinutrients and phytochemicals found in foods have been categorized as having both adverse and beneficial health effects in humans. These concentration-dependent effects may be manipulated in such away that advantage is taken from their health-related benefits so that management of chronic diseases becomes possible. For this purpose, current recommendations suggest that consumption of plant such as grains, fruits and vegetables be increased in order to prevent or possibly cure cardiovascular disease, diabetes and cancer (*1, 2*). Nonetheless, there is a concern about high intake of foods that are rich in antinutrients due to their increased burden on the body's tolerance to potentially harmful compounds (*3, 4*). For example, phytic acid, lectins, phenolic compounds and tannins, saponins, enzyme inhibitors, cyanogenic glycosides and glucosinolates have shown to reduce the availability of certain nutrients and impair growth. Some compounds such as phytoestrogens and lignans have also been linked to induction of infertility in humans. Therefore, it is prudent to examine all aspects related to food antinutrients, including their potential health benefits and methods of analyses.

When used at low levels, phytic acid, lectins and phenolic compounds as well as enzyme inhibitors and saponins have been shown to reduce the blood glucose and/or plasma cholesterol and triacylglycerols. Meanwhile, phenolic compounds from plant sources, phytic acid, protease inhibitors, saponins, lignans and phytoestrogens have been demonstrated to reduce cancer risks. This monograph intends to cover the occurrence and problems associated with the use of plant food antinutrients, their removal from foods in order to detoxify them, and changes that occur in them during processing. Potential beneficial health effects of food antinutrients and phytochemicals will also be discussed.

Phenolic Compounds and their Antioxidant Activity in Foods and Biological Systems

Phenolic compounds are found in reasonably large quantities and in a variety of chemical forms in plant foods and serve as secondary metabolites that protect plant tissues against injuries and insect and animal attack. Phenolic compounds in plant foods belong to the families of phenolic acids, flavonoids, isoflavonoids, and tocopherols, among others.

Phenolic compounds found in foods generally contribute to their astringency and may also reduce the availability of certain minerals such as zinc. During thermal processing, phenolic compounds may undergo oxidation and oxidized phenolics so formed, such as quinones, may combine with amino acids, thus making them nutritionally unavailable. Furthermore, such reaction products are generally highly colored in nature and impart undesirable dark colors to foods. Therefore, in some cases, it might be beneficial to remove phenolics in foods by devising novel processing techniques. The isolated phenolics may then be used in different applications in order to control oxidation of food lipids.

The most widely distributed phenolics in plant foods are tocopherols. Both tocopherols (alpha, beta, gamma and delta) and tocotrienols (alpha, beta and gamma) may be present. While the vitamin E activity of α-tocopherol exceeds that of other tocopherols/tocotrienols, its *in-vitro* activity as an antioxidant is not as good as the other tocopherols (*5*) and follows the trend given below.

$$\alpha\text{-tocopherol} < \beta\text{-tocopherol} \approx \gamma\text{-tocopherol} < \delta\text{-tocopherol}$$

In general, grains and particularly their oil and oil-rich fractions such as those from the germ provide a good source of compounds with vitamin E activity (*6*). It should also be noted that grains are a unique source of tocotrienols which are known to inhibit cholesterol synthesis.

Phenolic acids are another group of antioxidants found abundantly in whole grains and oilseeds, particularly in the bran layer (*7*). Phenolic acids in foods, such as those of benzoic and cinnamic acid derivatives occur in the free, esterified/etherified and insoluble-bound forms. The antioxidant activity of phenolic acids in a meat model system was recently reported (*8*). It has been shown that the *in-vitro* activity of phenolic acids as antioxidants was in the following order *(9)*.

Protocatechuic > Chlorogenic > Caffeic > Vannilic > Syringic > p-Coumaric

Flavonoids and isoflavonoids are other groups of phenolic antioxidants found in foods. Green tea is a rich source (25%) of flavonoids of the catechin type. Tea catechins as well as myricetin, another flavonoid found in plat foods, are among the strongest natural antioxidants found in nature (*10*). In addition, isoflavonoids, such as those found in soybean in relatively large amounts, also exhibit potent antioxidant activity. The skin of onions contains up to 6% flavonoids, mainly quercetin (*11*).

Since lipid peroxidation is implicated as being a cause of atherosclerosis and many degenerative diseases such as cataract and the aging process, it is anticipated that use of natural antioxidants not only protects foods against rancidity development, it might also augment the body for its antioxidant defense mechanism, particularly in the elderly. Figure 1 shows possible contribution of free radicals in diseases and potential effect of antioxidants in preventing/controlling free radical formation and their deleterious effects.

Phytates

Phytic acid (*myo*-inositol-1,2,3,4,5,6-hexakis dihydrogen phosphate; PA) is present in foods in varying concentrations of 0.1-6.0% (*12, 13*). Phytates are found as crystalline globoids inside protein bodies in the cotyledon of legumes and oilseeds or in the bran region of cereal grains (*4, 12*). Phytic acid, with its highly negatively charged structure, is a very reactive compound and particularly attracts positively-charged ions such as those of zinc and calcium. Therefore, it is considered as a food antinutrient and its removal during food processing has been considered (*14, 15*). In addition, PA may also react with charged groups of proteins, either directly or indirectly, via negatively charged groups of proteins mediated by a positively-charged metal ion such as calcium. Interaction of PA with starch molecules, directly via hydrogen bonding with phosphate groups or indirectly through proteins to which it is attached, is also possible. Such bindings may reduce the solubility and digestibility of protein and starch components of food. However, literature data in this area of research are often contradictory (*16-18*).

Potential beneficial effects of PA relate to its ability to lower blood glucose response to starchy foods (*19*). Removal of PA from navy beans was reported to cause an increase in blood glucose response while addition of PA back to the beans flattened the response (*20*). The effect of reducing the blood glucose response may be exerted by influencing the rate of starch digestion. Antinutrients such as PA and tannins may lower the rate of starch digestion by the same mechanism that makes them antinutrients (*21*). Such compounds can bind directly with the amylase enzyme, thus inactivating it. Indirectly, PA and tannins may also bind with calcium which is required for stabilization of amylase activity or possibly with starch in order to affect its gelatinization or accessibility to digestive enzymes (*4*).

Phytic acid has also been implicated as having a significant effect on reducing plasma cholesterol and levels of triacylglycerols (*22*). The effect is thought to be related to the ability of PA to bind to zinc and thus lower the ratio of plasma zinc to copper which is known to dispose humans to cardiovascular disease (*22*). The effect may also be related to the ability of PA to reduce the plasma glucose and insulin concentrations which may, in turn, lead to reduced stimulus for hepatic lipid synthesis (*21, 24*).

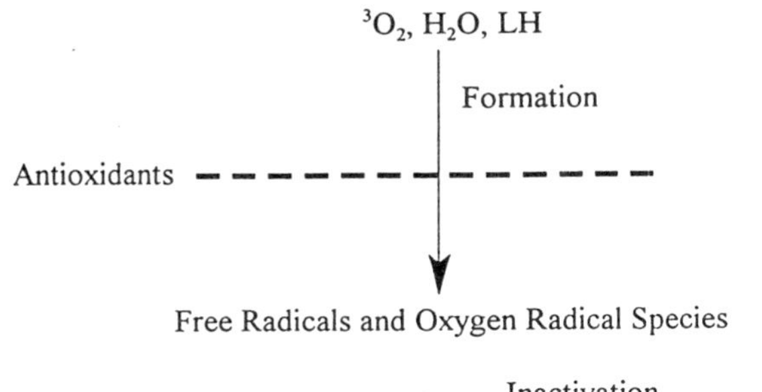
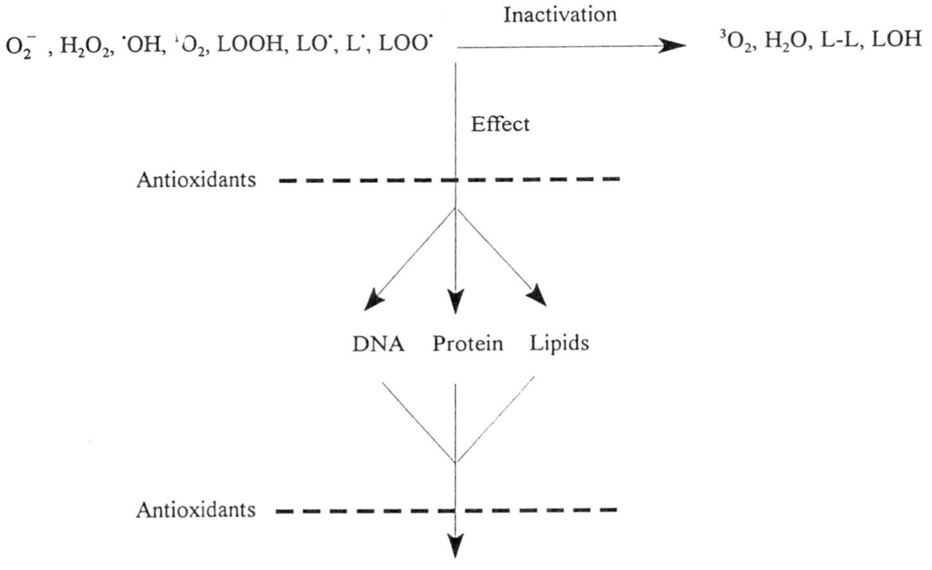

Figure 1. Formation of free radicals, their effects, and neutralization by antioxidants.

Another potential beneficial health effect of phytates relates to their protective influence against cancer. When PA was added to rat diet at 0.6-2.0%, a negative relationship was observed between PA concentration and the epithelial cell proliferation in the ascending and the descending colon (*25*). Furthermore, when PA was added at 1-2% in the drinking water, either one week prior or 2 to 21 weeks after the induction of carcinogen, a significant reduction in the number and size of tumors in the colon of test animals was noted (*26, 27*). Further details on the beneficial effects of PA on reducing carcinogenesis are provided by Thompson (*4*).

The effect of PA in reducing cancer risk may be exerted by a number of mechanisms. PA may bind iron, a catalyst of lipid peroxidation, and thus may reduce formation of free radicals and subsequently breakdown of cellular membranes which encourage cell proliferation (*28*). Alternatively, PA may bind to zinc which is required for DNA synthesis, thus reducing cell proliferation. Furthermore, PA may interact with bacterial enzymes such as β-glucosidase and mucinase in the colon, therefore reducing the conversion of primary to secondary bile acids which are considered as promoters of tumorogenesis. Involvement of inositol phosphate, produced upon PA hydrolysis, also enhances the activity of the baseline killer cells (*29*).

Enzyme Inhibitors

Protease inhibitors. Protease inhibitors are found in abundance in raw cereals and legumes. Due to their particular protein nature, protease inhibitors may be easily denatured by heat treatment. This class of compounds is abundant in raw cereals and legumes, especially soybean. Due to their particular protein nature, protease inhibitors may be easily denatured by heat processing although some residual activity may still remain in the commercially produced products (*30*). The antinutrient activity of protease inhibitors is associated with growth inhibition and pancreatic hypertrophy (*30*). Trypsin inhibitors in soybean give rise to inactivation and loss of trypsin in the small intestine, thus trigger the release of cholecystokinin and induce pancreatic synthesis of excess trypsin and burden on sulfur-containing amino acids in requirement of the body (*31*). Potential beneficial effects of protease inhibitors remain unclear, although lower incidences of pancreatic cancer has been observed in populations where the intake of soybean and its products is high (*32*). This topic is thoroughly discussed in a later chapter.

Amylase inhibitors. Amylase inhibitors are also very heat labile and have been reported as having hypoglycemic effects (*33*). However, instability of this inhibitor under the conditions of the gastrointestinal tract resulted in failure to reduce insulin responses and increase the caloric output of food by using them as starch blocker tablets (*34*).

While protease inhibitors have been linked with pancreatic cancer in animal studies, they may also act as anticarcinogenic agents. The Bowman-Birk inhibitors derived from soybean have been shown to inhibit or prevent the development of chemically-induced cancer of the liver, lung, colon, oral and oesophagus (*35-38*).

Saponins

Saponins are made of a steroid (or triterpene) group attached to a sugar moiety. These surface-active compounds are found in legumes as well as certain spices and herbs. Saponins are able to lyse erthrocytes due to their interaction with cholesterol in the erythrocyte membrane (*39, 40*). Saponins, which have a bitter taste, are toxic in high concentrations and may also affect nutrient absorption by inhibition of metabolic and digestive enzymes as well as binding with nutrients such as zinc. Due to their strong surface activity, saponins are of interest for their beneficial biological effects. The hypocholesterolemic effect of saponins is quite strong, especially when fed in the presence of cholesterol (*41-43*). These topics are discussed in detail in later chapters.

Lignans, Phytoestrogen and Other Related Compounds

Lignans and phytoestrogens are present in higher plants including cereals, legumes, oilseeds, fruits and vegetables. Lignans are a group of diphenolic compounds with a dibenzylbutane backbone. This group of compound is also present in biological fluids in man and animals produced by the bacterial flora in the colon from plant precursors. Meanwhile, phytoestrogens are present mainly as isoflavones and the coumestans. The most abundant isoflavones are glycosides of genestein and diadzein the 4-methyl ether derivatives formononetin and Biochanin A. The important coumestans include coumestrol, 4′-methoxy coumestrol, sativol, tri-foliol and repensol.

Dietary phytoestrogens may cause infertility and liver disease (*44-46*). Similarly, lignans are thought to have oestrogenic and antifertility effects. however, epidemiological data and biological properties of phytoestrogens and lignans suggest that they may serve as important compounds in the prevention and control of cancer, particularly the hormone-dependent ones. Therefore, vegetarians were found to have a higher urinary excretion of lignans and isoflavonic phytoestrogens than breast cancer patients and omnivores (*47*). Japanese women on traditional diets were found to have a lower risk of cancer and excreted 14% higher amounts of phytoestrogen than the Finnish women of high risk cancer (*48*). Strong support for anticarcinogenicity of phytoestrogen comes from studies which showed excellent relationship between cancer risk and soybean intake in the Japanese population as well as between soybean intake and urinary excretion of phytoestrogen (*48*). Induced mammary tumors by chemicals were also inhibited when soybeans were fed to test subjects (*49*). A reduction in cell proliferation as well as colon and mammary cancer risks was also observed when diets of rats were supplemented with flaxseed, a known source of mammalian lignan precursors (*50-52*). The exact mechanism by which these anticarcinogenic activities are exerted is not yet fully understood but several mechanisms have been postulated and detailed in the literature (*46, 53-55*).

Lectins and Haemagglutinins

Lectins and haemagglutinins are sugar-binding proteins which may bind and agglutinate red blood cells. They occur in most plant foods including those that are often consumed in the raw form (*56*). Although lectins in soybeans and peanuts are not toxic, those from jack beans, winged beans, kidney beans, mung beans, lima beans, and castor beans are all toxic when taken orally (*57*). The toxicity of lecins arises from their biding with the specific receptor sites on the epithelial cells of the intestinal mucosa with subsequent lesion and abnormal development of microvillae (*57*). The consumption of lectin-containing foods may lead to endogenous loss of nitrogen and protein utilization. The carbohydrates and proteins that are undigested and unabsorbed in the small intestines reach the colon where they are fermented by the bacterial flora to short-chain fatty acids and gases. These may in turn contribute to some of the gastrointestinal symptoms associated with the intake of raw beans or purified lectins. The lectin-induced disruption of the intestinal mucosa may allow entrance of the bacteria and their endotoxins to the blood stream and cause toxic response (*58*). Lectins may also be internalized directly and cause systemic effects such as increased protein catabolism and breakdown of stored fat and glycogen, and disturbance in mineral metabolism (*57*).

Literature Cited

1. Committee on Diet and Health. Diet and Health. Implications for Reducing Chronic Disease Risk. National Academy Press, Washington, DC, USA, **1989**.
2. Scientific Review Committee. Nutrition Recommendations. Health and Welfare Canada, Supply and Services Canada, Ottawa, Canada, **1990**.
3. Liener, I.E. *J. Nutr.* **1986,** *116*, 920-923.
4. Thompson, L.U. *Food Res. Intern.* **1993,** *26*, 131-149.
5. Jung, M.Y.; Choe, E.; Min, D.B. *J. Food Sci.* **1991,** *56*, 807-815.
6. McLaughlin, P.J.; Weihrauch, J.L. *J. Am. Diet. Assoc.* **1979,** *75*, 647-651.
7. Onyeneho, S.N.; Hettiarachchy, N.S. *J. Agric. Food Chem.* **1992,** *49*, 1496-1500.
8. Shahidi, F.; Wanasundara, P.K.J.P.D. In *Phenolic Compounds in Food and Their Effects on Health. 1*; Ho, C.-T.; Lee, C.Y.; Huang, M.-T., Eds.; ACS Symposium Series 506; American Chemical Society: Washington, DC, **1992,** pp. 214-222.
9. Cuveller, M.-E.; Richard, H.; Berset, C. *Biosci. Biotech. Biochem.* **1992,** *56*, 324-326.
10. Wanasundara, U.N.; Shahidi, F. *Food Chem.* **1994,** *50*, 393-396.
11. Herrmann, K. *J. Food Technol.* **1996,** *11*, 433-450.
12. Reddy, N.R.; Sathe, S.K.; Salunkhe, D.K. *Adv. Food Res.* **1982,** *28*, 1-92.
13. Harland, B.; Oberleas, D. *World Rev. Nutr. Diet.* **1987,** *52*, 235-258.
14. Reddy, N.R.; Sathe, S.; Pierson, M. *J. Food Sci.* **1988,** *53*, 107-110.
15. Tzeng, Y.-M.; Diosady, L.L.; Rubin, L.J. *J. Food Sci.* **1990,** *55*, 1147-1152.

16. Atwal, A.; Eskin, N.; McDonald, B.; Vaisey-Genser, M. *Nutr. Rep. Int.* **1980,** *21,* 257-267.
17. Thompson, L.U.; Sarraino, M. *J. Agric. Food Chem.* **1986,** 34, 468-469.
18. Reinhold, J.; Nasr, K.; Lahimgarzadeh, A.; Hedayati, H. *Lancet,* **1973,** *1,* 283-288.
19. Yoon, J.H.; Thompson, L.U.; Jenkins, D.J.A. *Am. J. Clin. Nutr.* **1983,** *38,* 835-842.
20. Thompson, L.U.; Button, C.L.; Jenkins, D.J.A. *Am. J. Clin. Nutr.* **1987,** *38,* 481-488.
21. Thompson, L.U. *Food Technol.* **1988,** *42 (4),* 123-132.
22. Jariwalls, R.J.; Sabin, R.; Lawson, S.; Herman, Z.S. *J. Appl. Nutr.* **1990,** *42,* 18-28.
23. Klevay, L.M. *Nutr. Rep. Int.* **1977,** *15,* 587-593.
24. Nielson, B.K.; Thomson, L.U.; Bird, R.P. *Cancer Lett.* **1987,** *37,* 317-325.
25. Wolever, T.M.S. *World Res. Nutr. Diet.* **1990,** *62,* 120-125.
26. Shamsuddin, A.M.; Elsayed, A.M.; Ullah, A. *Carcinogenesis* **1988,** *9,* 577-580.
27. Shamsuddin, A.M.; Ullah, A.; Chakvarthy, A. *Carcinogenesis* **1989,** *10,* 1461-1463.
28. Fraf, E.; Eaton, J.W. *Free Rad. Biol. Med.* **1990,** *8,* 61-69.
29. Baten, A.; Ullah, A.; Tomazic, V.J.; Shamsuddin, A.M. *Carcinogenesis* **1989,** *10,* 1595-1599.
30. Hathcock, J.N. In *Nutrient and Toxicological Consequences of Food Processing*; Friedman, M., Ed.; Plenum Press: New York, USA, **1991,** pp. 273-279.
31. Liener, I.E.; Kakade, M.L. In *Toxic Constituents of Plant Foodstuffs*; Liener, I.E., Ed.; Academic Press: New York, **1980,** pp. 7-71.
32. Kennedy, A.R.; Billings, P.C. In *Anticarcinogenesis and Radiation Protection*; Corutti, P.A.; Nygerard, D.F.; Simic, M.G., Eds.; Plenum Press: New York, USA, **1987,** pp. 285-295.
33. Plus, W.; Kemp, N. *Diabeblogia* **1973,** *9,* 97-101.
34. Carlson, G.L.; Li, B., Ban, P.; Olsen, W. *Science* **1983,** *219,* 393-399.
35. St. Clair, W.H.; Billings, P.C.; Kennedy, A.R. *Cancer Lett.* **1990,** *52,* 145-152.
36. Witschi, H.; Kennedy, A.R. *Carcinogenesis* **1989,** *10,* 2275-2277.
37. Messachi, D.V.; Billings, P.; Shklar, G.; Kennedy, A.R. *J. Nat. Cancer Inst.* **1986,** *76,* 447-452.
38. Messina, M.; Barnes, S. *J. Nat. Cancer Inst.* **1991,** *83,* 541-546.
39. Birk, Y.; Peri, I. In *Toxic Constituents of Plant Foodstuffs*; Liener, I.E., Ed.; Academic Press: New York, USA, **1980,** pp. 161-182.
40. Scott, M.T.; Gross-Sampson, M.; Bomford, R. 1985. *Int. Arch. Allergy, Appl. Immunol.* **1985,** *77,* 409-412.
41. Gestener, B.; Assa, Y.; Henis, Y.; Tencer, Y.; Royman, M.; Birk, Y.; Bondi, A. *Biochem. Biophys. Acta* **1972,** *270,*181-187.
42. Sidhu, G.S.; Oakenful, D.G. *Br. J. Nutr.* **1986,** *55,* 643-649.

43. Oakenful, D.; Sidhu, G.S. *Eur. J. Clin. Nutr.* **1990,** *44*, 79-88.
44. Bonnetts, H.W.; Underwood, E.J.; Shier, F.L. *Aust. J. Agric. Res.* **1946,** *22*, 131-138.
45. Lindsay, D.R.; Kelly, R.W. *Aust. Vet. J.* **1946,** *46*, 219-222.
46. Setchell, K.D.R.; Adlercreutz, H. In *Role of the Gut Flora in Toxicology and Cancer*; Rowland, I.R., Ed.; Academic Press: London, UK, **1988,** pp. 315-395.
47. Adlercreutz, H.; Fotsis, T.; Bannwart, C.; Wahala, K.; Makela, T.; Brunow, G.; Haase, T. *J. Steroid Biochem.* **1988,** *25*, 791-797.
48. Adlercreutz, H.; Honjo, H.; Higashi, A.; Fotsis, T.; Hamalanien, E.; Hasegawa, T.; Okada, H. *Scand. J. Clin. Lab. Invest.* **1988,** 48 (suppl. 190), 190.
49. Barnes, S.; Grubbs, C.; Setchell, K.D.R.; Carlson, J. In *Mutagens and Carcinogens in the Diet*; Pariza, M.W.; Aeschbacher, H.; Felton, J.S.; Sato, S., Eds.; Wiley-Liss: New York, **1990,** pp. 239-253.
50. Thompson, L.V.; Robb, P.; Serraino, M.; Cheung, F. *Nutr. Cancer* **1991,** *16*, 43-52.
51. Serraino, M.; Thompson, L.U. *Cancer Lett.* **1991,** *60*, 135-142.
52. Serraino, M.; Thompson, L.U. *Cancer Lett.* **1991,** *63*, 159-165.
53. Adlercreutz, H. *Scand. J. Clin. Lab Invest.* **1990,** 50 (Supp. 201), 3-23.
54. Adlercreutz, H. In *Nutrition, Toxicology, and Cancer*; Rowland, I.R., Ed.; CRC Press: Boca Raton, FL, **1991,** pp. 131-193.
55. Adlercreutz, H.; Mousavi, Y.; Loukpvaara, M.; Hamalanien, E. In *The New Biology of Steroid Hormones*; Hochberg, R.; Naftolin, F., Eds.; Raven Press: New York, **1991,** pp. 145-154.
56. Nachbar, M.S.; Oppenheim, J.D. *Am. J. Clin. Nutr.* **1980,** *33*, 2238-2345.
57. Liener, I.E. In *Food Proteins*; Kinsella, J.E.; Soucie, W.G.; AOCS Press: Champaign, IL, **1989,** pp. 329-353.
58. Banwell, J.G.; Howard, R.; Cooper, D.; Costerton, J.W. 1985. *Appl. Environ. Microbiol.* **1985,** *50,* 68-80.

Chapter 2

Protease and α-Amylase Inhibitors of Higher Plants

John R. Whitaker

Department of Food Science and Technology, University of California, Davis, CA 95616

Numerous proteins in higher plants inhibit enzymes found in plants, animals (including humans) and microorganisms. Some of these inhibitors have endogenous physiological functions in the plants; others appear to have a protective role. Some enzyme inhibitors have nutritional implications in human diets. Protease inhibitors are known for all four types of proteases. The best studied are the serpins (inhibit serine-type proteases) and the cystatins (inhibit the sulfhydryl-type proteases). Based on disulfide bond arrangement, primary amino acid sequences and other factors, the groups can be further subdivided. Some of these protease inhibitors have been bioengineered into plants to protect them (and their seeds) against pests. The α-amylase inhibitors are under intense study. Those from beans have sequence homology with lectins and arcelins also found in beans. Some enzyme inhibitors are multi-headed, inhibiting more than one molecule of the same enzyme, same class of enzyme or different classes of enzymes. The structural diversity, mechanism of inhibition, and importance of these inhibitors are discussed.

Naturally occurring proteins in wheat that inhibit α-amylases were first reported in 1933 (*1, 2*), rediscovered in 1943 (*3, 4*) and again in 1973 (*5, 6*). Proteinaceous α-amylase inhibitors of common beans (*Phaseolus vulgaris*) were first reported in 1945 (*7*) and again in 1968 (*8*) and detailed investigations began in 1975 (*9*). α-Amylase inhibitors have now been reported in a variety of plants, including maize (*10*), barley (*11*), sorghum (*12*), millet (*13*), rye (*14*), bajra (*15*), peanuts (*16*), acorns (*17, 18*), taro roots (*19, 20*) and mangoes (*21*).

Protease inhibitors were detected first in soybeans and purified from them by Kunitz (*22*), followed by the discovery and isolation of the Bowman-Birk inhibitors (*7*). About 20,000 papers dealing with protease inhibitors have appeared since 1945 (*23*). Trypsin inhibitors, the best studied, are ubiquitous, being present in plants, animals and microorganisms. Inhibitors of other proteases have received much less attention.

Investigation of the structure and function of naturally occurring inhibitors got off to a slow start because they were often considered to be only of academic interest.

But with the realization that they may be important in the control of plant pests (*23-26*), they are now receiving much attention from plant scientists, molecular biologists and DNA-protein engineers (*23*). Some inhibitors, especially from microorganisms (*27*) have received much attention for use in medical applications. Nutritionists and biochemists have been interested in the potential and/or real effect on digestion of foods in the GI tract (*28-34*). They are also quite useful in analytical purposes, including purification of enzymes and distinguishing among different enzymes (*35-37*).

There are a large number of general reviews on proteinaeous enzyme inhibitors (*23, 38-56*).

Protease Inhibitors

This group of proteinaeous inhibitors are the best studied and thousands of articles have been written on the subject. The trypsin inhibitors are ubiquitous and they, and others, often occur in several isoinhibitor forms. For example, pink Brazilian beans (*61*) and red kidney beans (*62*) contain three and four isoinhibitors, respectively, that have been purified. In this chapter, I will deal only with the classification, diversity, mechanism of action and structure of a few examples of protease inhibitors.

Classification. Initially, classification was based on the protease inhibited, such as trypsin. Classification now includes the class of protease inhibitor, i.e., a serine protease, a cysteine protease, an aspartic acid protease or a metalloprotease inhibitor, recognizing the mechanism-based nature of binding of inhibitor to the enzyme (Table I). More recently, classification has been made also on the basis of similarities

Table I. FAMILIES OF PLANT PROTEASE INHIBITORS

Serine Protease Inhibitors (Serfins)

1. Bowman-Birk (trypsin/chymotrypsin)[a]
2. Kunitz (trypsin; others)[a]
3. Potato I (chymotrypsin; trypsin)[a]
4. Potato II (trypsin; chymotrypsin)[a]
5. Cucurbit (trypsin)
6. Cereal superfamily (amylase, trypsin)[b]
7. Ragi I-2 family (amylase, protease)[b]
8. Maize 22 kDa/thaumatin/PR (amylase, trypsin)[b]

Cysteine Protease Inhibitors (Cystatins and Stefins)

1. Cystatin super family
2. Cystatin family
3. Stefin family
4. Fitocystatin family

Metallo Protease Inhibitors

1. Carboxypeptidase

[a]Second enzyme listed binds less tightly.
[b]Double headed inhibitor.

```
 1   SOYBEAN CII                                SDHSSSDDE
 2   PEANUT AII                                    EA*SSS
 3   ADZUKI BEAN                             SGHHE*TTDEPS*
 4   MACROTYLOMA DE3                           DHHHSTDEPS*
 5   VICIA ANGUSTIFOLIA                                GDD
 6   MUNG BEAN                               SSHHH*S*DEPS*
 7   ALFALFA LEAF
 8   RICE BRAN                                          M*
 9   WHEAT GERM II-4                                    AT
10   WHEAT GERM I-2b                                  AAKK
11   COIX                                             GDEK

                      20      ↓ 30       40
 1               -SSKPCCDLCMCTASM--PP--QCHCADIRLN
 2               -DDNV**NG*L*DRRA--**YXE*V*Y*TFDH
 3               -*******Q*-**K**--**--K*R*S****D
 4               -*******E*A**K*I--**--**R*T*V***
 5               -VKSA***T*L**R*Q--**--T*R*V*VGER
 6               -**E****S*R**K*I--**--**********
 7                 TTA**NF*P**R*I--**--**R*T**GET
 8               RPW*-***NIKRLPTKPD**--*WR*N*ELEP
 9               RPWK-***RAI**K*F--**--M*R*M*MVE-
10               RPW*-**DRAI**R*F--**--I*R*M*QVFE
11               RPWE-***IA***R*I--**--I*R*V*KVDR

                            ↓
                   50       60        70
 1               -S*HSA*DR*A*TRSM--P---G*R*LDTTDF
 2               --*PAS*NS*V****N--*P--Q***R*K*QG
 3               -******KS****Y*I--*A--K*F*T*IN**
 4               -******SS*V**F*I--*A--Q*V*V*MK**
 5               --*****NH*V*NY*N--*P--Q*Q*F*8HK*
 6               -******KS*M*****--*G--K******D**
 7               --*****KT*L**K*I--*P--Q***T*I*N*
 8               SQ*TA**KS*REAPGPFGKLI--*EDIYWGAD
 9                   -Q*AAT*KK*GPAT*DSSRR--V*EDXY
10                  --*P*T*KA*GPSVGDPSRR--V*QDQYV
11               --*SD**KD*EE*ED--NRH--V*FDTYIG*P

                  80       90
 1               -*YKP-*KSSDEDDD
 2               R*PVTE*R*
 3               -**E*-****RD**WDN
 4               -**A*-****HD*
 5               -***A-*H**EKEEVIKN
 6               -****-*E*M*K***
 7               -**PK-*N
 8               P--G*F*TP
11               ---G*T*HDD
```

Figure 1. Amino acid sequences of some legume and cereal protease inhibitors that belong to the Bowman-Birk family. The - is included for alignment purposes and the * indicates the amino acid is identical to the one in the reference protein (No. 1). The arrows indicate the binding sites for trypsin and/or chymotrypsin. Adapted from Ref. *23*.

in the primary amino acid sequences (*23*) (Figure 1) and/or the disulfide bond location (*41*) (Figure 2). Table I shows the families of plant protease inhibitors that have received much attention. The best studied are the serine protease inhibitors. The cysteine protease inhibitors have been studied mostly in animals, but more recent studies on some cysteine protease inhibitors in plants indicate they belong to the cystatin family. These include inhibitors from rice (*63*) and corn (*64*). Cysteine protease inhibitors have been reported in cowpeas (*65*), pumpkin seeds (*66*) and potato tubers (*67*), but further classification is not yet possible.

Figure 1 shows the amino acid sequence alignment of 11 Bowman-Birk inhibitors from a number of plant sources. The most prominent alignments are the cysteine (C) residues. Proline (P) residues at positions 25 and 26 (soybean inhibitor) and aspartic acid (D) residue at position 32 (soybean inhibitor) are the only other amino acid residues preserved in all nine inhibitors. As shown, the active site amino acid residues immediately adjacent to the arrows are not conserved. In general, they are arginine (R) or lysine (K) on the carbonyl side of the peptide bond and serine (S) on the amino side for trypsin inhibitors and a hydrophobic amino acid residue (alanine (A), tyrosine (Y), phenylalanine (F) or valine (V)) on the carbonyl side of the peptide bond and a serine (S) on the amino side for chymotrypsin inhibitors. But there are other variations.

Figure 2 shows the disulfide bond locations and arrangements in inhibitors from a fungus and from five plants (*41*); all are unique. The arrows indicate the primary specificity site for binding with the enzymes. The Bowman-Birk inhibitors have two binding sites. Turkey ovomucoid has three binding sites (*68, 69*) and avian egg white ovoinhibitor has up to six binding sites.

Amino Acid Diversity at the Active Site of Inhibitors

Diversity at the Active Site. As shown in Figure 1, the amino acids surrounding the site can be quite variable. This is best shown by data for 112 avian egg white ovomucoids in Figure 3 (*70*). The six 1/2 Cys residues (8, 16, 24, 35, 38 and 56, as shown by the vertical row of numbers on left side of Figure 3) are strictly conserved, as well as a few other residues. But around the binding site (positions 18 and 19) there is great diversity. In most biologically active proteins the replacement of an active site amino acid residue with another, even with small differences, leads to loss of activity. But this is not true of the protease inhibitors, antibodies (immunoglobulins) and lectins, which have in common that they bind with a variety of proteins or glycoproteins with diverse amino acid sequences (such as trypsins from different sources). Therefore, diversity is a positive factor (*70*) in the amino acid sequences of these three groups of proteins. These variant proteins probably have differences in secondary and tertiary structures, even if subtle in some cases, that permit them to distinguish among closely related compounds.

Mechanism of Action. The initial binding of protease inhibitors with enzymes appears to occur at the active site in the same manner as any protein substrate. For example, trypsin binds to the inhibitor at an arginine or lysine residue via the positively charged guanidine or α-amino group interacting with the negatively charged aspartate residue at the bottom of the specificity "pocket" and by hydrogen bonds via the peptide bonds of residues along the peptide chain forming part of the active site. However, because the inhibitor binds so tightly ($K_i \sim 10^{-9}$ to 10^{-11} M) and/or because of constraints of conformation of the inhibitor at the binding site, the potential scissile bond of the inhibitor is not hydrolyzed. If the pH is decreased from 8 to 3-4, K_i is very large (complex is loosely bound) and slow hydrolysis can occur (*41*). This is described by Eqn. 1 where E is the enzyme, I is the inhibitor, E•I is the loosely bound complex ($K_i \sim 10^{-2}$-10^{-5} M) formed in step 1, E-I is the tightly bound complex

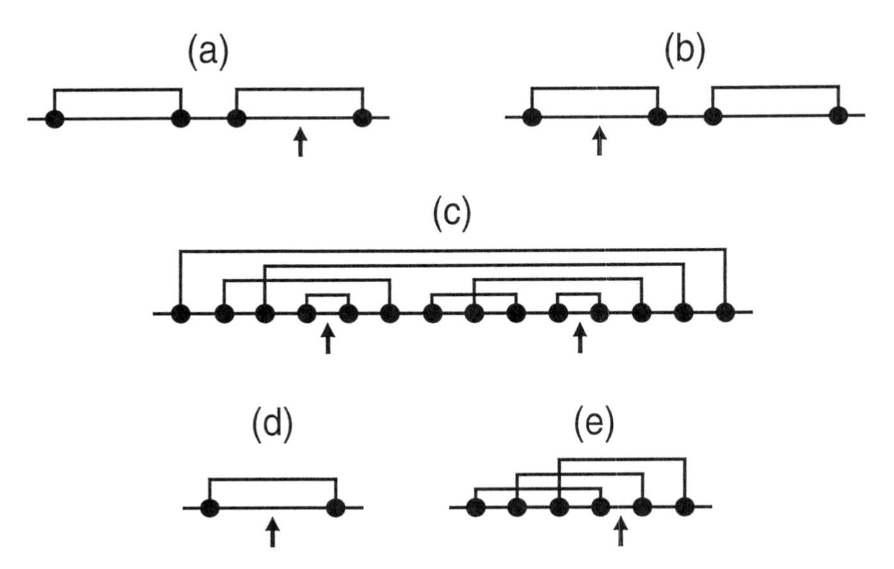

Figure 2. Location and arrangement of disulfide bonds in some plant and microbial protease inhibitors. Sources and groups (with 1/2 Cys residues numbered from left to right): a) *Streptomyces albogriseelus* (35, 50, 71, 101); b) soybean, Kunitz (35, 82, 121, 130); c) lima bean, Bowman-Birk (18, 19, 22, 24, 32, 34, 42, 46, 49, 51, 59, 61, 68, 72); d) potato, I (5, 49); e) potato, II (34, 38, 46, 58, 61, 62, 68, 79, 91, 95, 104, 118, 121, 122, 128, 139). Reproduced with permission from Ref. *41*. Copyright 1980 Annual Reviews.

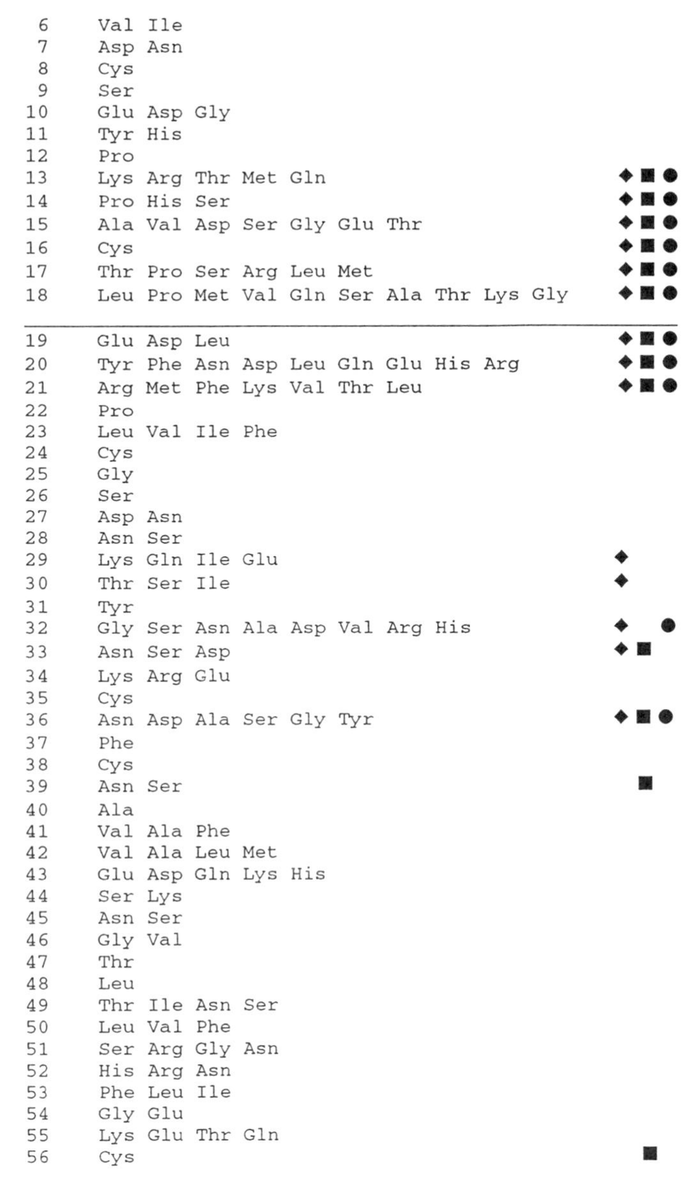

Position	Residues	◆	■	●
6	Val Ile			
7	Asp Asn			
8	Cys			
9	Ser			
10	Glu Asp Gly			
11	Tyr His			
12	Pro			
13	Lys Arg Thr Met Gln	◆	■	●
14	Pro His Ser	◆	■	●
15	Ala Val Asp Ser Gly Glu Thr	◆	■	●
16	Cys	◆	■	●
17	Thr Pro Ser Arg Leu Met	◆	■	●
18	Leu Pro Met Val Gln Ser Ala Thr Lys Gly	◆	■	●
19	Glu Asp Leu	◆	■	●
20	Tyr Phe Asn Asp Leu Gln Glu His Arg	◆	■	●
21	Arg Met Phe Lys Val Thr Leu	◆	■	●
22	Pro			
23	Leu Val Ile Phe			
24	Cys			
25	Gly			
26	Ser			
27	Asp Asn			
28	Asn Ser			
29	Lys Gln Ile Glu	◆		
30	Thr Ser Ile	◆		
31	Tyr			
32	Gly Ser Asn Ala Asp Val Arg His	◆		●
33	Asn Ser Asp	◆	■	
34	Lys Arg Glu			
35	Cys			
36	Asn Asp Ala Ser Gly Tyr	◆	■	●
37	Phe			
38	Cys			
39	Asn Ser		■	
40	Ala			
41	Val Ala Phe			
42	Val Ala Leu Met			
43	Glu Asp Gln Lys His			
44	Ser Lys			
45	Asn Ser			
46	Gly Val			
47	Thr			
48	Leu			
49	Thr Ile Asn Ser			
50	Leu Val Phe			
51	Ser Arg Gly Asn			
52	His Arg Asn			
53	Phe Leu Ile			
54	Gly Glu			
55	Lys Glu Thr Gln			
56	Cys		■	

Figure 3. Amino acid sequence of turkey ovomucoid third domain (left column of amino acids shown vertically) and changes found in one or more of 111 other avian ovomucoids for each amino acid position (shown horizontally) (*69*). The primary specificity binding site is shown by the horizontal line between positions 18 and 19. Residues that bind with human leukocyte elastase (◆) (*71*), bovine chymotrypsin A (■) (*72*) and *Streptomyces griseus* proteinase 8 (●) (*73*). Reproduced with permission from Ref. *70*. Copyright 1987 Cold Spring Harbor Laboratory.

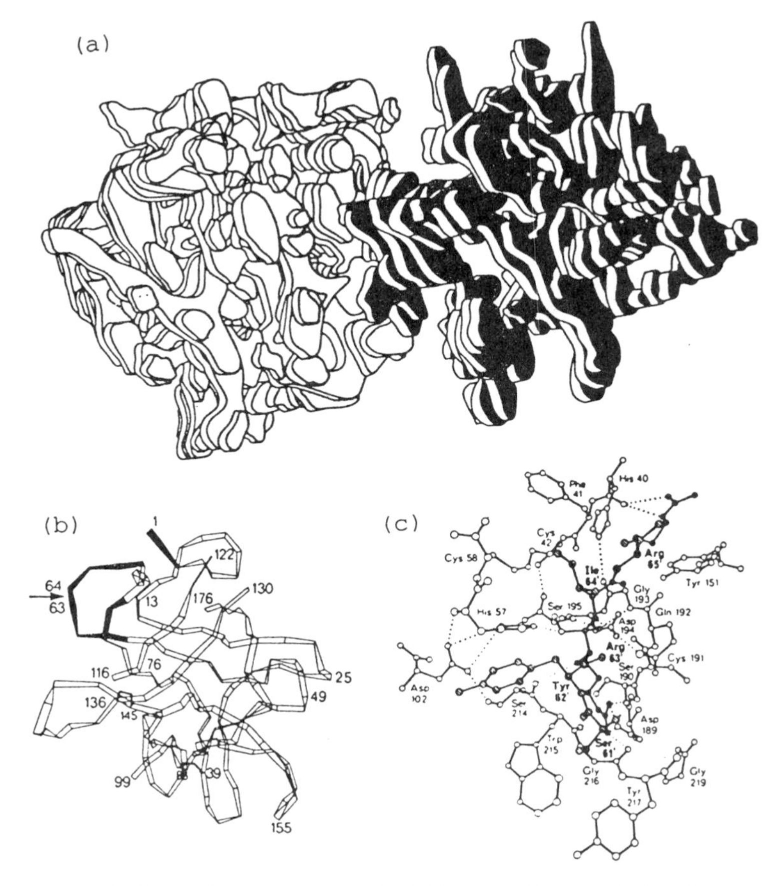

Figure 4. Binding of soybean trypsin inhibitor (STI; Kunitz) with porcine trypsin. a) Model of the STI-trypsin complex, made from an electron density map at 5 Å-resolution. The lightly shaded part is trypsin, the dark part STI. b) Ribbon diagram of STI showing the contact regions (black) with trypsin. c) Active site regions of trypsin (thin lines) and STI (heavy lines) contact in complex formation. Reproduced with permission from Ref. *75*. Copyright 1974 American Chemical Society.

(K_i~10^{-9}-10^{-11} M) formed at pH 8 in the subsequent second step 2 and E•I* is the modified inhibitor (peptide bond at the binding site slowly hydrolyzed) as found at pH 3-4. Enzymes in which a catalytic group has been removed (trypsin vs

$$E + I \underset{}{\overset{1}{\rightleftharpoons}} E{\cdot}I \overset{2}{\rightleftharpoons} E\text{-}I \qquad (1)$$
$$\downarrow 3$$
$$E{\cdot}I^*$$

anhydrotrypsin for example) but the binding site is intact may bind inhibitor as tightly in step 2 as the original enzyme but step 3 does not occur. Step 3 is not essential for inhibition.

Enzyme•Inhibitor Complex. We described the enzyme•inhibitor complex mechanistically above. E•I has been crystallized and its structure determined in the case of pancreatic trypsin inhibitor (*74*), Kunitz soybean trypsin inhibitor (*75*) and the subtilisin inhibitor (*76, 77*). Figure 4 shows the complex of porcine trypsin with soybean trypsin inhibitor (STI) (Kunitz) at 2.6 Å resolution. The model of the E•I complex at 5 Å resolution (Figure 4a) indicates the general area of contact between the two proteins (STI with MW of 20,400 and trypsin of 24,000). Figure 4b shows the chain folding of STI and the amino acid residues (Asp1, Asn13, Pro60, Ser61, Tyr62, Arg63, Ile64, Arg65 and His71) in contact with trypsin residues (Ser195, His57, Gly193, Asp189, Ser214 and Gly216). Details of the contact between the two proteins at the atomic level is shown in Figure 4c. These interactions are very similar to those between trypsin and a normal polypeptide substrate.

Figure 5 shows the α-carbon atom structure of human stefin B with the cysteine protease papain (*78*). The inhibitor, at top, binds to the enzyme at the lip of the crevice between the two domains where the normal substrate binds to the enzyme. Again, there is limited surface area contact between the two and it is at the active site.

α-Amylase Inhibitors

There are two types of proteinaeous inhibitors of α-amylases, the microbial peptides obtained from various strains of *Streptomyces* (*79*) and the higher plant proteins primarily derived from legumes and cereals (*56*). Data on the α-amylase inhibitors are not as complete as for many of the protease inhibitors, but rapid progress is being made on their primary sequences, more recently by cDNA analyses (*79a*).

Amino Acid Sequences of Selected Inhibitors. Figure 6 compares the amino acid sequences of five of the microbial peptide inhibitors. They range in molecular weight from 3936 for AI-3688 to 8500 for Haim II. The four 1/2 Cys residues (11, 27, 45 and 73) are conserved. There are several other primary amino acid sequences that are conserved among the five inhibitors. Of particular interest are the residues -Trp-Arg-Tyr- at positions 18-20 and residues 59-70 known to be involved in binding to α-amylase (*80, 82, 83*). The four 1/2 Cys residues (see above) are located at identical positions in the aligned inhibitors, as are Val12, 31, 33 and 35, Asn25, Tyr20, 37 and 60, Gly51 and 59 and Thr55. The tertiary structure of Hoe-467A (tendamistat) indicates the -Trp-Arg-Tyr- sequence forms a prominent projection on the surface of the molecule and that residues 59-70 are aligned in proximity to the -Trp-Arg-Tyr- sequence (see below).

The primary amino acid sequences of three wheat α-amylase inhibitors (WAI) are shown in Figure 7 (*84*). All three have 123 to 124 amino acids. WAI-0.19 and WAI-0.53 have 96% homology, WAI-0.19 and WAI-0.28 have 62% homology and WAI-0.28 and WAI-0.53 have 52% homology. All are monomers of ~14,000 MW.

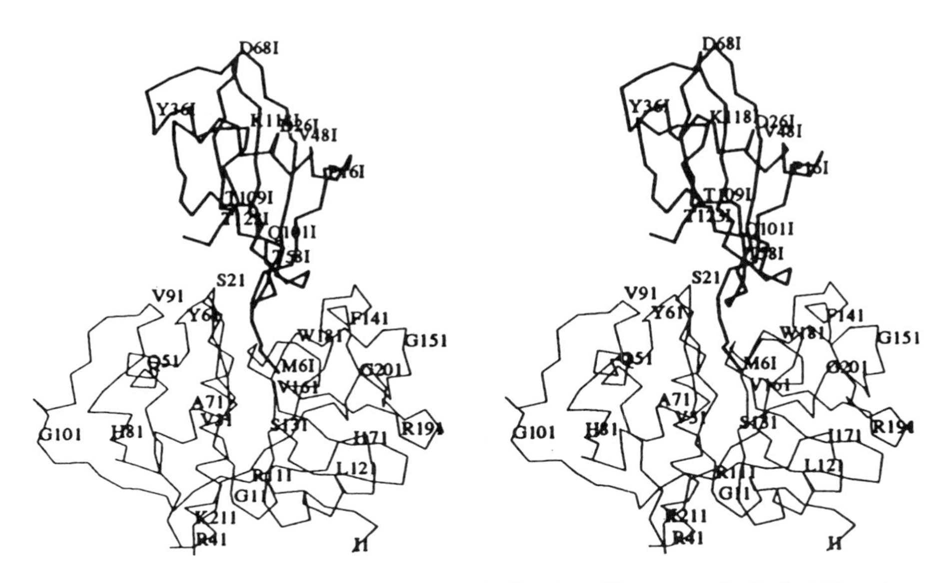

Figure 5. Tertiary structures of papain (thin lines) and human stefin B (bold lines) in the enzyme-inhibitor complex. Reproduced with permission from Ref. *78*. Copyright 1994 Japan Scientific Societies Press and Karger.

```
                    10        20        30        40        50
Hoe-467A  DTTESEPAPSCVTLYQSWRYSQADNGCAETVTVKVVYEDDTEGLCYAVAP 50
Z-2685       ATGS*VAE**GYF*****TDVH****DA*S*T*E*THGQWAP*RVIE* 48
Haim II         IA**A**HFTAD***TFVT***SIDYS*T*A*G*G*DVP*RSAN* 45
Paim I        A*****A**VM*E****TT*A****D**S*S*A*Q*GAT*P*ATLP* 47
AI-3688      ATGS***D**ESF*****TDVR***SDA***V*Q**              36

         51        60        70
Hoe-467A  GQITTVGDGYIGSHGHARYLARCL                          74
Z-2685    *GWA*FA-**GTDGDYVTG*HT*DPATPS                     76
Haim II   *D*L*FP-**GTRRDEVLGAVL*ATDGSALPVT                 77
Paim I    *ATV***E**L*E***PDH**L*PSS                        73
AI-3688                                                     36
```

Figure 6. Comparison of the amino acid sequences of five *Streptomyces* polypeptide α-amylase inhibitors. The numbers on the right side sum the number of amino acid residues in a particular protein to that point. See Figure 1 legend for explanation of - and *. Adapted from Refs. *80 and 81* .

There are other inhibitors in wheat that appear to be dimers and trimers of one or more of the monomeric inhibitors. Nothing is known of their secondary and tertiary structures and of the amino acids that might be in the active site. One tryptophan, four arginines and one tyrosine residues are conserved, indicating that these residues may be important parts of the active site.

The cDNA derived sequences of four α-amylase inhibitors from common beans (*Phaseolos vulgaris*) have been determined recently ((*79a* , *85, 86,* (Figure 8)). The α-amylase inhibitors are synthesized as pre-pro-proteins of about 266 amino acids. The 23 amino acid signal peptide is removed by a specific protease that hydrolyzes a specific -Ser-Ala- peptide bond (position 1), giving a 243 amino acid pro-protein which has no inhibitory activity until it is processed by a protease at the Asn85-Ser86 peptide bond (*86a*), to give α- (77 aa) and β- (146 aa) peptide chains that fold and assemble into a mature tetrameric protein (see below) with inhibitory activity. Sequences 1-4 of Figure 8 are cDNA derived sequences of α-amylase inhibitors and sequence 5 is an α-amylase inhibitor-like protein with no inhibitory activity following cloning and expression of the gene into tobacco (Chrispeels, M.S., University of California, San Diego, personal communication, 1996). All five pro-proteins are 239 to 261 amino acids in length. All begin with an alanyl residue at the N-terminal end (position #1). All have a protease processing site at -Asn85-Ser86- that must be split to give activity (*86a*). There is high amino acid sequence homology among them (Table II). α-Amylase inhibitor (αAI-5) from black beans has 97.5% homology with

Table II. Amino Acid Sequence Homologies Among α-Amylase Inhibitors, Arcelin-1 and Lectins[a]

	aAI-4	αAI-5	αAI-1	αAI-2	αAI-3	ARL-1	PHA-E[c]
αAI-4	--						
αAI-5	95.1	--					
αAI-1	93.4	97.5	--				
αAI-2	74.9	77.4	77.4	--			
αAI-3	55.6	57.6	57.6	57.7	--		
ARL-1	51.9	55.1	55.1	55.6	59.0	--	
PHA-E	40.3	43.2	43.2	42.7	49.4	53.0	--
PHA-L	41.2	43.6	43.6	42.3	50.2	55.5	82.5
Lectin[b]	33.7	35.0	35.8	38.1	36.4	35.6	40.1

[a]Data from Figure 8 expressed in percent.
[b]Pea lectin.
[c]The relatedness to PHA-L is 38.7%.

αAI-1 (LLP, *88*), 95.1% homology with αAI-4 from white kidney beans, 77.4% homology with αAI-2 from a wild common bean (*86*) and 58% with αAI-3 from a wild common bean (*85*) (with no inhibitory activity (*86a*)). About 5% of the heterogeneity among these five αAIs arises from the amino acid sequence between residues 192 and 203 (block area).

Lectins, arcelins and α-amylase inhibitors appear to have evolved from a common ancestral gene (*85, 89*; Figure 8). The genes for the arcelins and lectins are about the same size as those for the α-amylase inhibitors, the signal peptide is very similar and they are processed at a -Ser-Ala- or -Ser-Ser- peptide bond (arrow (1); Figure 8). But arcelin-1, PHA-E and PHA-L proteins are not processed further,

```
                    10        20        30        40        50
WAI-0.19   SGPWM-CYPGQAFQVPALPACRPLLRLQCNGSQVPEAVLRDCCQQLAHIS  49
WAI-0.53   *****-*************G*****K*********************D**  49
                  D
WAI-0.28   ****SW*N*ATGYK*S**TG**AMVK***V*****************D*N  50

           51       60        70        80        90       100
WAI-0.19   -EWCRCGALYSMLDSMYKEHGAQEGQAGTGAFPRCRRFVVKLTAASITAV  98
WAI-0.53   -**P*****************VS**********S****************  98
                         * A
WAI-0.28   N******D*S***RAV*Q*L*VR**K---EVI*G**K**M******VPE*  97

          101      110       120
WAI-0.19   CRLPIVVDASGDGAYVCK-DVAAYPDA                        124
WAI-0.53   ******************-********                        124
           GP                   *
WAI-0.28   *KV**P-NP***R*G**YG*WC****V                        123
```

Figure 7. Amino acid sequences of wheat α-amylase inhibitors WAI-0.19, WAI-0.28 and WAI-0.53. *, Same as amino acid in WAI-0.19; -, used for alignment. WAI-0.28 has several isoinhibitor forms, as indicated by the variable amino acids at five positions (listed above the line). For explanation of numbers on right side, see Figure 6 legend. Adapted from Ref. *84*.

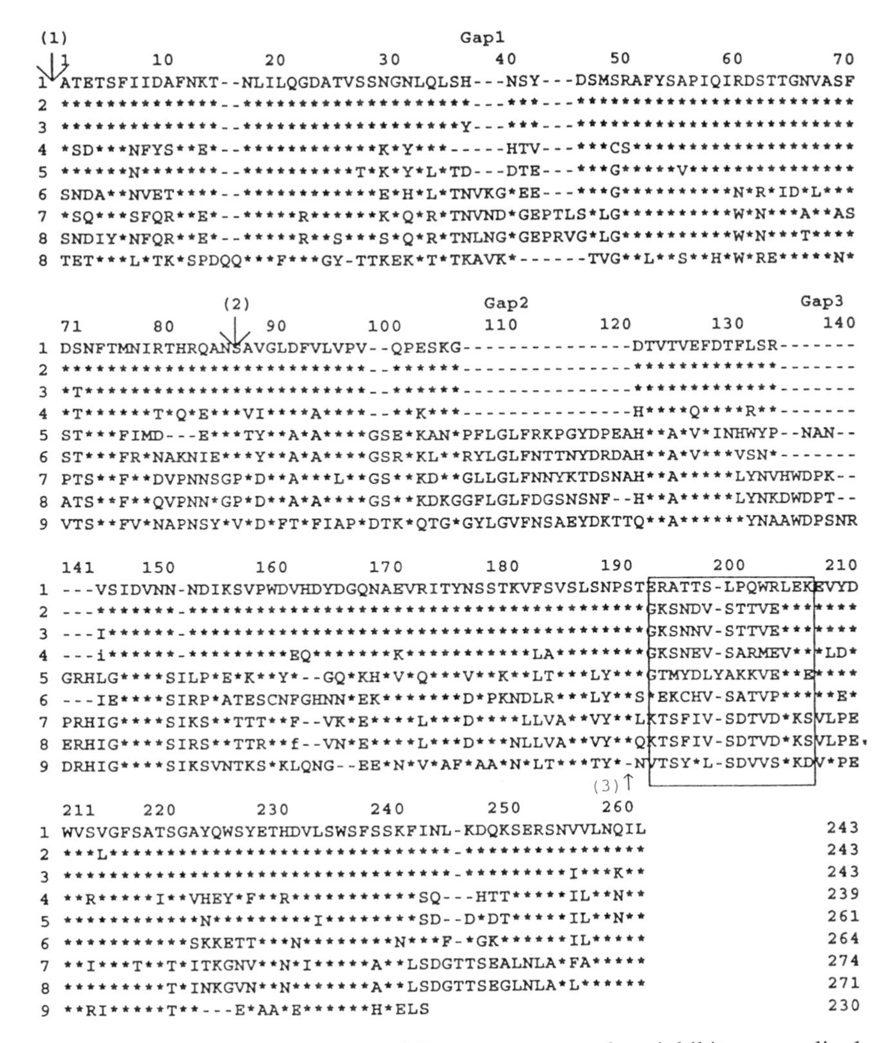

Figure 8. Amino acid sequences of five mature α-amylase inhibitors, arcelin-1 and three lectins. 1) White kidney bean 858A αAI-4; Lee, S. C. and Whitaker, J. R., University of California, Davis, unpublished data; 2) black bean αAI-5; Lee, S. C. and Whitaker, J. R., University of California, Davis, unpublished data; 3) greensleeves bean αAI-1; Refs. *87* and *88* ; 4) wild common bean αAI-2; Ref. *86* ; 5) wild common bean αAI-3-like protein (has no inhibitory activity); Ref. *85;* 6) arcelin-1; Ref. *89* and *90;* 7, PHA-E; Ref. *91*; 8) PHA-L; Ref.*91;* 9) pea lectin; Ref. *92*. *, Same as the amino acid in αAI-4; -, gap inserted for alignment; ↓(1) or ↓(2) sites of proteolytic processing. The boxed segments indicate the less homogeneous regions and ↑(3) indicates processing of pea lectin.

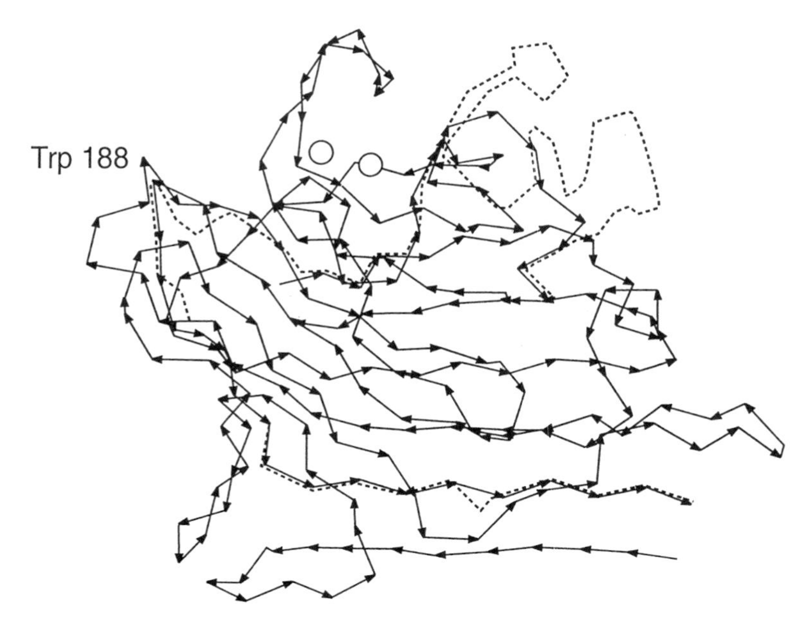

Figure 9. Tertiary structures of pea lectin (*95*). ---, by x-ray; —, α-amylase inhibitor-1 by modeling (*96*). Trp188 may be important in binding to α-amylase. Reproduced with permission from Ref. *96*. Copyright 1995 Oxford University Press.

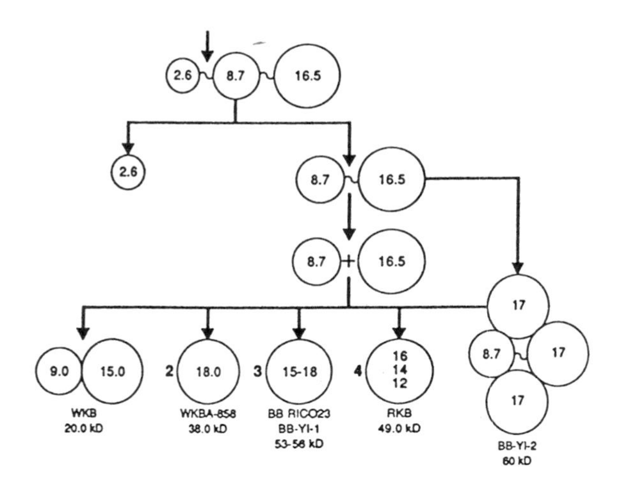

Figure 10. Hypothetical quaternary structures of some α-amylase inhibitors from common beans (*Phaseolus vulgaris*). Reproduced with permission from Ref. *56*. The numbers in the circles indicate molecular weight (in kD) of the unit. WKB, white kidney bean; WKBA-858, experimental cultivar (University of California, Davis) of white kidney bean; BB RICO23, black bean var. Rico 23 (Brazilian); BB-YI-1 and BB-YI-2, isoinhibitors 1 and 2 isolated by Yin (*93*) from black bean; RKB, red kidney bean. Copyright 1994 Blackie Academic & Professional.

unlike the α-AI pro-protein. Pea lectin (#9 in Figure 8) is further processed to α- and β-subunits at arrow (3). A nine amino acid sequence may also be removed from the C-terminal end of the β-subunit (large one) of pea lectin (*95*).

The deduced amino acid sequences of αAI-1, αAI-4 and αAI-5 are collaborated by Edman degradation sequencing of the N-terminal ends of the α- and β-subunits of the isolated mature proteins (*79a*; *87, 93, 94*, Finardi-Filho, F. and Lajolo, F. M., University of Sao Paulo, unpublished data).

Secondary, Tertiary and Quaternary Structure of α-Amylase Inhibitors. Nothing is known about the secondary and tertiary structures of the α-amylase inhibitors. Based on the amino acid sequence homology between αAI-1 and the pea lectin (Figure 8), and the crystal structure of pea lectin (*95*), Mirkov *et al.* (*96*) proposed, by model building, a tertiary structure for αAI-1 (Figure 9). The tertiary structures are very similar, differing only in three loops (at the top of figure) found in pea lectin, but not in αAI-1 or the other amylase inhibitors. These loops are formed by amino acid residues in the regions 37-45, 106-120 and 133-143 (Figure 8). A very interesting finding is that a tryptophan (shown as 188), an arginine and a tyrosine residue are in proximity in the folded αAI-1, the same three residues responsible for α-amylase inhibitory activity in tendamistat (*81-83*); these same groups were suggested to be important based on chemical modification of α-amylase inhibitor (*94, 97, 98*). Mutations of cDNA for αAI-1 in which changes in Arg74Asn and Trp-Ser-Tyr188-190Gly-Asn-Val were made were inactive (*96*). (The mutated residues as numbered in Figure 8 are Arg82Asn and Trp-Ser-Tyr225-227Gly-Asn-Val.) Arg82 is the fourth residue from the C-terminal end of the α-chain, while the Trp-Ser-Tyr225-227 sequence is in the β-chain. These are brought into close proximity in the folded protein (Figure 9).

Quaternary Structure. Based on the subunit model of Moreno and Chrispeels (*87*) from their comparison of the cDNA derived amino acid sequences for Greensleeves α-amylase inhibitor, using the lectin-like protein (LLP, Ref. *88*), Ho *et al.* (*56*) proposed a hypothetical quaternary structure based on the size of the αAI genes and electrophoretic and gel filtration methods of separating the subunits (Figure 10). The model used various numbers of the α- (8700 MW) and β- (16,500 MW) subunits to account for molecular weights of inhibitors ranging from 20,000 to 60,000 (*56*). If Arg82 of the α-subunit is part of the active site, as suggested above, then the models that do not include the α-subunit or show it as part of the unprocessed pro-inhibitor cannot be correct. The amino acid sequences for αAI-1, αAI-2, αAI-4 and αAI-5, given an α-subunit of 77 amino acids (8700 MW) and β-subunit of 146 amino acids (16,500 MW), would account for a molecular weight of 25,200; with 8 to 15% by weight of carbohydrate reported for various inhibitors, the MW could range from 27,200 to 28,900. No reported mature active inhibitors have MWs in this range. An $\alpha_2\beta_2$ tetrameric model would give MWs of 54,400 to 57,800; most of the characterized α-amylase inhibitors have molecular weights close to this range (*56*). However, the 20,000 and 40,000 MW inhibitors from white kidney beans (WKB858A and WKB858B (*94, 99*)) do not appear to fit this model. Further research must also determine the nature of an active site that might accommodate two sets of active site residues Arg82 and Trp-Ser-Tyr225-227, since the inhibitor is known to form 1:1 complexes with α-amylase in most cases.

Mechanism of Action. Kinetic data indicate that the α-amylase and inhibitor first form a loose 1:1 complex ($K_i \sim 10^{-5}$ M) by a diffusion-controlled reaction which then undergoes a slow conformational change to form a tight complex ($K_i \sim 10^{-10}$ to 10^{-11} M) (*97, 98, 100*), (Eqn. 2), where E is enzyme, I is inhibitor, E•I is loose complex and E-I* is tight complex. The complex is much less soluble than the separate

```
(a)
                  10        20        30        40        50
                                    AR A
RATI      SVG--TSCIPGMAIPHNPLDSCRWYVSTRTCGVGPRLATQEMKARCCRQL     48
BTI        F*--D**A**D*L*****RA**T**VSQI*HQ****L*SD**R***DE*     47
WAI        S*PWSW*N*ATGYKVSA*TG**AM*KLQ-*V-*SQVPEAVL-RD**Q**     46

          51        60        70        80        90       100
                             S
RATI      EAIPA-YCRCEAVRILMDGVVTPSGQHEGRLLQD-LPGCPRQVQRAFAPK     96
BTI       S****-******L**I*Q****WQ*AF**AYFK*S-*N***ER*TSY*AN     95
WAI       AD*NNEW***GDLSSMLRA*YQEL*VR**---KEV****RKE*MKLT*AS     93

         101       110       120       130
RATI      LVTEVECNIATI----HGGPFCLSLL-GAGE                       122
BTI       ***PQ****G**----**SAY*PE*QP*Y*                        121
WAI       -*P**-*KVPIPNPSGDRAGV*YGDWCAYPDV                      123

(b)
                  10        20        30        40        50
BASI      ADPPPVHDTDGHELRADANYYVLSANRAHGGGLTMAPGHGRHCPLFVSQD     50
SBTI         DF*L*NE*NP*ENGGT**I**DIT*F-**IRA**TGNER***T*V*S     46
WBTI         E*LL*SE*ELV*NGGT**L*PDRW*L***IEA*ATGTET***T*VRS     47

          51        60        70        80        90       100
BASI      PNGQHDGFPVRITPYGVAPSDKIIRLSTDVRISFRAYTTCLQS-TEWHID     99
SBTI      R*ELDK*IGTI*SSPYRIRFIAEGHPLSLKFD**AVIML*YGIP***SVV     96
WBTI      **EVSV*E*L**SSQLRSGFIPD---YSL***G*ANPPK*AP*P-W*TVV     93

         101       110       120       130       140       150
BASI      SELAAGRRHVITGPVKDPSPSGRENAFRIEKYHGAEVSEYKLMSC-----    144
SBTI      ED*PE*PAVK*-*EN**AM-D*WFRLE*VSDDEFNN---***VF*PQQAE    141
WBTI      EDQPQQPSVKL-SEL*STKFDYLFKFEKVTS-KF---*S***KY*AKR--    136

         151       160       170       180       190
BASI      GDWQQDLGVFRDLKGGAWFLGATEPYHVVV-FKKAPPA                181
SBTI      D*K*G*I*ISI*HDD*TRR*VVSKNKPL**Q*Q*LDKESL              181
WBTI      -*T*K*I*IY**Q**Y*R-*VV*DENPL**I***--YESS              171

(c)
                  10        20        30        40        50
RAI                                                       AI      2
BAPI                                  MARAQVLLMAAALVLMLTAAPRAAV*L     25
MATI      AVFTVVNQCPFTVWAASVPVGGGRRLNRGESWRITAPAGTTAA-RI-W*R     48

          51        60        70        80        90       100
RAI       S-GCQVS-SAIGPCLAYARGAGAAPSASCQSGVRSLNAAARTTADRRAAC     50
BAPI      N-****D-*KMK***T*VQ*-*PG**GE*CN***D*HNQ*QSSG**QTV*     72
MATI      TG*-*FDA*GR*S*RTGDC*G----VVQ*-T*YGRAPNTLAEY*LKQFN-     91

         101       110       120       130       140       150
RAI       NCSLKSAASRVSGLNAGKASSIPGRCGVRLPYAISAS--IDCSRVNN       95
BAPI      *-**GI*RGIH***LNN*A***SK*N*NV**T**PD--*****IY         117
MATI      ------------N*DFFD-I**LDG--FNV**SFLPDGGSG***GPRCAV    126

         151       160       170       180       190       200
MATI      DVNARCPAELRQDGVCNNACPVFKKDEYCCVGSAANNCHPTNYSRYFKGQ    176

         201       210       220       230
MATI      CPDAYSYPKDDATSTFTCPAGTNYKVVFCP                        206
```

Figure 11. Amino acid sequences of several plant double-headed inhibitors and comparison with other plant single-headed inhibitors. a) RATI, ragi α-amylase/trypsin inhibitor; BTI, barley trypsin inhibitor; and WAI-0.28, wheat α-amylase inhibitor. Adapted from Ref. *103*. b) BASI, barley α-amylase/subtilisin inhibitor; SBTI, Kunitz soybean trypsin inhibitor; and WBTI, winged bean trypsin inhibitor. Adapted from Ref. *104*. c) RAI, ragi α-amylase inhibitor; BAPI, barley α-amylase/trypsin inhibitor; and MATI, maize α-amylase/trypsin inhibitor. See Figures 1 and 6 for explanation of-, * and numbers at right side of amino acid sequence. AR A located at amino acid residues 27, 28 and 30 (at top of figure) are for a second isoinhibitor of RATI. Adapted from Ref. *105*.

$$E + I \overset{1}{\rightleftharpoons} E{\cdot}I \overset{2}{\rightleftharpoons} E\text{-}I^* \qquad (2)$$

proteins. Mirkov *et al.* (*96*) proposed that the inhibitor might bind at the active site of α-amylase, in part by Arg82 electrostatically interacting with the two catalytic Asp residues of α-amylase (*101*) while the flanking Trp225 and Tyr227 may bind to neighboring subsites (occupied by starch when it binds). This suggestion is contrary to data showing that the complex retains ~5% α-amylase activity on small substrates (*97, 102*), and binds to starch and to maltose (a competitive inhibitor; *98, 100*). We have suggested that binding of inhibitor is at another site on α-amylase, the "glycogen binding site" and that the large conformational change (E_a~39 kcal/mol) that occurs decreases, but does not eliminate, the catalytic site (*97, 98, 100*). Binding is not via the carbohydrate groups of the inhibitor (*98*), as shown by mutation of the five potential carbohydrate sites (*96*), with retention of full activity.

Multidomain-Multisite Inhibitors

We already discussed above the two binding sites of the Bowman-Birk type inhibitors (Figure 1). Generally, one site binds trypsin tightly (K_i ~10^{-10} to 10^{-11} M) and the other site binds chymotrypsin (or elastase or another protease in some cases) less tightly (K_i ~10^{-9} M). In the absence of chymotrypsin, trypsin can also bind (K_i ~10^{-8} to 10^{-9} M) into the chymotrypsin site (*62*). These are called double-headed inhibitors. Some ovomucoids can bind two or three molecules of enzyme simultaneously and independently; the avian egg white ovoinhibitor can bind up to six enzyme molecules simultaneously. This is because these inhibitors have several independent binding domains that appear to have evolved by gene duplication (*41*).

There are several cereal and corn proteins that bind with a protease and an α-amylase. Examples of these inhibitors are shown in Figure 11. Note the marked similarities among amino acid sequences of the double-headed inhibitors and the single-headed protease inhibitors. Much work needs to be done on these inhibitors to determine if the two binding sites are independent or overlap; the secondary and tertiary structures and the active sites need to be elucidated also.

Recently, the primary, secondary and tertiary structures of a double-headed protease/α-amylase inhibitor from ragi were reported ((ragi bifunctional inhibitor (RBI and RATI are synonymous); Figure 12)) (*106*). The inhibitor contains 122 amino acids (MW ~13,800) and five disulfide bonds arranged in the same manner as in wheat α-amylase inhibitors WAI-0.28 and WAI-0.53. It contains a globular four-helix motif. Val67-Ser69 and Gln73- Glu75 form an antiparallel β-sheet. There are 33% α-helix and about 7% β-sheet in the protein. The trypsin binding loop consists of Gly32, Pro33, Arg34, Leu35, Ala36 and Thr37. The primary specificity site (P1) to trypsin is Arg34. A similar sequence of amino acid residues is conserved in other protease inhibitors. The binding site for α-amylase to RBI was not obvious to Strobl *et al.* (*106*). The suggestion of Alagiri and Singh (*107*) that Trp22 and Tyr23 residues are involved in the binding site appears to be ruled out, since these residues are mainly involved in stabilization of the trypsin binding loop (*106*)).

The primary amino acid sequence of RATI (synonym RBI) is shown in Figure 11a, and compared with that of the barley trypsin inhibitor (BTI) and the wheat α-amylase inhibitor (WAI-0.18). Leaving out the amino-acid sequence 32-37 for the trypsin binding site, there are conserved arginine residues at positions 21 and 56 (using RATI sequence numbering; Figure 12), tyrosine residues at positions 23 and 54 and a tryptophan residue at position 22. Residues Arg-Trp-Tyr21-23 are located in one of the α-helices, some distance away from the trypsin binding site and adjacent to a disulfide bond (which would provide a rigid binding site). The Arg-Trp-Tyr21-23 sequence appears to be the most likely specificity site for α-amylase

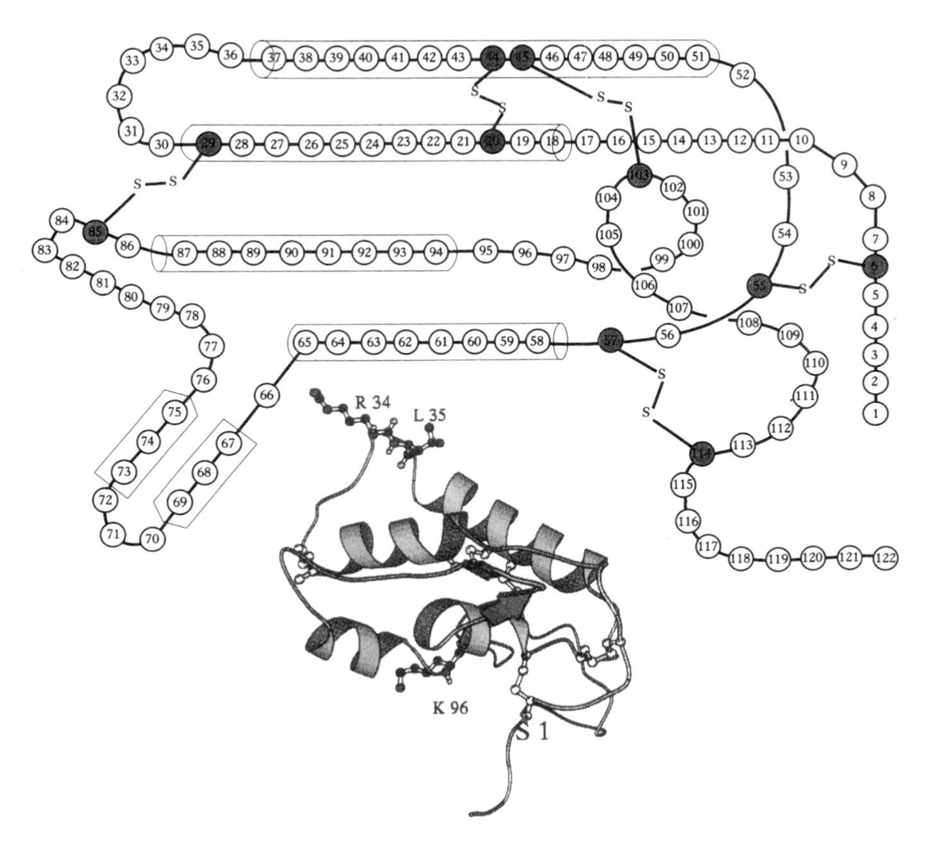

Figure 12. Primary, secondary and tertiary structures of ragi α-amylase/trypsin inhibitor (RATI). The numbered circles, from 1 to 122, represent the primary sequence; the cylinders represent α-helices of the secondary structure and the ten ● indicate 1/2 Cys residues that form the five disulfide bonds as shown. The arrow indicates the site for recognition and binding of trypsin. The small drawing at bottom (center) represents the tertiary structure of RATI, with the α-helices and random coils. Reproduced with permission from Ref. *106.* Copyright 1995 American Chemical Society.

binding. Tyr54 is in a random coil loop that appears to be too far from the Arg21 and Trp22 residues to be a part of the active site. The final answer of location of the α-amylase binding site requires additional research.

Literature Cited

1. Chrzaszcz, T.; Janicki, J. *Biochem. Z.* **1933,** *260,* 354-368.
2. Chrzaszcz, T.; Janicki, J. *Biochem. J.* **1934,** *28,* 296-304.
3. Kneen, E.; Sandstedt, R. M. *J. Am. Chem. Soc.* **1943,** *65,* 1247-1252.
4. Kneen, E.; Sandstedt, R. M. *Arch. Biochem. Biophys.* **1946,** *9,* 235-249.
5. Saunders, R. M.; Lang, J. A. *Phytochemistry.* **1973,** *12,* 1237-1241.
6. Silano, V.; Pocchiari, F.; Kasarda, D. D. *Biochim. Biophys. Acta.* **1973,** *317,* 139-148.
7. Bowman, D. E. *Science.* **1945,** *102,* 358-359.
8. Hernandez, A.; Jaffe', W. G. *Acta Cient. Venez.* **1968,** *19,* 183-185.
9. Marshall, J. J.; Lauda, C. M. *J. Biol. Chem.* **1975,** *250,* 8030-8037.
10. Blanco-Labra, A.; Iturbe-Chinas, F. A. *J. Food Biochem.* **1981,** *5,* 1-17.
11. Mundy, J.; Rogers, J. C. *Planta.* **1986,** *169,* 51-63.
12. Kutty, A. V. M.; Pattabiraman, T. N. *Biochem. Arch.* **1986,** *2,* 203-208.
13. Shivaraj, B.; Pattabiraman, T. N. *Indian J. Biochem. Biophys.* **1980,** *17,* 181-185.
14. Granum, P. E. *J. Food Biochem.* **1978,** *2,* 103-120.
15. Chandrasekher, G.; Pattabiraman, T. N. *Indian J. Biochem. Biophys.* **1985,** *20,* 241-245.
16. Irshad, M.; Sharma, C. B. *Biochim. Biophys. Acta.* **1981,** *659,* 326-333.
17. Stankovic, S. C.; Markovic, N. D. *Glasnik Hem. Drustva Beograd.* **1960-1961,** *25-26,* 519-525.
18. Stankovic, S. C.; Martovic, N. D. *Chem. Abstr.* **1963,** *59,* 3084d.
19. Narayana Rao, M.; Shurpalekar, K. S.; Sundaravalli, O. E. *Indian J. Biochem.* **1967,** *4,* 185.
20. Narayana Rao, M.; Shurpalekar, K. S.; Sundaravalli, O. E. *Indian J. Biochem.* **1970,** *7,* 241-243.
21. Mattoo, A. K.; Modi, V. V. *Enzymologia* **1970,** *39,* 237-247.
22. Kunitz, M. *Science.* **1945,** *101,* 668-669.
23. Richardson, M. *Methods Plant Biochem.* **1991,** *5,* 259-305.
24. Green, T. R.; Ryan, C. A. *Science.* **1972,** *175,* 776-777.
25. Ryan, C. A. *Trends Biochem. Sci.* **1978,** *5,* 148-150.
26. Brown, W. E.; Graham, J. S.; Lee, J. S.; Ryan, C. A. In: *Nutritional and Toxicological Significance of Enzyme Inhibitors in Foods;* Friedman, M., Ed.; Plenum Press: New York, N. Y., 1986; pp. 281-290.
27. Tajiri, T.; Koba, Y.; Ueda, S. *Agr. Biol. Chem.* **1983,** *47,* 671-679.
28. Savaiano, D. A.; Powers, J. R.; Costello, M. J.; Whitaker, J. R.; Clifford, A. J. *Nutr. Repts. Intl.* **1977,** *15,* 443-449.
29. Bo-Linn, G. W.; Santa Ana, C. A.; Morawski, S. G.; Fordtran, J. S. *New Engl. J. Med.* **1982,** *307,* 1413-1416.
30. Carlson, G. L.; Li, B. U. K.; Bass, P.; Olsen, W. A. *Science.* **1983**, *219,* 393-395.
31. Granum, P. E.; Holm, H., Wilcox, E.; Whitaker, J. R. *Nutr. Repts. Intl.* **1983,** *28,* 1233-1244.
32. Lajolo, F. M.; Nancini-Filho, J.; Menezes, E. W. *Nutr. Repts. Intl.* **1984,** *30,* 45-54.
33. Liener, I. E.; Donatucci, D. A.; Tarcza, J. C. *Am. J. Clin. Nutr.* ***1984,*** *39,* 196-200.

34. Layer, P.; Carlson, G. L.; Dimagno, P. *Gastroenterol.* **1985,** *88,* 1895-1902.
35. Buonocore, V.; de Biasi, M.-G.; Giardina, P.; Poerio, E.; Silano, V. *Biochim. Biophys. Acta.* **1985,** *831,* 40-48.
36. Courtois, P.; Franckson, J.R.M. *J. Clin. Chem. Clin. Biochem.* **1985,** *23,* 733-737.
37. Harmoinen, A.; Jokela, H.; Koivula, T.; Poppe, W. *J. Clin. Chem. Clin. Biochem.* **1986,** *24,* 903-905.
38. Ryan, C. A. In *The Biochemistry of Plants;* Marcus, A., Ed.; Academic Press: New York, N. Y., 1981, Vol. 6; pp. 351-370.
39. Richardson, M. *Food Chem.* **1981,** *6,* 235-253.
40. Richardson, M. *J. Biol. Educ.* **1981,** *15,* 178-182.
41. Laskowski, M. Jr.; Kato, I. *Ann. Rev. Biochem.* **1980,** *49,* 593-626.
42. Liener, I. E.; Kakade, M. L. In: *Toxic Constituents of Plant Foodstuffs;* Academic Press: London, 1980, 2nd edn.; pp. 7-71.
43. Ryan, C. A.; Walker-Simmons, M. In: *Biochemistry of Plants;* Marcus, E., Ed.; Academic Press: New York, N. Y., 1981, Vol. 6; pp. 321-347.
44. Whitaker, J. R. In: *Impact of Toxicology on Food Processing;* Ayers, J. C. and Kirshman, J. C., Eds; Avi Publishing Co., Inc.: Westport, CT., 1981; pp. 57-104.
45. Whitaker, J. R. In *Xenobiotics in Foods and Feeds;* Finley, J. W. and Schwass, D. E., Eds; ACS Symposium Series 234; American Chemical Society: Washington, D. C., 1983; pp. 15-46.
46. Warchalewski, J. R. *Nahrung* **1983,** *27,* 103-117.
47. Gatehouse, A.M.R. In *Developments in Food Proteins-3;* Hudson, P.J.F., Ed.; Elsevier: Amsterdam, 1984; pp. 245-294.
48. Xavier-Filho, J.; Campos, F.A.P. *Arg. Biol. Tecnol.* **1984,** *27,* 407-418.
49. Xavier-Filho, J.; Campos, F.A.P. In *Toxicants of Plant Origin;* Cheeke, P. R., Ed.; *Proteins and Amino Acids;* CRC Press Inc.: Boca Raton, FL., 1989, Vol. 3; pp. 1-27.
50. Buonocore, V.; Silano, V. In: *Nutritional and Toxicological Significance of Enzyme Inhibitors in Foods;* Friedman, M., Ed.; Plenum Press: New York, N. Y., 1986; pp. 483-507.
51. Rackis, J. J.; Wolf, W. J.; Baker, E. C. In: *Nutritional and Toxicological Significance of Enzyme Inhibitors in Foods;* Friedman, M., Ed.; Plenum Press: New York, N. Y., 1986; pp. 299-347.
52. Silano, V. In: *Enzymes and Their Role in Cereal Technlology;* Kruger, J. A. *et al.,* Eds.; Am. Assoc. Cereal Chem., Minneapolis, MN, 1987; pp. 141-199.
53. Weder, J. K. D. In: *Nutritional and Toxicological Significance of Enzyme Inhibitors in Plant Foods;* Friedman, M., Ed.; Plenum Press: New York, N. Y., 1986; pp. 239-279.
54. Garcia-Olmedo, F.; Salcedo, G.; Sanchez-Monge, R.; Gomez, L; Royo, T.; Carbonero, P. *Oxford Surveys Plant Mol. Cell. Biol.* **1987,** *4,* 275-334.
55. Xavier-Filho, J.; Ventura, M. M. *Comments Agr. Food Chem.* **1988,** *1,* 239-264.
56. Ho, M. F.; Yin, X.; Filho, F. F.; Lajolo, F.; Whitaker, J. R. In: *Protein Structure-Function Relationships in Foods;* Yada, R. Y., Jackman, R. L. and Smith, J. L., Eds.; Blackie Academic & Professional: London, 1994; pp. 89-119.
57. *Proteinase Action: Proc. Intl. Workshop,* August 29-31, 1983; Elodi, P., Ed.; Akademiac Kiado': Budapest, 1984.
58. *Proteinase Inhibitors;* Barrett, A. J. and Salvesen, G., Eds.; *Research Monograph in Cell and Tissue Physiology*, Vol. 12, Amsterdam, 1986.

59. *Proteases in Biological Control and Biotechnology;* Cunningham, D. D. and Long, G. L., Eds.; Alan R. Liss, Inc.: New York, N. Y., 1987.
60. *Proteinase Inhibitors: Medical and Biological Aspects;* Katunuma, N., Umezawa, H. and Holzer, H., Eds.; Japan Scientific Societies Press: Japan; Springer-Verlag: Berlin, 1983.
61. Whitaker, J. R.; Sgarbieri, V. C. *J. Food Biochem.* **1981,** *5,* 197-213.
62. Wu, C.; Whitaker, J. R. *J. Agr. Food Chem.* **1990,** *38,* 1523-1529.
63. Abe, K.; Kondo, H.; Arai, S. *Agr. Biol. Chem.* **1987,** *51,* 2763-2768.
64. Abe, M.; Whitaker, J. R. *Agr. Biol. Chem.* **1988,** *51,* 1583-1584.
65. Xavier-Filho, J.; Campos, F. A. P.; Ary, M. A.; Peres Silva, C.; Carvalho, M. M. M.; Macedo, M. L. R. *et al. J. Agr. Food Chem.* **1989,** *37,* 1139-1143.
66. Zimacheva, A. V.; Ievleva, E. V.; Mosolov, V. V. *Biochemistry* (English translation), **1989,** *53,* 640-645.
67. Heinrikson, R. L.; Kezdy, F. S. *Methods Enzymol.* **1976,** *45,* 740-751.
68. Kato, I.; Schrode, J.; Kohr, W. J.; Laskowski, M., Jr. *Biochemistry.* **1987,** *26,* 193-201.
69. Laskowski, M., Jr.; Kato, I.; Ardelt, W.; Cook, J.; Denton, A.; Empie, M. W.; Kohr, W. I.; Park, S. J.; Parks, K.; Schatzley, B. L.; Schoenberger, O. L.; Tashiro, M.; Vichot, G.; Whatley, H. E.; Wieczorek, A.; Wieczorek, M. *Biochemistry.* **1987,** *26,* 202-221.
70. Laskowski, M., Jr.; Kato, I.; Kohr, W. J.; Park, S. J.; Tashiro, M.; Whatley, H. E. *Cold Spring Harbor Symposia on Quantitative Biology.* **1987,** *52,* 545-553.
71. Bode, W.; Wei, A.-Z.; Huber, R.; Meyer, E.; Travis, J.; Neumann, S. *EMBO J.* **1986,** *5,* 2453-2458.
72. Fujinaga, M.; Sielecki, A. R.; Read, R. J.; Ardelt, W.; Laskowski, M., Jr.; James, M.N.G. *J. Mol. Biol.* **1987,** *195,* 397-418.
73. Read, R. J.; Fujinaga, M.; Sielecki, A. R.; James, M.N.G. *Biochemistry.* **1983,** *22,* 4420-4433.
74. Deisenhofer, M.; Steigemann, W. *Acta Cryst. B.* **1975,** *31,* 238-250.
75. Sweet, R. M.; Wright, H. T.; Janin, J.; Chothia, C. H.; Blow, D. M. *Biochemistry.* **1974,** *13,* 4212-4228.
76. Satow, Y.; Mitsui, Y.; Iitaka, Y. *J. Biochem. Tokyo.* **1978,** *84,* 897-906.
77. Mitsui, Y.; Satow, Y.; Watanabe, Y.; Hirone, S.; Iitaka, Y. *Nature.* **1979,** *277,* 447-452.
78. Turk, V.; Bode, W. In: *Biological Functions of Protease and Inhibitors;* Katunuma, N., Suzuki, K., Travis, J. and Fritz, H., Eds.; Japan Scientific Societies Press: Tokyo; Kargen: Basel, 1994; pp. 47-59.
79. Tajiri, T.; Koba, Y.; Veda, A. *Agr. Biol. Chem.* **1983,** *47,* 671-679.

79a. Lee, S.-C. *Comparative Amino Acid Sequences, and Deduced Tertiary and Quaternary Structures of White Kidney Bean and Black Bean α-Amylase Inhibitors;* Ph. D. Dissertation; University of California, Davis, 1996.

80. Hirayama, K.; Takahashi, R.; Akashi, S.; Fukuhara, K.; Oouchi, N.; Murai, A.; Arai, M.; Murao, S.; Tanaka, K.; Nojima, I. *Biochemistry .* **1987,** *26,* 6483-6488.
81. Vertesy, L.; Tripier, D. *FEBS Lett.* **1985,** *185,* 187-190.
82. Arai, M.; Oouchi, N.; Murao, S. *Agr. Biol. Chem.* **1985,** *49,* 987-991.
83. Arai, M.; Oouchi, N.; Goto, A.; Ogura, S.; Murao, A. *Agr. Biol. Chem.* **1985,** *49,* 1523-1524.
84. Maeda, K.; Kakabayashi, S.; Matsubara, H. *Biochim. Biophys. Acta.* **1985,** *828,* 213-221.
85. Mirkov, T. E.; Wahlstrom, J. M.; Hagiwara, K.; Finardi-Filho, F., Kjemtrup, S.; Chrispeels, M. J. *Plant Mol. Biol.* **1994,** *26,* 103-1113.

86. Suzuki, K.; Ishimoto, M.; Kitamura, K. *Biochim. Biophys. Acta.* **1994,** *1206,* 289-291.
86a. Pueyo, J. J.; Hunt, D. C.; Chrispeels, M. J. *Plant Physiol.* **1993,** *101,* 1341-1348.
87. Moreno, J.; Chrispeels, M. J. *Proc. Natl. Acad. Sci. USA* . **1989,** *86,* 1-5.
88. Hoffman, L. M.; Ma, Y.; Barker, R. F. *Nucl. Acids Res.* **1982,** *10,* 7819-7828.
89. Osborn, T. C.; Alexander, D. C.; Sun, S.S.M.; Cardona, C.; Bliss, F. A. *Science* . **1988,** *240,* 207-210.
90. Hartweek, L. M.; Vogelzang, R. D.; Osborn, T. C. *Plant Physiol.* **1991,** *97,* 204-211.
91. Hoffman, L. M.; Donaldson, D. D. *EMBO J.* **1985,** *4,* 883-889.
92. Higgins, T.J.V.; Chandler, P. M.; Zurawski, G.; Button, S. C.; Spencer, D. *J. Biol. Chem.* **1983,** *258,* 9544-9549.
93. Yin, X. *α-Amylase Inhibitors of Black Kidney Bean (Phaseolus vulgaris): Subunit Composition and Other Molecular Properties;* Ph.D. Dissertation; University of California, Davis, 1990.
94. Ho, M. F.; Whitaker, J. R. *J. Food Biochem.* **1993,** *17,* 35-52.
95. Einspahr, H.; Parks, E. H.; Suguna, K.; Subramanian, E.; Suddath, F. L. *J. Biol. Chem.* **1986,** *261,* 16518-16527.
96. Mirkov, T. E.; Evans, S. V.; Wahlstrom, J.; Gomez, L.; Young, N. M.; Chrispeels, M. J. *Glycobiology.* **1995,** *5,* 45-50.
97. Wilcox, E. R.; Whitaker, J. R. *Biochemistry* . **1984,** *23,* 1783-1791.
98. Wilcox, E. R.; Whitaker, J. R. *J. Food Biochem.* **1984,** *8,* 189-213.
99. Ho, M. F.; Whitaker, J. R. *J. Food Biochem.* **1993,** *17,* 15-33.
100. Powers, J. R.; Whitaker, J. R. *J. Food Biochem.* **1977,** *1,* 239-260.
101. Qian, M.; Haser, R.; Payan, F. *J. Mol. Biol.* **1993,** *231,* 785-799.
102. Whitaker, J. R.; Finardi-Filho, F.; Lajolo, F. M. *Biochimie.* **1988,** *70,* 1153-1161.
103. Odani, S.; Koide, T.; Ono, T. *J. Biol. Chem.* **1983,** *258,* 7998-8003.
104. Svendsen, I.; Hejgaard, J.; Mundy, J. *Carlsberg Res. Commun.* **1986,** *51,* 43-50.
105. Richardson, M.; Valdes-Rodriquez, S.; Blanco-Labra, A. *Nature.* **1987,** *327,* 432-434.
106. Strobl, S.; Muhlhahn, P.; Bernstein, R.; Wiltscheck, R.; Maskos, K.; Wunderlich, M.; Huber, R.; Glockshuber, R.; Holak, T. A. *Biochemistry.* **1995,** *34,* 8281-8293.
107. Alagiri, S.; Singh, T. P. *Biochim. Biophys. Acta.* **1993,** *1203,* 77-84.

Chapter 3

Plant Lectins: Properties, Nutritional Significance, and Function

Irvin E. Liener

Department of Biochemistry, University of Minnesota, St. Paul, MN 55108

Lectins constitute a ubiquitous class of proteins which are widely distributed in Nature, including plants which are commonly consumed in the diet of man and animals. The most characteristic property of the lectins is their ability to combine reversibly with sugars and glycoconjugates in a highly specific fashion. By virtue of their ability to bind glycoproteins on the surface of microvilli lining the small intestine, lectins lead to an interference with the absorption of nutrients. Other toxic effects may be attributed to their entrance into the circulatory system. In plants they are believed to be involved in the symbiotic relationship between legumes and N-fixing bacteria, and as part of their defense mechanism against predators. Possible beneficial applications of the lectins will be pointed out.

That a toxic substance must be present in the seeds of certain plants was recognized as early as 1888 when Stillmark observed that the extreme toxicity of the castor bean could be attributed to a protein fraction which was capable of agglutinating red blood cells (1). He coined the name *ricin* for this particular substance since it was derived from *Ricinus communis*. It is now known that similar substances, often referred to as hemagglutinins, are very widely distributed in Nature even in such edible legumes as kidney beans, soybeans, lentils, peas, and many other plants which are commonly consumed as foods in the human diet (see Table I). Of primary concern to those who are interested in food safety is the question as to whether these substances are present in significant amounts so as to pose a risk to human health.

Biochemical Properties

Before addressing this question, however, a few basic facts regarding the biochemical properties of this unique class of proteins will be presented. In addition to their ability to agglutinate red blood cells with a specificity towards different blood groups, their most characteristic feature is their ability to bind with specific sugars or glycoconjugates. It was this high degree of specificity that led Boyd and Shapleigh (3) to suggest the term "lectin" (taken from the Latin word, *legere* = to choose) for this class of compounds. Examples of the properties of some of these lectins (molecular weights, subunit structure, and sugar specificity) are likewise presented in Table I.

Table I. Properties and Sugar Specificity of Some Lectins[a]

Botanical name	Common name	Molecular weight	Number of subunits	Sugar specificity[b]
Arachis hypogeae	Peanut	110,000	4	GalNAc
Canavalia ensiformis	Jack bean	105,000	4	Man
Dolichos biflorus	Horse gram	110,000	4	GalNAc
Glycine max	Soybean	120,000	4	GalNAc
Lathyrus odoratus	Sweet pea	52,000	4	Man
Lathyrus sativus	Chickling vetch	49,000	4	Man
Lens esculenta	Lentil	46,000	4	Man
Phaseolus lunatus	Lima bean	60,000	2	GalNAc
Phaseolus vulgaris	Kidney bean	126,000	4	GalNAc
Pisum sativum	Pea	49,000	4	Man
Psophocarpus tetragonolobus	Winged bean	58,000	2	GalNAc
Vicia faba	Fava bean	52,500	4	Man

[a]Based on information taken from Goldstein and Poretz (2).
[b]GalNAc, N-acetylgalactosamine; Man, mannose.

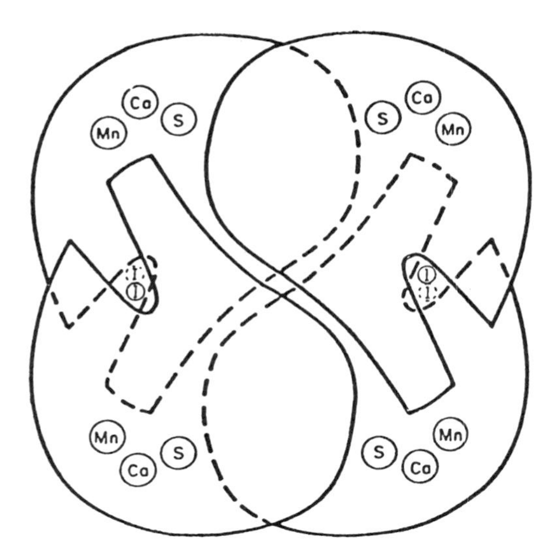

Fig. 1. Schematic representation of the tetrameric structure of concanavalin A. Manganese and calcium sites are indicated by Mn and Ca, respectively. The saccharide binding site is indicated by S and the hydrophobic site in the cavity by I. (Reproduced with permission from ref. 5. Copyright 1976 Macmillan Magazines Limited.).

The most important structural feature of these lectins is the fact that they contain 2 or 4 subunits, each of which has a sugar binding site. It is this feature of multivalency which accounts for the ability of lectins to agglutinate red blood cells by non-covalent binding to carbohydrate moieties located on the surface of the cell membrane or to precipitate glycoproteins, a property which is lost if the molecule is dissociated into subunits. This interaction between lectins and sugar-containing receptors can also be reversed or inhibited by the introduction of competing sugars. Although lectins are similar to antibodies in their ability to bind specific antigens, they differ in that they are not products of the immune system, their structures are diverse, and their specificity is restricted to carbohydrates.

The structural features of lectins are best illustrated by the lectins from the jack bean, also known as concanavalin A (con A), and the family of lectins from the kidney bean. The kidney bean lectins are sometimes referred to as PHA or phytohemagglutinins, a term which is really a misnomer and should be reserved as generic term for plant lectins.

Concanavalin A. Con A was first isolated and crystallized from the jack bean by Sumner and Howell in 1936 (4) and has been the most intensely investigated plant lectin. The main structural features of con A are shown in Figure 1. It is a tetramer comprised of four identical subunits, each of which has a molecular weight of 26,000. Each subunit has binding sites for Ca and Mn and a site directed primarily to mannose and to a lesser extent to glucose or to glycoproteins containing these sugars at their non-reducing ends.

Kidney Bean Lectin. Unlike con A which is comprised of four identical isomers, a somewhat more complicated situation arises when these subunits are not identical as illustrated by the family of lectins present in the kidney bean. As shown in Figure 2 there exists the possibility of five different types of lectins, the so-called isolectins, each of which is comprised of four subunits designated as E or L. These two different subunits, although having similar molecular weights, about 30,000, confer either erythroagglutinating activity (E subunit) or leukoagglutinating and mitogenic activity (L subunit) to the parent tetramer. Thus the isolectins referred to as L4 and E4 would have exclusively leukoagglutinating or hemagglutinating activity, respectively, whereas the other three isolectins (E_3L_1, E_2L_2, and E_1L_3) would display both activities depending on the relative proportion of these two subunits.

Many possible variations in this theme exists among the lectins which have been isolated from a large number of legumes depending on the number and size of the individual subunits and the type of linkage which connects them.

Physiological Effects in Animals

Soybean Lectin. The first indication that lectins might play a significant role as an antinutrient came from experiments with soybeans. The well known effect of heat in producing an improvement in the nutritive value of soybeans had been generally attributed in the main to the trypsin inhibitors (7). However, the addition of trypsin inhibitor to heated soybeans at a level equivalent to that of the raw bean failed to reduce its nutritive value to that of the raw bean (8). This indicated that some other factor could also be contributing to the poor nutritive value of the raw bean. In an attempt to identify the factor that might be responsible for this additional negative effect, it was noted that there was a positive relationship between the amount of heat treatment necessary to destroy hemagglutinating activity and the growth response of chicks (9). Since this association did not necessarily prove a cause and effect relationship, it was important to isolate this hemagglutinin in sufficient quantity to permit its incorporation

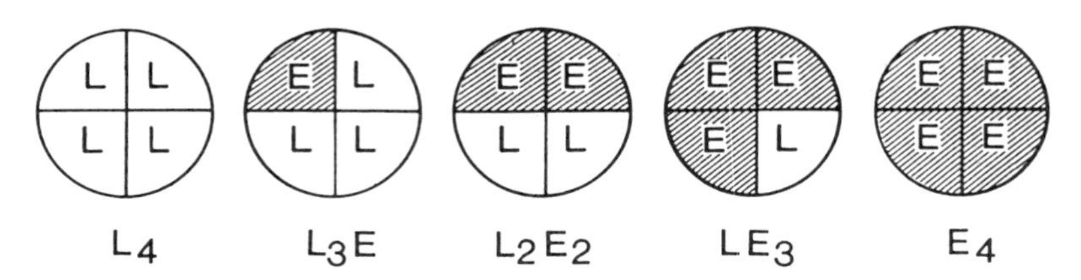

Fig. 2. Schematic representation of the tetrameric structure of the five isolectins in *Phseolus vulgaris*. See text for the properties of the isolectins consisting of the subunits denoted by L and E. (Adapted from ref. 6. Copyright National Academy of Science.).

Table II. Nutritional Value of "Lectin-free" Soybean Cultivar (T102) Compared with Commercial Variety of Soybeans (Amsoy) as Measured in Rats[a]

Soybean	Weight gain (g/21 days)	PER[b]	Lectin activity (HU/mg protein)[c]
Raw Amsoy	9.1	0.64	14×10^3
Raw T102	18.2	1.21	120
Heated Amsoy	68.9	2.58	0
Heated T102	73.4	2.73	0

[a]Data taken from Donatucci, PhD thesis, U. Minnesota, 1983.
[b]Protein Efficiency Ratio.
[c]HU = hemagglutinating units.

into diets fed to animals. When the purified lectin was added to the diet of rats receiving heated soybeans at a level equivalent to that present in raw soybeans, the growth response was reduced to about one-half that of the diet with heated soybeans (10). The availability of a soybean strain which lacked the gene for the lectin (courtesy of T. Hymowitz, University of Illinois) permitted a means for validating this conclusion. As shown in Table II the result of feeding this particular strain of soybeans, which had less than 0.05% of the lectin activity of a commercial variety of soybeans (Amsoy), produced an improvement in nutritive value, as measured by the Protein Efficiency Ratio in rats, that was about half of that produced by heat alone. This then confirmed the previous conclusion based on experiments in which the soybean lectin had been added directly to diets fed to rats (10). It is important to point out that the trypsin inhibitor content of both soybean varieties was essentially the same, thus ruling out any effect due to the trypsin inhibitors.

Kidney Bean Lectin. It had long been recognized that many varieties of the *Phaseolus vulgaris* including the kidney bean are poorly tolerated by animals unless they have been subjected to heat treatment. The presence of hemagglutinins in *P. vulgaris* had been reported as early as 1908 (11), but direct evidence that the hemagglutinin was the toxic agent depended on the subsequent isolation in 1962 of the lectins from black beans and kidney beans and its feeding to animals as part of the diet (12). The kidney bean lectin proved to be much more toxic than the soybean lectin as evidenced by the fact that, unlike the soybean lectin which simply retarded growth, the kidney bean lectin actually caused a loss in weight to the point where the animal eventually succumbed. At levels in excess of 1% of the diet not only was the growth of rats markedly inhibited but death followed in 7 to 10 days, similar to what is observed with the raw beans. More recent comprehensive studies by Pusztai and his group have greatly extended our knowledge concerning the mechanism whereby the kidney bean lectin exerts their toxic effect (see below).

Interaction with Intestines. Jaffe′ was the first to suggest that the toxicity of the lectins from *P. vulgaris* could be attributed to their ability to bind to specific receptor sites on the surface of the epithelial cells lining the intestinal tract (13). Subsequent studies by Pusztai and his group (14) have provided further evidence for the *in vivo* binding of various plant lectins to the intestinal mucosa and the physiological effects that ensue (16). Shown in Figure 3 is a immunofluorescence micrograph of part of a transverse section through the duodenum of a rat fed a diet containing kidney beans. Incubation with rabbit anti-lectin immunoglobulins and fluorescein isothiocyanate-conjugated anti-rabbit IgG shows immunofluorescence in the brush border region. Accompanying the binding of the lectin to the intestinal wall is the appearance of lesions and severe disruption and abnormal development of the microvilli as exemplified by Figure 4 which shows electron micrographs of sections through the apical regions from rats fed diets containing 5% raw kidney beans compared to a control with 5% casein. The precise nature of the receptor site on the microvilli remains uncertain, but is believed to be identical, or at least similar, to the binding site on erythrocytes for the E4 lectin (17). This conclusion is based on the fact that a glycopeptide isolated from the membrane of erythrocytes inhibited the binding of E4 to the microvillus membrane vesicle of the intestinal mucosa.

Effect of Absorption of Nutrients. One of the major consequences of the damage inflicted to the intestinal mucosa by lectins is severe impairment in the absorption of nutrients across the intestinal wall. By using a technique known as vascular intestinal perfusion, it is possible to demonstrate that lectins can interfere with the absorption of a nutrient such as glucose (18). In this experiment radioactive

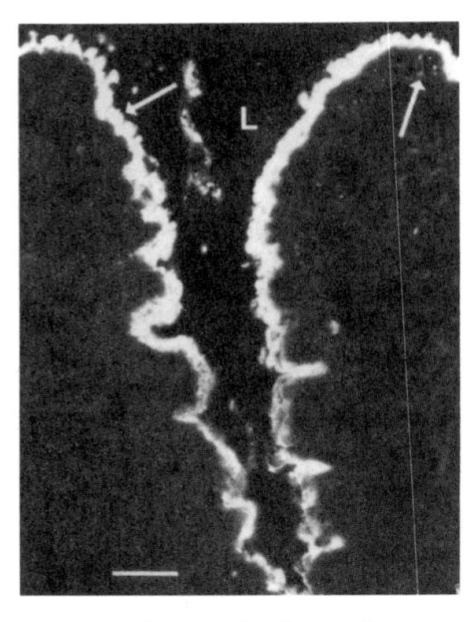

Fig. 3. Immunofluorescence micrograph of part of a transverse section of the duodenum of a rat fed a diet containing raw kidney beans. See text for further details. (Reproduced with permission from ref. 15. Copyright 1980 Chapman Hall Ltd.).

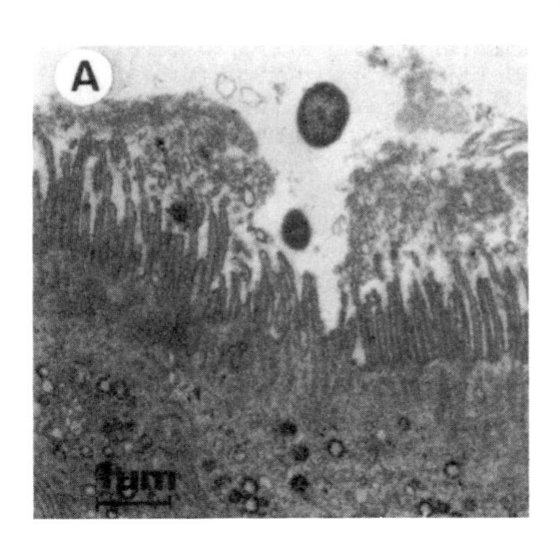

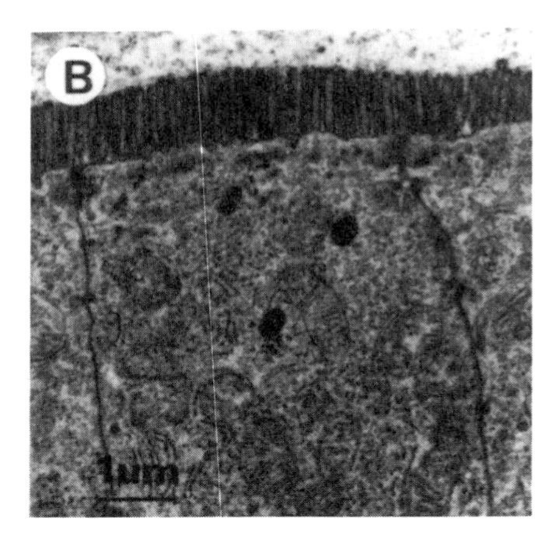

Fig. 4. Electron micrograph of sections of the small intestine of rats fed diets containing (A) 5% kidney beans and 5% casein with (B) 10% casein. (Reproduced with permission from ref. 16. Copyright 1979 Society of Chemical Industry.).

glucose was injected into the lumen of the isolated intestines taken from rats that had been fed a casein diet with and without the lectin isolated from the navy bean. The absorption of glucose was then monitored by the appearance of radioactive glucose in the perfusate. As shown in Figure 5 the rate of absorption of glucose across the intestinal wall of rats fed the lectin-containing diet was less than half that of controls fed the same diet but lacking the lectin.

The ability of lectins to interfere with the absorptive capacity of the intestine is not restricted to sugars, however. Other workers have shown that the inclusion of lectins into the diet of rats caused a malabsorption of amino acids, lipids, vitamin B_{12}, and an interference with ion transport (19-21). This non-specific interference with the absorption of nutrients is no doubt one of the reasons that the protein of raw legumes is so poorly utilized. One cannot, however, exclude the possibility that lectins can also inhibit the activity of certain intestinal enzymes which participate in the digestion of proteins. For example, it has been reported that lectins can inhibit enterokinase (22), the enzyme that activates trypsinogen, as well as the peptidases derived from the brush border membrane of the intestinal mucosa (23,24). In addition to N losses due to malabsorption and an inhibition of protein digestion, the kidney bean lectin has been shown to cause a rapid increase in the turnover and shedding of the epithelial cells lining the small intestines (25). This endogenous loss of nitrogen would further exacerbate the toxic effects of the lectin with respect to protein utilization.

Bacterial Colonization. Another factor contributing to the toxic effects of lectins is the observation that germ-free animals are better able to tolerate raw legumes in their diet than conventional animals (26-28). This observation is accompanied by the fact that an overgrowth or colonization of coliform bacteria has been noted in the small intestines of rats (29,30) and chicks (31) fed diets containing raw beans or the purified lectins derived therefrom. The explanation for this phenomenon is not entirely clear, but a most likely explanation is that the lectins, because of their polyvalent property, can bind to the receptor sites on the brush border as well as receptor sites on the surface of the bacteria. The lectin thus acts as a glue or sandwich between the intestines and the bacteria. The toxic effect produced by this bacterial overgrowth has been explained on the basis that the permeability of the small intestine has been altered to permit normally innocuous bacteria, or the endotoxins which they produce, to gain entrance into the bloodstream with frequently fatal consequences (27).

Systemic Effect. Lectins and their antibodies which they elicit can be detected immunochemically in the blood of rats and pigs fed raw kidney beans (32,33). This would indicate that the lectins, either intact or partially digested, may be absorbed and thus enter the circulatory system. In addition to a direct toxic effect on the internal organs, other systemic effects which have been observed include an increase in protein and lipid catabolism, a depletion of glycogen in muscle tissue, and an elevation in blood insulin levels (14).

Significance in the Human Diet

The fact that lectins are so widely distributed in food items commonly consumed by humans raises the important question as to whether they pose any significant risk to human health. In general, most lectins are inactivated by heat treatment such as that involved in commercial processing or household cooking. For example, Table III shows the effectiveness of processing on the lectin content of food items containing soybeans as an ingredient. It is doubtful whether such low levels of lectin activity pose a risk to human health. Of greater concern is the toxic effects which are associated with

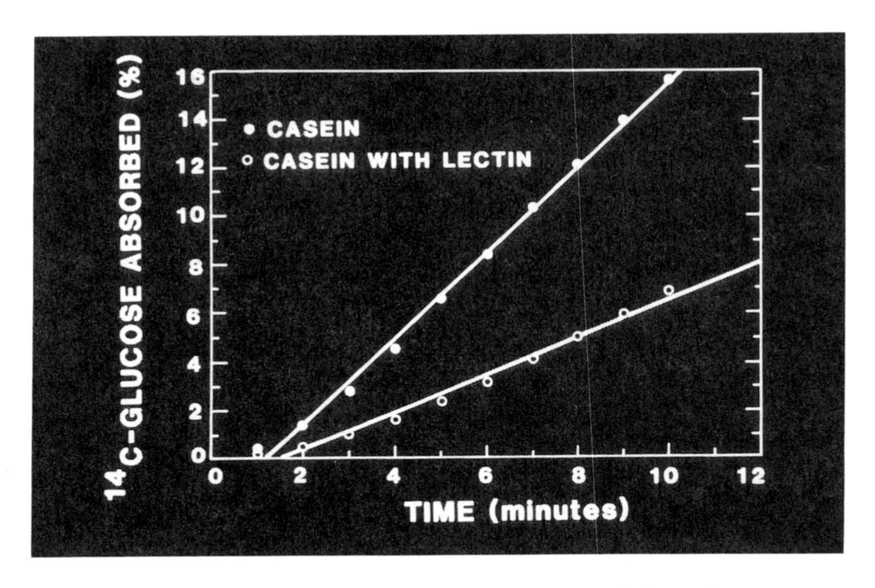

Fig. 5. Rate of absorption of glucose by intestines of rats fed navy bean (*Phaseolus vulgaris*) lectin compared to controls without lectin. (Reproduced with permission from ref. 18. Copyright 1987 American Institute of Nutrition.).

Table III. Lectin Content of Processed Foods Containing Soybeans as an Ingredient[a]

Product	% of Raw soy
Raw soybeans	100.0
Unprocessed soy flour	47.0
Defatted soy flour	4.3
Textured meat analog	0.4
Breakfast cereal	5.2
Soy milk	0.5
Cookie	0.7

[a]Adapted from ref. 34.

the inadvertent consumption of beans and dishes derived therefrom which have been improperly cooked or processed.

There are a number of reports in the literature of human intoxication in which lectins appear to have been the causative agent (35,36). In 1948 a severe outbreak of gastroenteritis occurred among the population of West Berlin due to the consumption of partially cooked beans that had been air-lifted into the city during the blockade by the Russians (37). In Tanzania mothers feed their infants a porridge consisting of a mixture of beans and corn. This gruel was found to contain fairly high levels of lectin and was thus considered potentially harmful to young infants (38). This problem presumably arises from the fact that primitive cooking is often done in earthenware pots on a wood fire, so that a thick viscous mass like beans and corn may impede heat transfer. Thus, in the absence of vigorous stirring, or perhaps in attempt to conserve firewood, significant amounts of the lectin may escape destruction. A reduction in the boiling point of water, such as would be encountered in certain mountainous regions of the world, might also contribute to the incomplete destruction of the lectin (39).

An outbreak of what appeared to be food poisoning occurred in England in 1976 (40). A party of school boys on holiday ate kidney beans that had been soaked in water but had not been cooked. All nine of the boys who ate the beans became acutely nauseated with 1 to 1 1/2 hours and began to vomit followed by diarrhea. As few as 4 to 5 beans were sufficient to produce these reactions. Two of the boys were admitted to the hospital and needed intravenous infusion. However, recovery was rapid in all cases. This incident prompted a television program by BBC in which this problem was brought to the attention of the public by the Ministry of Health. The public was requested to report any similar experiences they might have had. Rather surprisingly the response revealed that 870 individuals reported that they had become ill following the consumption of raw beans as part of a salad, or in dishes such as a stew, casserole, or chili concarne, all of these having been prepared in a slow-cooker ("Crockpot"). It turns out that the times and temperatures involved in the slow cooking of kidney beans under household conditions may not be sufficient to destroy all of the lectin activity (41,42), even though the beans cooked under these conditions were considered acceptable in terms of texture and palatability by trained test panels (43). This would support the documented toxicity of kidney beans cooked in slow cookers involving relatively low temperatures for long periods of time (40). Prompted by these reports warning labels may now be found on labels of dried kidney beans sold in the retail food markets in England (Figure 6). This label recommends that the beans be boiled for at least 10 min.

Physiological Role in the Plant

At least two hypotheses have been advanced to explain the role that lectins play in plants: (a) as mediators or recognition molecules involved in the symbiotic relationship between legumes and nitrogen-fixing bacteria, and (b) as a defense mechanism against insects and microbial pathogens [see review by Chrispeels and Raikhel (44)].

Interaction of Legumes with N-fixing Bacteria. The association between legumes and N-fixing bacteria, mainly those of genus *Rhizobium*, and legumes is highly specific, and the lectins are believed to play a crucial role in this symbiotic relationship [see review by Kijne *et al.* (45)]. That lectins are responsible for this specific interaction is based on the finding that the lectin from a particular legume (host) such as soybeans binds only to the rhizobial species which nodulates the soybean but not to bacteria which are symbiotic to other legumes. It is sometimes possible to overcome this specificity barrier as exemplified by the introduction of the pea lectin gene into the roots of white clover (46). The latter then becomes susceptible

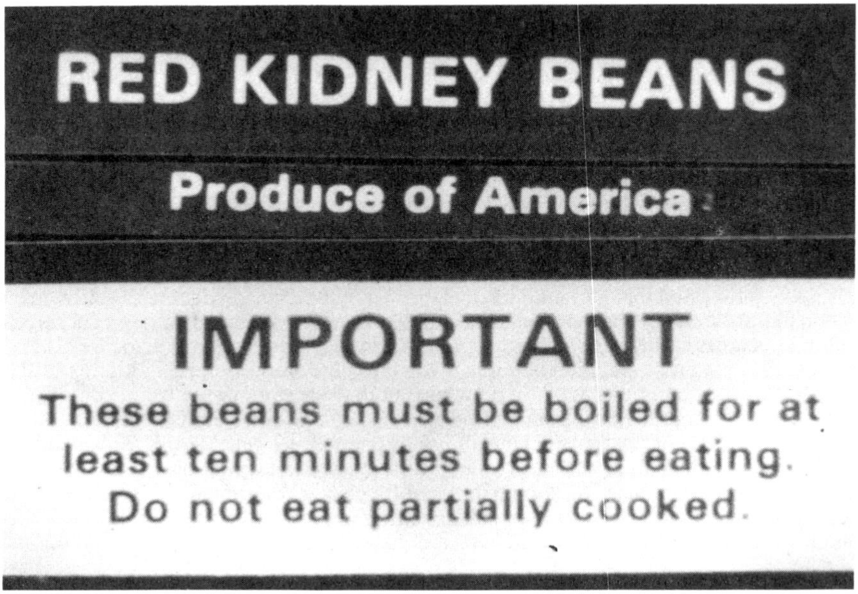

Fig. 6. Warning label that has been placed on packets of dry kidney beans sold in the retail market in England. (Reproduced with permission from ref. 2. Copyright 1986 Academic Press.).

to infection by a rhizobium which is specific for the pea but to which it is normally resistant. Increasingly more attention is being paid to the isolation and characterization of the root lectins which appear to be quite different from the seed lectins (47). It would appear that the binding of host lectins may send a signal to the bacteria that turns on a gene which leads to the synthesis of a specific polysaccharide on the bacterial cell wall that serves as a receptor site for the lectin. One of the aims of current research in this area is to identify the specific carbohydrate ligand on the bacterial surface that binds to the root lectins. One can perhaps contemplate that it might become possible in the future to introduce specific lectin genes known to bind to nodulating bacteria into non-leguminous plants, such as cereal grains, which are incapable of fixing N. The agricultural importance of such an achievement is obvious since it would reduce the need for artificial fertilizer with its attendant environmental problems.

Defense Mechanism Against Predators. Various lines of evidence suggest that lectins may be involved in the defense mechanism of plants against insects and pathogenic microorganisms (44). There is some question, however, as to whether the toxic effect against insects is actually due to the lectin or due at least in part to a contamination with an amylase inhibitor which is much more toxic to the bruchid beetle (48). Another protein named *arcelin*, similar in many respects to the phytohemagglutinin of the common bean (PHA) but not identical to it, has been isolated from a wild variant of *P. vulgaris*. The latter is resistant to the bean weevil which is a natural predator of the domesticated cultivars (49,50). Unlike the common varieties of *P. vulgaris* which may contain toxic levels of PHA, a line of beans which was especially rich in arcelin and low in PHA was found to be relatively non-toxic to rats (51). Transfer of the arcelin allele to bean cultivars that were non-resistant to this insect, by back crossing or by the addition of purified arcelin to artificial seeds, resulted in high levels of insect resistance. All three of these proteins (PHA, amylase inhibitor, and arcelin) display extensive homology with respect to amino acid and nucleotide sequences (52). It is believed that they evolved from a common ancestral gene, and the resulting proteins acquired somewhat different biological properties which are, nevertheless, related to a role which they play in the defensive mechanisms of plants.

The ability to be able to incorporate lectin genes into transgenic plants to enhance their resistance to a wide variety of predators has already been demonstrated in a number of cases (53). If such transgenic plants containing significant levels of lectin were to be used as a source of food, it is highly unlikely that they would be toxic since lectins can be effectively destroyed by the heat treatment that such foods would ordinarily receive upon cooking.

Other Applications of Lectins

Because of their unique property of being able to bind in a specific fashion to sugars and other glycoconjugates, the lectins have broad application in research and medical laboratories [see review by Lis and Sharon (54)]. Some of these applications are briefly described below.

Lectins, either in solution or in an immobilized form, have proved extremely useful for the detection and identification of many diverse glycoconjugates. The identification of blood group substances, membrane receptors, and the detection of malignant cells are but a few examples of the application of lectins in this regard. Immobilized lectins in particular have permitted the isolation of large quantities of specific glycoconjugates from complex biological extracts.

One of the most interesting application of lectins to medical research was the finding that lectins could be used to prevent graft rejection in bone marrow transplantation (55). This derives from the fact that the soybean lectin can be used to remove the mature T-cells responsible for graft rejection. This procedure has proved successful in the treatment of immune deficiency in children (known as "bubble children" because they have to be kept in absolute isolation in plastic chambers to prevent infection) and in the treatment of several of the irradiated victims of the Chernobyl accident in Russia.

Over 40 years ago, following the isolation of the soybean lectin, it was reported (56) that the injection of this lectin into rats retarded the growth of a transplantable tumor (Walker carcinoma). It is only recently, however, that this ability to inhibit tumor growth has also been demonstrated with the kidney bean lectin (57). The initial growth of the Krebs II ascites tumors injected into mice was retarded when the animals were fed a diet containing this lectin. To what extent these promising results with animals will lead to the development of effective anticarcinogenic agents in humans remains to be seen.

Literature Cited

1. Stillmark, **1888,** Dissertation, Dorpat University (now Tartu Estonia).
2. Goldstein, I.J.; Poretz, R.D. In *The Lectins, Properties, Function, and Applications in Biology and Medicine,* Liener, I.E.; Sharon, N.; Goldstein, I.J.; Eds; Academic Press Inc.: San Diego, CA, 1986, 33-247.
3. Boyd, W.C.; Shapleigh, E. *Science* **1954,** *119,* 419.
4. Sumner, J.B.; Howell, I.F. *J. Bacteriol.* **1932,** *32,* 227-237.
5. Becker, J.W.; Reeke, G.N.; Cunningham, B.A.; Edelman, G.M. *Nature* **1976,** *259*, 406-409.
6. Miller, J.B.; Hsu, R.; Heinrickson, R.; Yachnin, S. *Proc. Natl. Acad. Sci. USA* **1975,** *72*, 1388-1391.
7. Liener, I.E.; Kakade, M.L. In *Toxic Constituents of Plant Foodstuffs,* Liener, I.E., Ed., Academic Press, New York, 1980, pp. 7-71.
8. Liener, I.E. *J. Nutr.* **1949,** *39,* 325-339.
9. Liener, I.E. *J. Nutr.* **1953,** *49,* 609-620.
10. Liener, I.E. *J. Nutr.* **1953,** *49,* 527-539.
11. Landsteiner, K.; Raubitschek, H. *Zentrabl. Bakteriol. Parasitekd. Infektionsk., Abt. 2: Orig.* **1908,** *45,* 660-667.
12. Honavar, P.M.; Shih, C.-V.; Liener, I.E. *J. Nutr.* **1962,** *77*, 109-114.
13. Jaffé, W.G.; Planchert, A.; Paez-Pumar, J.I.; Torrealba, R.; Francheschi, D.N. *Arch. Venez. Nutr.* **1955,** *6,* 195-205.
14. Pusztai, A. *Plant Lectins;* Cambridge University Press, Cambridge, UK, 1991.
15. King, T.P.; Pusztai, A.; Clark, E.M.W. *Histochem. J.* **1980,** *12*, 201-208.
16. Pusztai, A.; Clarke, E.M.W.; King, T.P.; Stewart, J.C. *J. Sci. Food Agric.* **1979,** *30,* 843-848.
17. Boldt, D.H.; Barnwell, J.G. *Biochim. Biophys. Acta* **1985,** *843,* 230-237.
18. Donatucci, D.A.; Liener, I.E.; Gross, C.J. *J. Nutr.* **1987,** *117,* 2154-2160.
19. Kawatra, B.L.; Bhatia, I.S. *Biochem. Physiol. Pflanz.* **1979,** *174,* 283-288.
20. Banwell, J.G.; Boldt, D.R.; Meyers, J.; Weber, F.L.; Miller, B.; Howard, R. *Gastroenterology* **1983,** *84,* 506-515.
21. Dobbins, J.W.; Laurensen, J.P.; Gorelick, F.S.; Banwell, J.G. *Gastroenterology* **1983,** *84,* 1138 (abstract).
22. Rouanet, J.-M.; Besancon, P.; LaFont, J. *Experientia* **1983,** *39,* 1356-1357.
23. Kim, S.S.; Brophy, E.J.; Nicholson, J.A. *J. Biol. Chem.* **1976,** *251,* 3206-3212.
24. Erickson, R.H.; Kim, Y.S. *Biochim. Biophys. Acta* **1983,** *743,* 37-42.
25. Pusztai, A.; Grant, G.; de Oliveira, J.T.A. *IRCS Med. Sci.* **1986,** *14,* 205-208.

26. Hewitt, D.; Coates, M.E.; Kakade, M.L.; Liener, I.E. *Br. J. Nutr.* **1973,** *29,* 423-435.
27. Jayne-Williams, D.J. *Nature (London) New Biol.* **1973,** *243,* 150-151.
28. Rattray, E.A.S.; Palmer, S.; Pusztai, A. *J. Sci. Food Agric.* **1974,** *25,* 1035-1040.
29. Wilson, A.B.; King, T.P.; Clarke, E.M.W.; Pusztai, A. *J. Comp. Pathol.* **1980,** *90,* 597-602.
30. Banwell, J.G.; Howard, R.; Cooper, D.; Costerton, J.W. *Appl. Environ. Microbiol.* **1985,** *50,* 68-80.
31. Untawale, G.G.; McGinnis, J. *Poultry Sci.* **1976,** *55,* 2101-2105.
32. Williams, P.E.V.; Pusztai, A.; Macdearmid, A.; Innes, G.M. *Anim. Feed Sci. Technol.* **1984,** *12,* 1-10.
33. Grant, G.; Greer, F.; McKenzie, N.H.; Pusztai, A. *J. Sci. Food Agric.* **1985,** *36,* 409-414.
34. Calderon de la Barca, A.M.; Vazquez-Moreno, L.; Robles-Burgueno, M.R. *Food Chemistry* **1991,** *39,* 321-327.
35. Faschingbauer, H.; Kofler, L. *Wien. Klin. Wochenschr.* **1929,** *42,* 1069-1071.
36. Haidvogl, M.; Fritsch, G.; Grubbauer, H.M. *Padiat. Padol.* **1979,** *14,* 293-296.
37. Griebel, C.Z. *Lebensm.-Unters.-Forsch.* **1950,** *90,* 191-197.
38. Korte, R. *Ecol. Food Nutr.* **1972,** *1,* 303-307.
39. Jaffé, W.G. In *Nutritional Aspects of Common Beans and Other Legume Seeds as Animal and Human Foods,* Jaffe, W.G., Ed., Archivos Latinoamericanos de Nutricion, Caracas, Venezuela, **1973,** p. 200.
40. Bender, A.E.; Readi, G.B. *J. Plant Foods* **1982,** *4,* 15-22.
41. Grant, G.; More, L.J.; McKenzie, N.H.; Stewart, J.C.; Pusztai, A. **1983,** *Br. J. Nutr.* **1983,** *50,* 207-214.
42. Löwgren, M.; Liener, I.E. *Qual. Plant. Plant Foods Hum. Nutr.* **1986,** *36,* 147-154.
43. Coffey, D.G.; Uebersax, M.A.; Hosfield, G.L.; Brunner, J.R. *J. Food Sci.* **1985,** *50,* 78-81.
44. Chrispeels, M.J.; Raikhel, N.V. *Plant Cell* **1991,** *1,* 1-9.
45. Kijne, J.; Diaz, C.; dePater, S.; Lugtenberg, B. In *Advances in Lectin Research;* Franz, H., Ed.; Ullstein Mosby GmbH Co., Berlin, 1992, pp. 15-50.
46. Diaz, C.L.; Melchers, L.S.; Hooykaas, P.J.J.; Lugtenberg, B.J.J.; Kijne, J.W. *Nature* **1989,** *338,* 579-581.
47. Nap, J.-P.; Bisseling, T. *Science* **1990,** *250,* 948-954.
48. Huessing, J.E.; Shade, R.E.; Chrispeels, M.J.; Murdock, L.L. *Plant Physiol.* **1991,** *96,* 993-996.
49. Osborn, T.C.; Alexander, D.C.; Sun, S.S.M.; Cardona, C.; Bliss, F.A. *Science* **1988,** *240,* 207-210.
50. Osborn, T.C.; Burow, M.; Bliss, F.A. *Plant Physiol.* **1988,** *86,* 399-405.
51. Pusztai, A.; Grant, G.; Stewart, J.C.; Bardocz, S.; Ewen, S.W.B.; Gatehouse, A.M.R. *J. Agric. Food Chem.* **1993,** *41,* 436-440.
52. Hartweck, L.M.; Vogelzang, R.D.; Osborn, T.C. *Plant Physiol.* **1991,** *97,* 204-211.
53. Gatehouse, A.M.R.; Gatehouse, J.A. In *Effects of Antinutrients on the Nutritional Value of Legume Diets;* Barcocz, S.; Gelencser, E.; Pusztai, A., Eds. Office for Official Publications for the European Communities, Luxembourg, 1996, Vol. 1, pp. 14-21.
54. Lis, H.; Sharon, N. In *Lectins, Properties, Functions, and Applications in Biology and Medicine;* Liener, I.E.; Sharon, N.; Goldstein, I.J., Eds.; Academic Press, San Diego, CA, 1986; pp. 293-370.
55. *Lectins.* Sharon, N.; Lis, H., Eds.; Chapman and Hall, London, 1989. pp. 89-90.
56. Liener, I.E.; Seto, T.A. *Cancer Res.* **1955,** *15,* 407-409.
57. Pryme, I.F.; Pusztai, A.J.; Bardocz, S. *Int. J. Oncol.* **1994,** *5,* 1105-1107.

Chapter 4

Antinutritional and Allergenic Proteins

H. Frøkiær, T. M. R Jørgensen, A. Rosendal, M. C. Tonsgaard, and V. Barkholt

Department of Biochemistry and Nutrition, Building 224, Technical University of Denmark, DK–2800 Lyngby, Denmark

Immunologic and immunochemical methods are valuable tools in the investigation of food allergens and antinutritial proteins. Protease inhibitors and lectins are present in a broad spectrum of foods. They can retain their activity during passage through the digestive tract and induce different biological reactions, beneficial as well as harmful. The potential nutritional importance of legume inhibitors and lectins is evaluated from analyses of their concentration, biological activity and their stability during food processing conditions. Food allergens comprise a very varied group of proteins and glycoproteins with different sensitization routes for children and adults that are currently discussed. Food allergens' induction of oral tolerance was investigated by oral immunization of mice with milk or β-lactoglobulin.

A major part of our common protein sources may give rise to adverse reactions. Food components with general adverse effects are described as toxic or antinutritional, whereas adverse reactions towards components in amounts that do not normally give rise to adverse effects are described as either intolerance or allergy.

Antinutritional Proteins

Protease inhibitors as well as lectins are widely spread in nature and have essential importance for the availability of nutrients upon ingestion. They occur in most, if not all, plants and protect them against attack from microorganisms and insects through various low-molecular-weight compounds.

Crops with a high protein content, e.g. members of the leguminosae family, contain especially high levels of antinutritional proteins. A common property of the antinutritional proteins is their resistance towards proteolytic degradation as well as a generally high stability towards heat treatment.

Lectins. Lectins occur in many plants and are characterized by their highly specific carbohydrate binding activity. However, lectins are not only present in plants but also in microorganisms and mammals, including humans. Although characterized by their common capability to bind carbohydrate, lectins form a very heterogeneous group of proteins with respect to their molecular structure, carbohydrate specificity and biological activity.

Lectins were earlier named phytohemagglutinins as the first plant lectins discovered were able to agglutinate red blood cells (*1*). As plant lectins are not always capable of hemagglutination, a new definition has been introduced. According to this, all proteins possessing at least one non-catalytic domain that specifically binds mono- or oligosaccharides reversibly are defined as lectins. Plant lectins can be divided into four groups: legume lectins, chitin-binding lectins, monocot mannose-binding lectins and type-2 ribosome-inactivating proteins (*2*).

The two groups of lectins described below are most important in connection with food antinutrients.

Legume Lectins. Proteins from the legume lectin group are structurally related lectins that occur exclusively in legumes. Most of these lectins are glycoproteins. In spite of their structural relationship, legume lectins show a broad variation in their specificity towards carbohydrates and thereby in their biological activities (*3*).

Type-2 Ribosome-inactivating Lectins. Type-2 ribosome-inactivating lectins which includes ricin from castor bean bind to the surface of most eucaryotic cells and are taken up by the cells. Part of the lectin is able to attach to the 60 S ribosomal subunit and thereby halting the protein synthesis (*3*).

Biological Effects of Lectins. Although the first recognized activity of lectins was the capability of these proteins to agglutinate red blood cells, this activity is not important with respect to most lectins found in plant seeds used for human consumption and animal feed.

Lectins bind with high affinity to oligosaccharides which are absent in plant but are abundant in bacteria as polysaccharides of bacterial cell walls and in animals especially as constituents of glycoproteins in cell membranes. Several membrane integrated glycoproteins function as receptors for hormones and cytokines or are directly involved in cell to cell contact. Binding of lectins to these glycoproteins may therefore trigger reactions such as cell division and growth (mitogenic effect), cell maturation and cell death (*3*).

Responses in Small Intestine to Lectins. The generally high resistance of lectins towards proteolytic degradation enables them to pass through the digestive tract of humans and animals in the intact form (*3*). Furthermore, when the ingested lectin reaches the small intestine it may bind to certain glycoproteins on the surface of intestinal cells according to the carbohydrate specificity of the lectin. Plant lectins, specific for N-acetylgalactosamine/galatose, e.g. soybean agglutinin (SBA) and wheat germ agglutinin (WGA), or specific for complex carbohydrate containing these residues (e.g. kidney bean lectin, PHA), bind strongly to receptors on epithelium cells in the digestive tract and stimulate the growth of the small intestine and the pancreas as well as accumulation of polyamines in these sites (*4*). Lectins that bind less avidly to the epithelium cells, e.g. mannose specific lectins such as pea lectin (PSA), induce less prominent growth of the intestine and pancreas.

As the glycosylation patterns of membranes of the epithelial cells differ with animal species and age of the animal, the effects of a lectin may vary widely. Except for a few incidences of accidental poisoning of humans with lectins from castor beans and red kidney beans, little is known about the toxicity of plant lectins to humans, either the acute toxicity or possible effects upon prolonged ingestion (*3*).

However, experiments with the binding of various lectins to membrane of the human intestinal cancer cell line Caco2 have provided information of the binding activity of several legume lectins (*5*). PHA and SBA bound strongly to the cells and resulted in a shortening of the microvilli of the cells. PSA bound less avidly but to more sites on the cells and did not result in any decrease in the length of the microvilli. Moreover, these studies showed that the binding of lectins with high avidity for carbohydrate on the cell surface caused increased growth of the cells and destruction of the cytoskeleton (*6*). The effects of these and other lectins may either impair the capacity of the epithelial cells to absorb nutrients from the gut or increase the permeability of the mucosal barrier to macromolecules.

Isolectins with different specificity. Many legumes contain two isolectins (e.g. kidney bean and castor bean) which show high sequence homology but differ in carbohydrate specificity and activity (*7*). Of the two different isoforms of kidney bean lectin, PHA-L and PHA-E, PHA-E binds to a larger extent to the surface of epithelial cells than PHA-L, whereas only PHA-L has been demonstrated to be mitogenic towards lymphocytes (*8*).

Effects of Lectins on Cells of the Immune System. Independent of the capability of lectins to induce growth of intestinal cells, lectins may interfere with other cells of animals and humans. As lectin binds to the epithelium of the small intestine, it is absorbed by epithelium cells by endocytosis and taken up into systemic circulation in the intact form (*2*).

Once the lectin reaches the circulatory system it may affect cells which are not present in the digestive tract. Animals fed soybean or red kidney bean develop a strong humoral antibody response towards the respective lectin (*9,10*), in contrast to other dietary proteins (*11*) which usually cause immunological tolerance when ingested. Furthermore, many lectins exhibit strong mitogenic effects on different subpopulations of lymphocytes. Lectins taken up systemically may therefore change the micro-environment in the lymphoid organs and thereby interfere with the immune status of the recipient.

We have investigated the mitogenic effects on mouse lymphocytes from various lymph organs of four lectins from legume seeds extensively used for human consumption and animal feed, PHA-L, SBA, PSA, and peanut lectin (PNA). Of these lectins only SBA has been demonstrated to cause damage upon binding to epithelial cells in the small intestine while the other lectins are known to bind with less avidity to intestinal cells without causing damage to the epithelium (*7,9*). As shown in Figure 1, SBA does not significantly stimulate the growth of mouse lymphocytes in any lymph organ except for spleen cells. PHA-L and PSA demonstrated mitogenic activity towards lymphocytes of lymph nodes, spleen and to a lesser extent lymphocytes from Peyers' patches. PNA stimulated, exclusively, the growth of lymphocytes from Peyer's patches.

The lymph nodes of the intestine, Peyer's patches (PP) are surrounded by specialized epithelial cells, M cells, with only slightly folded membranes compared to other enterocytes. M cells contain less lysosomes and are capable to transport dietary proteins, including lectins, to PP for presentation to the immune system (*12*).

PNA is known to bind to immature B cells in the germinal centers which are constantly abundant in PP in contrast to other lymphoid organs due to the constant presentation of dietary proteins and microorganisms.Our results indicate that PNA, in addition to its ability to bind to B cells in germinal centers, has the ability to stimulate proliferation of the B cells. Therefore, small amounts of lectins, apparently harmless upon digestion, may change the interaction between lymphocytes in PP as well as in other lymphoid tissues because of their mitogenic effects.

Long-term effects of lectins. As most of the observed effects of dietary lectins have been obtained through short-term experiments, it is not known whether exposure over a long period may have any effect. The fact that many apparently harmless dietary lectins are able to cross the epithelium of the intestine and exhibit mitogenic activity towards different cell types raises an important question as to whether even small amounts of lectins can induce uncontrolled cell growth in organs or tissues. It remains an important issue to investigate whether such effects could lead to induction of malignant or benign turmors or immune disorders.

Analysis of lectins. As indicated by their earlier name, hemagglutinins, lectins can be detected by their agglutinating activities on red blood cells. This agglutinating activity has been used to detect active lectins from winged bean and jack bean in gut content and feces (*3,14-16*), whereas SBA activity was not found in similar experiments (*17*). These different results may be explained either by different stabilities or binding activities of the lectins or by differences in sensitivity of the hemagglutination assays. Hemagglutination assay is performed by a dilution series of lectins added to a fixed concentration of blood cells. The lowest dilution of lectin agglutinating the blood cells is visually determined. The sensitivity of hemagglutination analyses varies strongly depending of the actual lectin, the source of blood cells and the length of time since the blood was drawn. Some lectins are not capable of agglutinating blood cells at all and the sensitivity of hemagglutination assays for lectins capable of agglutination varies at least 1000 fold, typically in the range of 50 ng/ml to 50 μg/ml.

Alternative to hemagglutination assays the most widespread methods to determine lectins are different immunochemical analyses (*18*). Among the immunochemical analyses, ELISA (enzyme linked immunosorbent assay) is the easiest

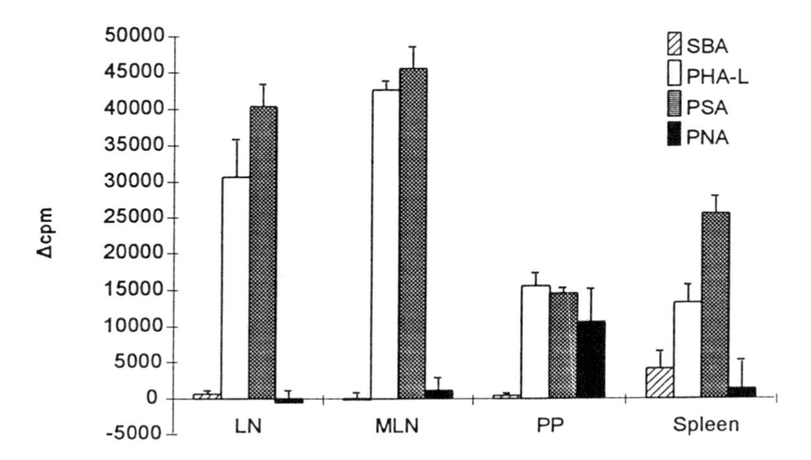

Figure 1. Mitogic activity of four legume lectins towards mouse lymphocytes from different lymph organs. Proliferation of lymphocytes was measured as incorporation of H^3-labelled thymidine: Lymphocytes ($2 \cdot 10^6$ cells/ml) were stimulated with lectins from soybean (SBA), kidney bean (PHA-L), pea (PSA) and peanut (PNA). LN: inguinal and popliteal lymph nodes, MLN: mesenteric lymph nodes; PP: Peyer's patches.

to perform and usually also the most sensitive. Using a sandwich ELISA with two monoclonal antibodies against PSA, we were able to detect as little as 0.5 ng/ml of lectin is shown in Figure 2. Another type of assay is FLIA (functional lectin immuno assay), where lectin in the sample reacts with glycoproteins attached to the solid phase and is detected by a specific antibody. FLIA is not as sensitive as the sandwich ELISA but ensures that only the active lectin is detected. The FLIA method can be used to follow inactivation of lectin during food processing as illustrated in Figure 2, whereas the sandwich ELISA is useful to determine the amount of lectin absorbed from the intestine in feeding trials as this method is the most sensitive (unpublished results).

Proteinase Inhibitors. Similar to lectins, inhibitors of proteolytic enzymes are able to resist proteolytic degradation and thus survive passage through the digestive tract. The inhibitors bind tightly to the proteolytic enzymes. Thereby the enzymes are inactivated and the resultant enzyme-inhibitor complex as well as unbound inhibitors pass undegraded through the digestive tract causing poor utility of the food and in addition loss of endogenous protein.

Inhibitors of trypsin and chymotrypsin are abundant in seeds of most plants, but as for lectins the highest levels of trypsin inhibitors are found in the protein rich seeds of legumes (*19*). Two classes of trypsin inhibitors are found to dominate in legume seeds, the Kunitz soybean trypsin inhibitor (KSTI) and the Bowman-Birk inhibitor (BBI). These two inhibitors are present in high levels in soybean from where they were first isolated (Table I).

Kunitz Soybean Inhibitor. KSTI is a 20.1 kD protein that binds very tightly to trypsin in a 1:1 molar ratio. This inhibitor consists of 180 amino acid residues, including 2 disulfide bonds, and is tightly folded due to a hydrophobic interior which makes the inhibitor very stable at pH values in the range of pH 1-12 (20). KSTI is only present in two species of the leguminosae family used for food and feed: soy bean and winged bean, but homologous proteins are present in many other plant seeds including other legumes that are not used for human consumption.

Bowman-Birk Inhibitors. BBI inhibitors are small and very compact proteins with a molecular weight of approximately 8 kD containing 7 disulfide bonds per molecule which makes the inhibitor extremely heat and acid stable (*21*). This inhibitor usually exists in various isoforms in the plant seed (*22,23*). BBI contains a binding site for trypsin and a separate site for chymotrypsin, both proteases are secreted by the pancreas into the lumen of the upper small intestine. The BBI type inhibitors are present in the seeds of most legumes. European pea breeding programs have resulted in production of pea cultivars with a very low content of the BBI-type trypsin inhibitor (Table I).

Trypsin inhibitors are generally rich in cysteine and, with sulfur-containing amino acids being nutritionally limiting amino acids in legume proteins, cultivars with a low content of BBI-type inhibitors may have a lower nutritional quality than cultivars with a high inhibitor content if they can be utilized by humans and animals, e.g. after heat denaturation.

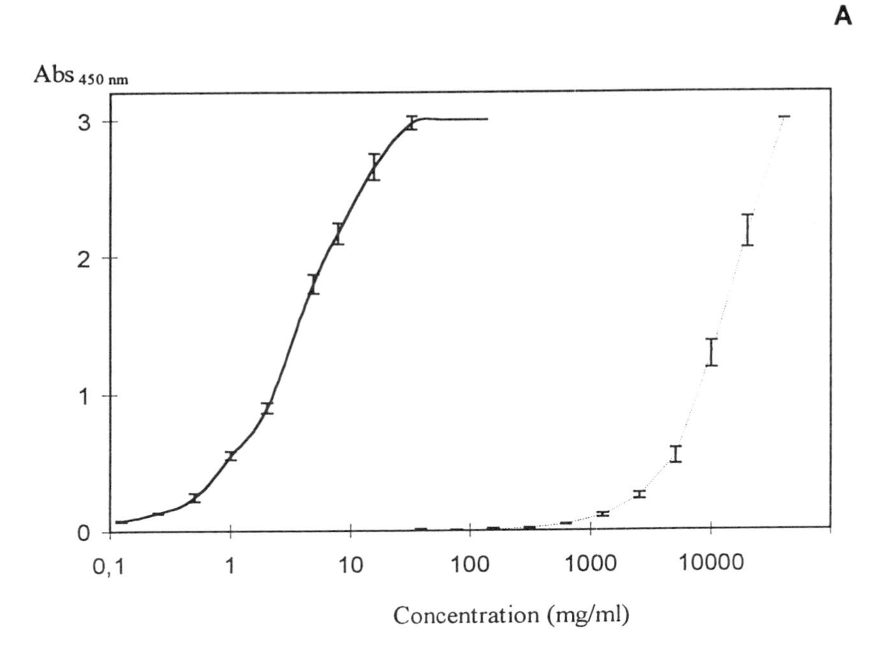

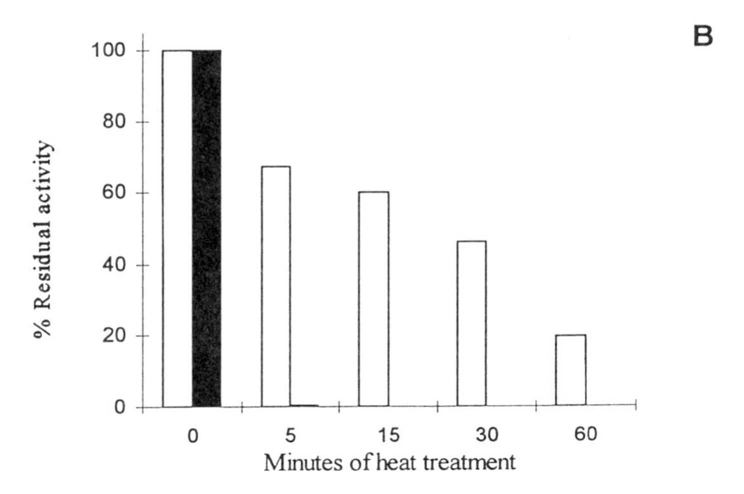

Figure 2. Immunochemical analysis of pea lectin (PSA) using monoclonal antibodies. A: Standard curves for two different types of ELISA. Solid line: sandwich ELISA; stippled line: FLIA. B: Inactivation of PSA in extracts of pea flour heat treated for various time at 72°C (white bars) and 100°C (black bars) and analyzed by FLIA.

Table I. Approximate inhibitor content per g flour measured in ELISA with monoclonal antibodies specific against the respective inhibitors in soybean, pea and chickpea

Inhibitor	*Soybean*	*Pea*	*Chickpea*
KSTI-type	20 mg/g	-	-
BBI-type	2-3 mg/g	0.05-3 mg/g[a]	2-3 mg/g

[a] special bred cultivars of pea contain a very low level of BBI type inhibitor.

Nutritional effects of trypsin inhibitors. The trypsin inhibitors (TI) are suspected to decrease the growth rate and cause enlarged pancreas in animals fed high levels of TI. The effects are most pronounced in young growing animals (*24*). To which extent these effects are due to the inhibitor content in the feed is still unclear in spite of many feeding experiments cited in the literature (for reviews, see ref. *25,26*). Most experiments have been performed with feeding of raw flour compared to heat treated flour. As many storage proteins from legume seeds are poorly degraded in their native form by proteolytic enzymes, it is not possible to differentiate between the role of protease inhibitors and other proteins, including lectins, in the feed. As lectins are by now known to cause enlargement of the pancreas, many of the effects earlier ascribed to trypsin inhibitors may at least partly be caused by the presence of lectins in the meal used for feeding trials.

Only few experiments have been performed with isolated inhibitors added to standard feeds. Affinity purified TI from cow peas added to the feed of growing rats led to a lower daily weight gain (*27*). As isolation of large amounts of TI is a laborious and expensive task, concentrates of antinutritional components including TI and lectins from pea have been used for feeding trials using pigs (*26*). As lectins and possibly also other proteins present in the TI concentrate are resistant to proteolytic degradation, it is difficult to evaluate the effects of TI per se in these experiments.

TI devoid of lectins and other antinutritional components was isolated from pea and fed to rats in balance trials (*28,29*). High levels of pea trypsin inhibitors corresponding to up to 29 % pea protein from a cultivar with a high level of TI in the diet had only a slightly lowering effect on the biological value and no effect on the true protein digestibility (*29*). In contrast, equivalent amounts of KSTI and BBI from soybean added to a standard feed showed a significant decrease in both the biological value and the true protein digestibility (unpublished). Similar results have been obtained when chick pea TI added to heated soybean flour was fed to young chickens (*30*). The added chick pea TI did not affect the daily weight gain, whereas raw soybean has been shown to decrease the daily weight gain of chickens and other growing animals (*26*). Likewise, groups of young chickens fed 30% peas showed the same performance whether the peas had a high or a low content of TI (*31*).

Among other factors which may lead to different results when investigating the nutritional effects of TI are the diet associated with the TI, animal species and age as well as the feeding strategy.

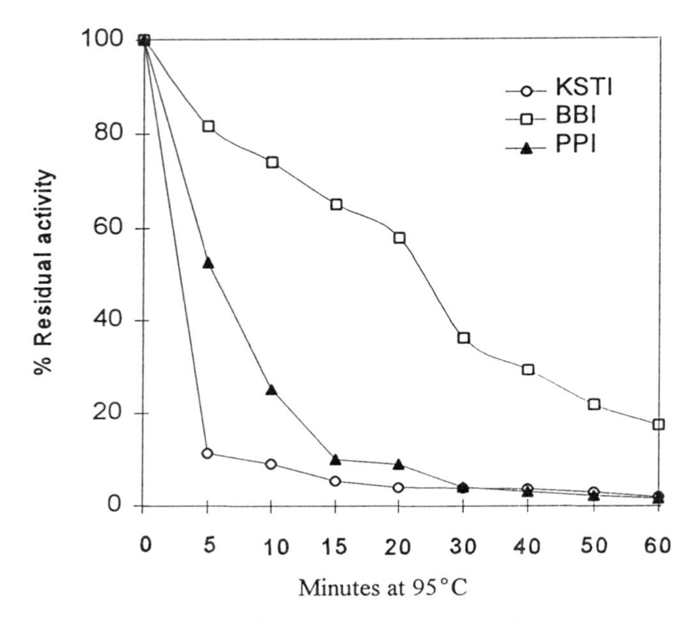

Figure 3. Heat treatment of KSTI and BBI from soybean and pea protease (PPI) at 95°C. To assure identical conditions, extracts of soybean and pea were mixed and subsequently heat treated. Samples were assayed for residual activity by competitive ELISA using monoclonal antibodies specific for KSTI, BBI and PPI, respectively.

Different Physio-chemical properties of TI. Homologous trypsin inhibitors from different legumes differ with respect to their heat resistance (Figure 3). Protease inhibitors may also differ in their resistance to low pH and pepsin, the milieu in the stomach, as well as gall in the upper part of the intestine. These factors may very well influence the activity of the ingested TI when it reaches the upper part of the intestine. Another important point is that most feeding experiments are performed using short-term feeding schedules. Long-term effects of the TI remain to be evaluated.

Long-term effects of TI. Long-term effects of TI do not necessarily have to be adverse. Studies on BBI type inhibitors have indicated that this group of inhibitors are involved in prevention of tumor development in vitro (*31,32*). Another inhibitor of trypsin and chymotrypsin from amaranth seed demonstrated growth depressing effects on the MCF7 breast cancer cells *in vitro* (26).

Epidemiological studies have shown that diets rich in legumes have beneficial anticarcinogenic effects on breast, colon and prostatic cancers. This indicates that legumes contain protective components. BBI type inhibitors may very well be one such agent.

Analysis of protease inhibitors. The content of trypsin and chymotrypsin inhibitors is traditionally determined by enzymatic methods, but these methods are very dependent on the concentration of protein, non-protein inhibitors as well as analytical performance of the assay system. Moreover, the enzymatic methods cannot distinguish different types of inhibitors from one another.

Immunochemical methods, especially when based on monoclonal antibodies, have proven very useful to determine the content of the two types of inhibitors in soybean, peas and chickpeas (*34-37*). ELISA based on monoclonal antibodies is also useful to characterize the heat stability of the two types of trypsin inhibitors (*38*). We have demonstrated that UHT treatment as well as traditional heat treatment of soymilk KSTI is quite easily inactivated, whereas BBI is very resistant towards heat. Trypsin inhibitors of the BBI type are the predominant TI in legumes.

In spite of the high sequential and structural homology between TI's from different legume crops, they show very different resistance towards heat. This is illustrated in (Figure 3), where extracts of soy and pea flours were mixed in a 1: 4 ratio and heat treated at 95°C for various periods. Monoclonal antibodies specific against native KSTI, BBI (*38*) and pea protease inhibitor (PPI) (*36*) were used to follow the inactivation of the three inhibitors in the protein mixture. BBI was demonstrated to be very heat stable, whereas PPI was only slightly more resistant towards heat than KSTI. Together with the above mentioned feeding trials with PPI, KSTI and BBI, this stresses the impact of investigation of individual inhibitors from different plant seeds before any adverse or beneficial nutritional effects are claimed.

Food Allergens

Adverse reactions to food are divided into toxic and non-toxic. Non-toxic reactions are further divided into allergy and intolerance although many of the symptoms are identical. Only reactions with an immunological basis should be called allergy, whereas non-immunological reactions should be called intolerance (*39*). This differentiation may be considered very academic, but it is justified by important characteristics of food allergy: a major part of these reactions are of the IgE-dependent immediate type and may give very severe symptoms within a short time after ingestion of the offending food. Furthermore, the lower limit of ingested material for symptoms to occur cannot be inferred from one incident to the next. In contrast to this, intolerance is most often characterized by a threshold (a lower limit) for the occurrence of adverse reactions.

Food Intolerance. The mechanisms of food intolerance cover a very wide spectrum (40). Some of these mechanisms are well known, e.g. metabolic disorders like lactose intolerance and effects of pharmacological agents such as serotonin in tomatoes. Symptoms from histamine in fish can mimic some of the initial symptoms of food allergy. Additives and dyes are often called allergenic, but the mechanism of their adverse effect is rarely an immunologic reaction (*41*). Although intolerance is generally caused by non-protein material, allergy is most often caused by proteins or glycoproteins. The food allergies discussed in the following are all IgE-dependent, but there are other adverse reactions with immunological basis such as celiac disease induced by the gliadin fraction of gluten (*42).*

Prevalence of Food Allergy. There is a large discrepancy between the number of people who report to have experienced food allergic reactions and the number whose symptoms can be confirmed by double blind placebo controlled food challenge (DBPCFC), the accepted method for diagnosis of food allergy. Furthermore, this diagnosis is time consuming and expensive, and it is therefore difficult to determine a generally accepted number for the true prevalence of food allergy. Skin prick test and *in vitro* determinations of specific IgE are commonly used for screening although their positive predictive accuracy is less than 50% (*43*). The prevalence in children up to three years was determined to 2.2% for cow's milk allergy (*44*) and to 8% for food allergy in general (*45*). The number of adults suffering from food allergy is currently estimated to be below 1% (*43*) .

Identification of Food Allergens. A list of common food allergens includes proteins from milk, egg, fish, shrimps, soybean, peanuts, rice, hazelnuts, apples and celery (*43,46,47*). Although specific IgE may be present without clinical symptoms many investigations rely on the analyses of specific IgE because binding of IgE is a major tool for identification of food components that may elicit an allergenic reaction. Many allergenic foods contain several allergenic components although a few are dominating. For example, analyses of serum from milk- and egg allergic patients show IgE-binding of most protein components in milk (*48*) and egg (*49*), respectively.

Characterization of Food Allergens. In spite of intense efforts through many years, it is not known what makes a food or a food component allergenic, and it has not been possible to demonstrate a distinction between allergenic epitopes and immunogenic epitopes in general. It is generally accepted that genetic predisposition in the recipient is an important factor for production of IgE. It is also assumed that a food allergen must be relatively resistant towards degradation as it must pass through the digestive tract and cross the intestinal barrier and still retain sufficient structure to give rise first to an immunological response and later to an allergic reaction.

Sensitization. Undegraded food proteins can indeed be detected in serum. As immunological reactions towards food are rare, the normal result of ingestion of a food component must be induction of tolerance. Thus, allergenicity may be considered as being a result of breaking of this tolerance or failure of induction of tolerance. The higher prevalence of food allergenicity among children could then be the result of, a not yet fully understood, intolerance because of an immature gastrointestinal immune system.

Sensitization of adults. For adults sensitization via inhalation has been proposed as the predominating route (50). This mechanism would account for the common cross reactivity of nuts, fruits and vegetables with well known pollen allergens, *Bet v* I and *Bet v* II from birch. *Bet v* II is homologous to the poly-proline binding protein profilin from yeast and mammals (*51*). It is found ubiquitously in plants and may be considered as a plant pan-allergen (*52*).

Carbohydrate Determinants on Allergenic Glycoproteins. Allergenic proteins from food are generally described as water soluble glycoproteins, resistant to heat, acid and proteases (*45)*, although there are many exceptions to this characteristic. The carbohydrate part of allergenic glycoproteins undoubtedly contributes to resistance towards proteolysis, but may also contribute to the immunogenicity. It is still an open, but very important, question to which extent the carbohydrate residues are part of the IgE binding epitopes. Some carbohydrate structures are found on many glycoproteins and such determinants would explain the observed cross-reactivity among a wide variety of vegetable glycoproteins (*53*). The participation of carbohydrate structure is also of importance for processing of potential allergens because carbohydrate and protein have different susceptibility to heat, acid or alkaline treatment and proteolysis.

Evaluation of Allergenicity of Novel Food. Traditional preparation of food and treatment of potential allergens like roasting of peanuts have resulted in the pattern of food allergenicity we know today. Changes in that pattern can be expected if food processing procedures are changed. Introduction of new protein sources or new combinations of food components e.g. by production of transgenic plants may create unexpected allergenic reactions. It is therefore urgent to evaluate the allergenicity of such products before it is manifested in patients as new allergies. As described recently (*54*). There is a manifold of information technology in our disposal that should be used to predict the allergenicity of such new products. Nonetheless, our knowledge about food allergenicity is still so insufficient that we cannot rely on existing information and more investigations are deemed necessary.

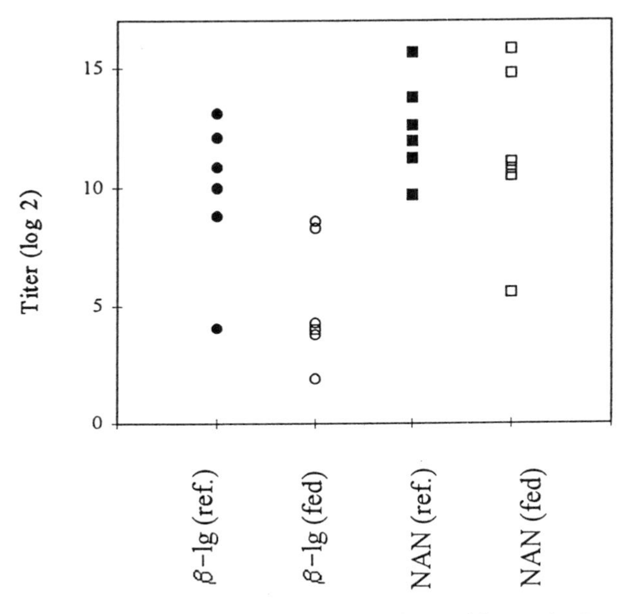

Figure 4. Analysis of induction of tolerance by oral immunisation of mice. The test groups contained 6 adult female mice (BALBcxCF1)f_1. They were fed β-lactoglobulin (β-lac(fed)) or NAN 1 ((NAN(fed)), normal infant formula) in their drinking water. After 17 weeks the fed mice and reference groups (ref.) were challenged with the product they had been drinking in Freunds Incomplete Adjuvant. The antibody titers were measured by ELISA in microtiter plates coated with β-lactoglobulin and detected by horse radish peroxidase coupled rabbit anti mouse antibodies. Mean titers within the groups: β-lac (fed): 5.15; β-lac (ref): 9.85; NAN(fed): 11.4; NAN(ref): 12.5.

***In vivo* Laboratory Investigations of Induction of Tolerance.** As it is not possible to use humans for these experiments, various animal models for investigation of food allergy are used (*55*). It is well established that feeding proteins to mice induces oral immunologic tolerance manifested by low or no responsiveness to the feed protein upon subsequent immunization with the same protein (*56*). As mentioned above, food allergy may be caused by a breakdown of oral tolerance or by a failure of its induction. We have therefore investigated various proteins for their capability to induce oral tolerance in mice. In the experiment presented here the animals were first generation cross over in order to obtain higher genetic diversity and a more varied response than could be obtained with inbred animals representing a narrow repertoire of MHC . The feeding period was longer than usual in immunisation experiments (*55*). The amount fed was realistic compared to the normal human intake.

Induction of Tolerance by Oral Immunization. The fact that many different proteins in a food item can be allergenic indicates that some compounds in a mixture can affect the allergenicity of the others, probably by affecting the induction of tolerance. The experiment described below compares the induction of tolerance towards β-lactoglobulin, pure and in combination with the other ingredients in a milk formula. (Figure 4)

Oral immunization of mice. In one group, each mouse was fed 1 mg β-lactoglobulin/day. In the other group, each mouse was fed 90 mg milk powder/day containing 1 mg β-lactoglobulin. The protein was added to the drinking water and the mice were fed rat pills ad libitum.

The feeding experiment continued for 17 weeks. After the end of the test feeding the mice were immunized with the product they had been drinking. The reference groups were immunized accordingly. Immunizations were made one and 14 days after the end of the test feeding. The antibody titers were measured by ELISA.

The pure β-lactoglobulin induced tolerance compared to the reference group ($p < 0.025$). This was also found by oral immunization with ovomucoid and KSTI (unpublished results). The more important result is that the same amount of the same protein did not induce tolerance when it was in a mixture with milk components compared to the reference group. There may be many reasons for this, either simply the effect of relative concentration or effects of a group of components e.g. lipids, or specific effects of single components. The type of assay presented here could be used to answer such questions. For legumes it would be interesting to analyze the effect of lectins because of their effect on the intestinal membrane as well as other antinutrients of low molecular weight suspected to interfere with digestion.

Conclusions

In spite of several common features within groups of lectins or inhibitors, the present knowledge illustrates high degrees of variation with respect to specificity, biological activity as well as resistance towards heat and proteolytic degradation. It is therefore of importance to further investigate the so called antinutritional proteins from different crops before making any generalizations on their nutritional effects. As most of the observed effects have been obtained by short-term studies, there is a need for studies on the long-term effects of a daily ingestion of small amounts of such material. Immunochemical methods such as ELISA based on monoclonal antibodies have proven to be a valuable tool to distinguish between different types of antinutional proteins as well as to detect very low concentrations in body fluids and tissues.

Food allergenicity is poorly understood although the problem has been the subject of intense research. The increased recognition of inhalation as a major route of sensitization of adults towards common structures may bring an important contribution to our understanding of this complex problem.

Acknowledgement

This work was supported by Center for Advanced Food Research, Denmark.

References

1. Elfstrand, M. In *Görbersdorfer Veröffentlichungen a;*. R. Kobert, Ed.; Enke: Stuttgart, GD, 1898, Band I, pp. 1-159.
2. Peumans, W.J.; Van Damme, E.J.M. *Trends in Food Sci. Technol.* **1996**, *71*, 132-138.
3. Putzai, A. *Plant Lectins*; Cambridge University Press: Cambridge, GB, 1991, pp 1-263.
4. Pusztai,A.; Grant, G.; Brown,D.S.; Bardocz,S.; Ewen, S.W.B.; Baintner, K.; Peunams,W.J.; Van Damme,E.J.M. In *Lectins. Biomedical Perspectives*; Pusztai,A; Bardocz, S.,Eds; Taylor and Francis Ltd.: London, G.B., 1995; pp. 141-154.
5. Koninkx, J.; Hendriks, H.; Van Rossum, J.; Van den Ingh, T.; Mouwen, J. *Gastroenterol.*, **1992**, *102*, 1516-1523.
6. Draaijer,M.; Koninkx, J.; Hendriks, H.; Kik, M.; Van Dijk, J.; Mouwen, J. *Biol. of the Cell*, **1989**, *65*, 29-35.
7. Hendriks, H.; Kik, M.; Koninkx, J.; Van en Ingh,T.; Mouwen, J. *Gut*, **1991**, *32*, 196-201.
8. Green, E.D.; Baenziger, J.U. *J. Biol.Chem.*, **1987**, *262*, 12018-12029.
9. Pusztai, A.; Ewen, S.W.B.; Grant, G.; Peumans, W.J.; van Damme, E.J.M.; Rubio, L.; Bardocz, S. *Digestion*, **1990**, *46* (suppl. 2), 308-316.
10. De Aizpurua, H.J.; Russell-Jones, G.J. *J.Exp.Med.*, **1988**, *167*,440-451.
11. Weiner, H.L. *Proc. Natl. Acad. Sci. USA,* **1994**, *91*, 10762-10765.
12. Neutra, M.R.; Kraehenbuhl, J.P. In *Handbook of Mucosal Immunology*; Ogra, P.L.; Mestecky, J.; Lamm, M.E.; Strober, W.; Mcghee, J.R.; Bienenstock, J., Eds.; Academic Press, Inc.: San Diego, CA, 1994, pp 27-40.
13. Kilpatrick, D.C. In *Lectins. Biomedical Perspectives.* Pusztai, A; Bardocz, S., Eds; Taylor and Francis Ltd.: London, G.B., 1995; pp. 155-182.
14. Nakata, S.; Kimura, T. *J. Nutr.*, **1985**, *115*, 1621-1629.

15. Higuchi,M.; Kawada, T.; Iwai, K. *J.Nutr.* **1989**, *119*, 490-495.
16. Jaffe, W.G.; Vega-Lette, C.L., *J.Nutr.*, **1968,** *94*, 203-210.
17. Grant, G.; van Driessche, E. In *Recent advances of research in antinutritional factors in legume seeds*, van der Poel, A.F.B.; Huisman, J.; Saini, H.S., Eds.; EAAP publication 70; Wageningen Pers: Wargeningen, NL, 1993; pp 219-233.
18. Gueguen, J.; van Oort, M.G.; Quillien, L.; Hessing, M. In *Recent advances of research in antinutritional factors in legume seeds*, van der Poel, A.F.B.; Huisman, J.; Saini, H.S., Eds.; EAAP publication 70; Wageningen Pers.
19. Wargeningen, NL, 1993; pp 9-30.
20. Richardson, M. *Phytochem.* **1977**, *16*, 159-169.
21. Wu, Y.V.; Scheraga, H.A. *Biochem.*, **1962**, *1*, 698-705.
22. Liener, I.E.; Kakade, M.L. In *Toxic Constituents in Plant foodstuffs*, Liener, I.E., Ed., 2. ed.; Academic Press: New York, NY, 1980, pp 7-77.
23. Ikenaka, T.; Noiroka, S. In *Proteinase Inhibitors. Research Monographs in Cells and Tissue Physiology*, Barrett, A.J.; Salvesen, G. Eds. Elsivier Science Publishers B.V., Cambridge, GB, 1986, pp 361-374.
24. Tan-Wilson, A.L.; Wilson, K.A. In *Nutritional and Toxilogical Significance of Enzyme Inhibitors in Foods, Adv.Exp.Med.Biol.*, Friedman, M. Ed.; 1986, Vol.199, pp 391-411.
25. Schneeman, B.O.; Gallaher, D. In *Nutritional and Toxilogical Significance of Enzyme Inhibitors in Foods, Adv.Exp.Med.Biol.*, Friedman, M. Ed.; 1986, Vol.199, pp 185-187.
26. Morgan,R.G.H.; Crass, R.A.; Oates, P.S. In In *Nutritional and Toxilogical Significance of Enzyme Inhibitors in Foods, Adv.Exp.Med.Biol.*, Friedman, M. Ed.; 1986, Vol.199, pp 81-90.
27. Le Guen, M.P.; Birk, Y. In *Recent advances of research in antinutritional factors in legume seeds*, van der Poel, A.F.B.; Huisman, J.; Saini, H.S., Eds.; EAAP publication 70; Wageningen Pers: Wargeningen, NL, 1993; pp 157-171.
28. Pusztai, A.; Grant, G.; Brown, D.J.; Stewart, J.C.; Bardocz, S. British J.Nutr. 1992, 68, 783-791.
29. Mortensen, K.; Olsen, L.R.; Sørensen, H.; Sørensen, S. In *Improving production and utilisation of grain legumes. Proceedings of 2nd European Conference on Grain Legumes.* Copenhagen, DK, 1995, pp.294-295.
30. Eggom, B.O.; Frøkiær, H.; Mortensen, K.; Olsen, L.R.; Sørensen, H.; Sørensen, S. In *Improving production and utilisation of grain legumes. Proceedings of 2nd European Conference on Grain Legumes.* Copenhagen, DK, 1995, pp. 270-271.
31. Birk, Y.; Smirnoff, P. In *Proceedings of the 1st European Conference on legume seeds*, Angers, Fr., 1991, pp 391-392.
32. Barrier-Guillot, B.; Castaing, J.; Peyronnet, C.; Lucbert, J. In In *Proceedings of the 1st European Conference on legume seeds*, Angers, Fr., 1991, pp 327-329.
33. Yavelow, J.; Finlay, T.H.; Kennedy, A.R.; Troll, W. *Cancer Research,* **1983**, *43*, 2454-2459.
34. Yavelow, J.; Collins, M.; Birk, Y.; Troll, W.; Kennedy,A.R. *Proc.Natl. Acad. Sci. USA,* **1985**, *82*, 5395-5399.
35. Brandon, D.L.; Bates, A.H.; Friedman, M. *J.Food Sci.,* **1988**, *53*, 102-106.
36. Brandon, D.L.; Bates, A.H.; Friedman, M. *J.Food Sci.*, **1989**, *37*, 1192-1196.

37. Frøkiær, H.; Hørlyck, L.; Barkholt, V.; Sørensen, H.; Sørensen, S. *Food Agric. Immunol.*, **1994**, *6*, 63-72.
38. Frøkiær, H.; Sørensen, S.; Henmar, H.; Savage, G.P. In *Improving production and utilisation of grain legumes. Proceedings of 2nd European Conference on Grain Legumes*. Copenhagen, DK, pp. 412-413.
39. Rouhana, A.; Adler-Nissen, J.; Cogan, U.; Frøkiær, H. *J. Food Sci.* **1996**, *61*, 265-269.
40. Ortolani, C.; Vighi, G. *Allergy* **1995**, *50*, 8-13.
41. Sampson, H.A., *J. Allergy Clin. Immunol.* **1986,** *78*, 212-219.
42. Bosso, J. V.; Simon, R. A. In *Food Allergy: Adverse Reactions to Food and Food Additives*; Metcalfe, D.D.; Sampson, H. A.; Simon, R. A., Eds; Blackwell Scientific Publications, Oxford, 1991, pp 288-300.
43. O'Mahony, S.; Ferguson, A. In *Food Allergy: Adverse Reactions to Food and Food Additives*; Metcalfe, D.D.; Sampson, H. A.; Simon, R. A., Eds; Blackwell Scientific Publications, Oxford, GB, 1991, pp 186-198.
44. Sampson, H.A. In *Allergy - Principles and Practice*; Middleton, E. Jr.; Reed, C. E.;Adkinson, N. F.; Yunginger, J. W.; Yunginger, J. W.; Busse, W. W. Eds.; Mosby: St. Louis, MO, 1993; pp 1661-1686.
45. Høst, A; Halken, S. *Allergy,* **1990**, *45*, 587-596.
46. Bock, S. A. *Pediatrics,* **1987**, *79*, 683-688.
47. Matsuda, T.; Nakamura, R. *Trends in Food Sci. and Techn.* **1993**, 4, 289-293.
48. Ebner, C.;Hirschwehr, R.; Bauer, L.; Breitneder, H.; Valenta, R.; Hoffmann, K.; Krebitz, M.; Kraft, D.; Scheiner, O. In *Highlights in Food Allergy*; Wütrich, B. Ortolani, C. Eds; Monographs in Allergy, Hanson, L Å, ; Shakib, F. Eds.; Karger: Basel, SH, 1996, Vol. 32; pp 73-77.
49. Gjesing, B.; Østerballe, O.; Schwartz, B. *Allergy* **1996**, *41*, 51-56.
50. Langeland, T. *Allergy* **1982**, *37*, 521-530.
51. Pastorello. E. A. ; Ispano, M.; Pravettoni, V.; Farioli, L.; Incorvaia, C.; Ansaloni, R.; Rotondo, F.; Vigano, G.; Ortolani, C. In Highlights in Food Allergy; Wütrich, B. Ortolani, C. Eds; *Monographs in Allergy*, Hanson, L Å,; Shakib, F. Eds.; Karger: Basel, SH, 1996, Vol. 32; pp 57-62.
52. Valenta, R.; Duchêne, M; Pettenburger, K.; Sillaber, C.; Valent, P.; Bettelheim, P.; Breitenbach, M.; Rumpold, H.; Kraft, D.; Scheiner, O. *Science* **1991**, *235*, 557-560.
53. Valenta, R.; Duchêne, M; Ebner, C.; Valent, P.; Sillaber, C.; Devailler, P.; Ferreira, F.; Tejkl, M.; Edelmann, H.; Kraft, D.; Scheiner, O. *J. Exp. Med.* **1992**, *2,* 377-385.
54. Aalberse, R.C.; van Ree, R. In *Highlights in Food Allergy*; Wütrich, B. Ortolani, C. Eds; Monographs in Allergy, Hanson, L Å,; Shakib, F. Eds.; Karger: Basel, SH, 1996, Vol. 32; pp 78-83.
55. Astwood, J. D.; Fuchs, R. L. *Trends in Food Sci. Techn.* **1996**, *7*, 219-226.
56. Pahud, J.J.; Schwarz, K.; Granato, D. *Food Allergy*; Schmidt, E., ed.; Nestlé Nutrition Workshop Series,; Nestec LTd. Vevey/Raven Press, Ltd.: New York, NY, 1988, Vol. 17, pp 199-207.
57. Mowat, A.M. In *Handbook of Mucosal Immunology*; ; Ogra, P.L.; Mestecky, J.; Lamm, M.E.; Strober, W.; Mcghee, J.R.; Bienenstock, J., Eds.; Academic Press Inc.: San Diego, CA, 1994, pp 185-201.

Chapter 5

Potato Polyphenols: Role in the Plant and in the Diet

Mendel Friedman

Western Regional Research Center, Agricultural Research Service, U.S. Department of Agriculture, 800 Buchanan Street, Albany, CA 94710

Potatoes and other plant foods accumulate a variety of secondary plant metabolites including phenolic compounds, phytoalexins, protease inhibitors, and glycoalkaloids, as a protection against adverse effects of bruising and injury by phytopathogens including bacteria, beetles, fungi, insects, and slugs. Since these phytochemicals are consumed by animals and humans as part of their normal diet, a need exists to develop a better understanding of the role of these compounds in both the plant and in the diet. To contribute to this effort, this integrated overview describes the biosynthesis and the role of phenolic compounds such as chlorogenic acid and tyrosine in host-plant resistance and their beneficial effects as antioxidants, antimutagens, anticarcinogens, and as antiglycemic agents. Also covered are analytical and compositional aspects of phenolic compounds in potatoes; ferrous ion- and heat-induced discolorations such as after-cooking blackening which seems to affect organoleptic but apparently not nutritional properties of potatoes; polyphenol oxidase-catalyzed browning reactions and their prevention; effects of baking, cooking, microwaving, light, and γ-radiation on the stability of chlorogenic acid; and recommendations for future research. The possibility that the net antioxidative potency of structurally different potato polyphenolic compounds is related to their net electrochemical oxidation-reduction (redox) potential merits validation. Understanding the multiple, overlapping roles of polyphenols in plant physiology and in food science and nutrition should stimulate interest in maximizing beneficial nutritional and health effects of polyphenols in the diet.

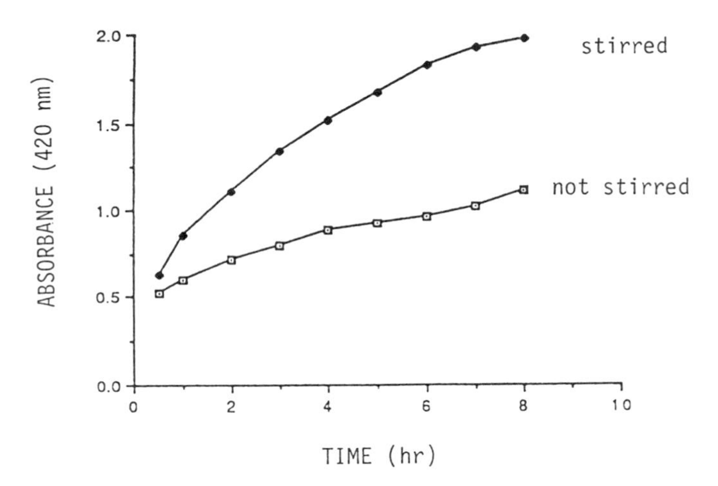

Figure 1. Effect of stirring on browning of a 5% slurry of Russet potato flesh measured at 420 nm (79).

Introduction

Polyphenolic compounds are secondary plant metabolites found in numerous plant species including potatoes (1). The oxidation products of phenolic compounds appear to be involved in defense of the plants against invading pathogens including bacteria, fungi, and viruses. Polymeric polyphenolic compunds seem to be more toxic to potential phytopathogens than the phenolic monomers such as chlorogenic acid from which they are derived. The polyphenol oxidase-catalyzed polymerization helps seal the injured plant surface and begins the healing process, analogous to the formation of fibrin blood clots in injured humans.

Enzyme-catalyzed browning reactions (2-5; Figure 1) of polyphenols continue after the food is harvested resulting in deterioration in flavor, color, and nutritional quality. For this reason, prevention of enzymatic browning in fruits and vegetables has been a major concern of food scientists.

Polyphenolic compounds have also been shown to possess antimutagenic, anticarcinogenic, antiglycemic, and antioxidative beneficial properties. These properties can be utilized in the prevention of rancidity and as health-promoting food ingredients.

In this overview, I attempt to integrate and correlate the widely scattered literature on the role of polyphenols in potatoes before and after harvest (1-140). Specifically covered are the following relevant aspects; analysis, biosynthesis, host-plant resistance, food browning and its prevention, and beneficial and adverse effects on food quality and safety. Suggestions for future research are also mentioned in order to catalyze progress in minimizing adverse effects and enhancing desirable ones of potato polyphenols.

Since chlorogenic acid constitutes up to 90% of the total phenolic content of potato tubers, most of the discussion centers around this compound. Figure 2 shows the structures of potato phenolics including chlorogenic acid and its isomers.

Analysis and Composition

Analytical methods for potato-polyphenols include gas-liquid chromatography (GLC), high-performance liquid chromatography (HPLC), thin-layer chromatography (TLC), and UV spectrophotometry, but surprisingly, apparently not immunoassays. Before analysis, the compounds have to be extracted and purified. Voigt and Noske (6) describe optimized extraction of chlorogenic acid from potatoes with the aid of acetone, ethanol, and methanol. The order of effectiveness was methanol > ethanol > acetone. They found that the chlorogenic acid content of stewed potatoes was the same as that of raw potatoes.

Reeve et al. (7) discovered in our laboratory that phenolic compounds are distributed mostly between the cortex and skin (peel) tissues of the potato (Figure 3). About 50% of the phenolic compounds were located in the potato

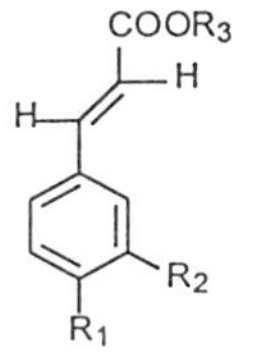

cinnamic acid: $R_1 = R_2 = R_3 = H$
p-coumaric acid: $R_1 = OH$, $R_2 = R_3 = H$
caffeic acid: $R_1 = R_2 = OH$, $R_3 = H$
ferulic acid: $R_1 = OH$, $R_2 = OCH_3$, $R_3 = H$
chlorogenic acid: $R_1 = R_2 = OH$, R_3 = quinic acid

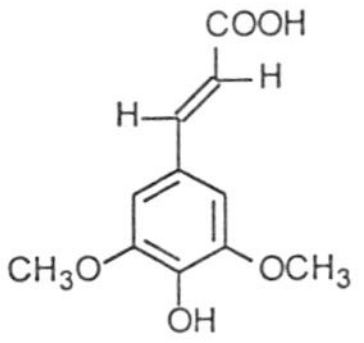

sinapic acid

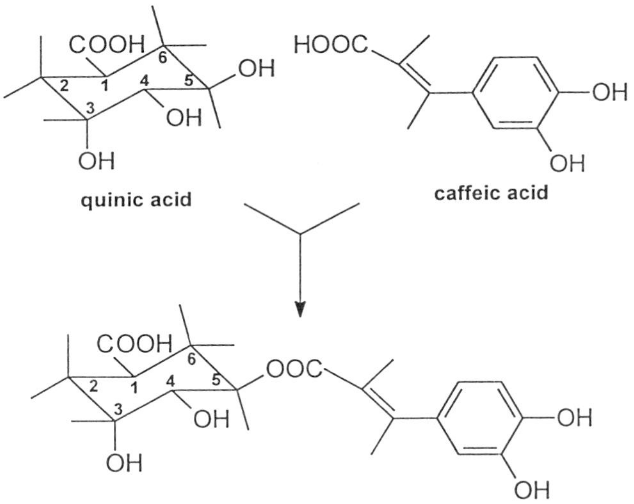

quinic acid

caffeic acid

chlorogenic acid (5-O-caffeylquinic acid)

cryptochlorogenic acid = 4-O-caffeoylquinic acid
neochlorogenic acid = 3-O-caffeoylquinic acid
isochlorogenic acid "a" = 4,5-di-O-caffeoylquinic acid
isochlorogenic acid "b" = 3,5-di-O-caffeoylquinic acid
isochlorogenic acid "c" = 3,4-di-O-caffeoylquinic acid

Figure 2. Structures of potato polyphenols and of chlorogenic acid isomers.

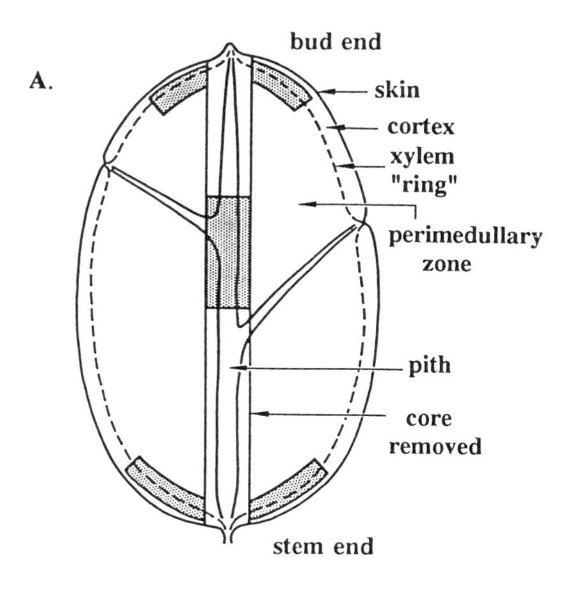

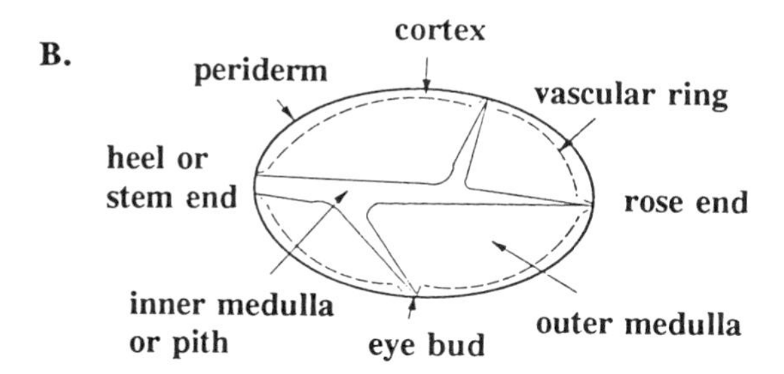

Figure 3. Cross section of a potato tuber as depicted by **A.** Reeve et al. (7); **B.** Olsson (140).

peel and adjoining tissues, while the remainder decreased in concentration from the outer towards the center of potato tubers (8).

Several isomeric chlorogenic acids have been found in potatoes. The major one, now designated as 5-caffeoylquinic acid according to the designation by the International Union of Pure and Applied Chemistry (IUPAC) is complemented by 3- and 4-caffeoylquinic acids. The IUPAC numbering system differs from that used by Brandl and Herrman (9). Whether these isomers occur naturally or are artifacts formed during extraction is an unresolved issue (10).

Tisza et al. (11) describe a new GC-MS method for the quantitation of chlorogenic, citric, malic, and caffeic acids and of fructose, glucose, and sucrose in freeze-dried potato samples. The chlorogenic acid content of freeze-dried potato tubers, but not sprouts, correlated with values obtained by UV spectroscopy.

Nagels et al. (12) and Brandl and Herrmann (9) describe HPLC separation of chlorogenic acid isomers in potatoes. Ten varieties of potato cultivars grown under the same conditions contained the following chlorogenic acid isomers: 3-0-caffeoylquinic (n-chlorogenic acid), 4-0-caffeoylquinic (cryptopchlorogenic acid), 5-0-caffeoylquinic (neochlorogenic acid), 3,4-dicaffeoylquinic, and 3,5-dicaffeoylquinic.

Sosulski et al. (13) report a total phenolic acid content of potato flour of 410 ppm, with free chlorogenic acid, contributing 83.2% to the total. Griffiths et al. (14) describe a colorimetric method for chlorogenic acid in freeze-dried potato tubers based on reaction of the acid with nitrous acid. Application of this method to 13 potato cultivars revealed that the chlorogenic acid content of the stolon end of tubers could in the majority of cases be related to the susceptibility to after-cooking-blackening. However, in two genetically related cultivars, the amount of blackening was greater than predicted solely on the basis of chlorogenic acid content.

Malmberg and Theander (15) found that spectrophotometric analysis of potato chlorogenic acid gave higher values than did analysis by HPLC or GLC. Spectrophotometry may give high values because chlorogenic-acid isomers contribute to the total absorbance. Since GLC requires derivatization, HPLC may be preferable as a general method. However, HPLC analysis may not always be satisfactory because chlorogenic acid undergoes a time-and light-dependent change. (See below).

Because of the unexpected problem we encountered with the high-performance liquid chromatographic (HPLC) method (16-19), we evaluated the effectiveness of the ultraviolet method for measuring chlorogenic acid in several varieties of fresh potatoes but also in parts of the potato plant, in processed potato products, and in weed seeds (Tables I-V).

Chlorogenic acid underwent a time- and light-dependent change in the methanolic and ethanolic extracts of potatoes used (Figure 4). The decrease of the chlorogenic acid peak on chromatograms was accompanied by a corresponding increase of a new peak. Use of ultraviolet spectrophotometry to estimate chlorgenic acid, was reproducible, apparently because the newly formed

Table I. Chlorogenic Acid Content (mg/100 g fresh weight) in Different Potato Varieties (18, 19)

Variety	Chlorogenic Acid
Idaho Russet (small tubers)	9.6 ± 0.5
Idaho Russet (large tubers)	14.2 ± 0.7
small red	13.3 ± 0.7
Simplot I	13.1 ± 0.9
Simplot II	16.5 ± 0.5
NDA 1725	17.4 ± 1.2
potato 3194	18.7 ± 2.1

Table II. Distribution of Chlorogenic Acid (mg/100 g fresh weight) in Parts of a Potato Plant (18, 19)

Sample	Chlorogenic Acid
tubers	17.4 ± 1.2
roots	26.3 ± 0.8
leaves	223.5 ± 0.9
sprouts	754.0 ± 25.2

Table III. Heat Stability of Chlorogenic Acid in Cooked Potatoes (mg/g freeze-dried weight) Determined by UV Spectrophotometry (18, 19)

Potato	Chlorogenic Acid
fresh	0.80 ± 0.05
baked	0.00
boiled	0.32 ± 0.01
microwaved	0.43 ± 0.02

Table IV. Stability of Chlorogenic Acid During Baking of Mixed Heavenly Blue Morning Glory and Wheat Flour (17)

Sample	% loss
unbaked mixed flour	0
convection oven baked muffin	
crust fraction	100
crumb fraction	65.1 ± 1.9
microwave oven baked muffin	77.0 ± 0.9

Table V. Effect of Growing Time in a Greenhouse on Chlorogenic Acid Content (mg/100 g fresh weight (n = 3) of Freeze-Dried NDA 1725 Potato Leaves (141) Determined by UV Spectrophotometry (Dao and Friedman, unpublished results)

Time	Chlorogenic Acid	
(Weeks)	Mean Value	95% C. I.(Coeffecients of Variation)
3	148.0	131.9, 164.1
4	152.1	135.9, 168.3
6	305.7	289.6, 321.9
7	423.1	406.9, 439.3
9	257.8	241.6, 274.0

compound(s) has an absorption maximum similar to that of chlorogenic acid (Figure 5). Nearly all of the chlorogenic acid in spiked potato powders was recovered. Thus, our results suggest that the UV method may have advantages over HPLC. Since HPLC can measure specific polyphenols in a mixture whereas UV spectroscopy cannot, extracts of plant, food, and animal tissues should be analyzed immediately to minimize formation of new products which may co-elute with known polyphenols on HPLC columns. Generally, HPLC, UV, and GC-MS methods need to be further compared, correlated, and validated.

Seven varieties of potatoes contained 754 mg/100 g fresh weight for sprouts, 224 mg/100g for leaves, 26 mg/100 g for roots, and 17 mg/100 g for tubers. Oven-baked potatoes contained 0% of the original amount of chlorogenic acid, boiled potatoes 35%, and microwaved potatoes 55%. Commercially processed french fried potatoes, mashed potato flakes, and potato skins contained no chlorogenic acid. The absence of chlorogenic acid was confirmed by thin-layer chromatograpy. Leaves contained high levels of chlorogenic acid (20; Table V). Figure 6 and Table IV illustrate the value of UV spectroscopy to measure changes in chlorogenic acid during baking.

These observations suggest that food processing conditions need to be defined and used to minimize destruction of chlorogenic acid. Moreover, the fate of chlorogenic acid and other polyphenols during food processing is largely unknown.

An analysis of total phenolic and chlorogenic acid content showed that immature potato tubers with undeveloped periderm had the same content as mature tubers (21). High temperature storage did not affect the total phenol and chlorogenic acid content. However, both total phenolic and chlorogenic acid levels increased on prolonged (5 week) storage at 0°C.

Pre-Harvest Events

Biosynthesis. Chlorogenic acid and related polyphenols are present in numerous plant species including those of the Solanaceae (11). Their main function in plants appears to be in defenses against phytopathogens, as described below for potatoes. Biosynthetically, chlorogenic acid is derived from phenylalanine, which in turn is formed via the shikimate pathways starting from phosphoenol puruvate and erythrose 4-phosphate, as described in detail by Herrmann (22) and Schmidt and Amrhein (23). Figure 7 shows the biosynthetic transformations of phenylalanine to chlorogenic acid (24). Chlorogenic acid biosynthesis has an absolute requirement for oxygen. This may explain why most of the chlorogenic acid is found in the outer periderm rather than in the cortex of the tuber (Figure 2).

Detailed discussion of the enzymes involved in the illustrated biosynthetic transformations is beyond the scope of this paper. A useful entry into the literature on this subject can be found in Moriguchi et al. (25) and Tanaka and Kojima (26).

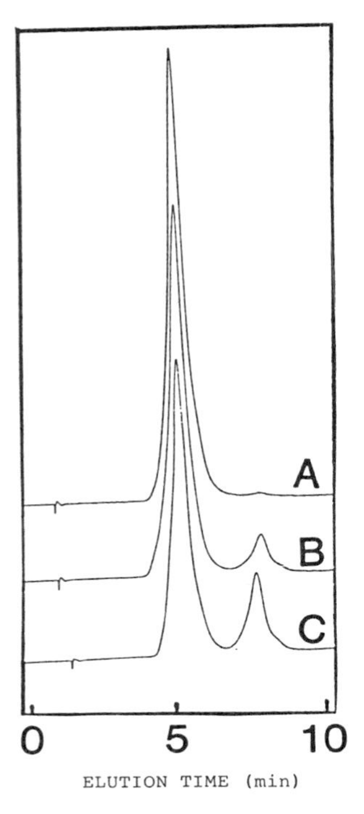

Figure 4. HPLC chromatogram of chlorogenic acid in methanol. **A**. freshly prepared; **B**. after 1 day; and **C**. after 7 days (18).

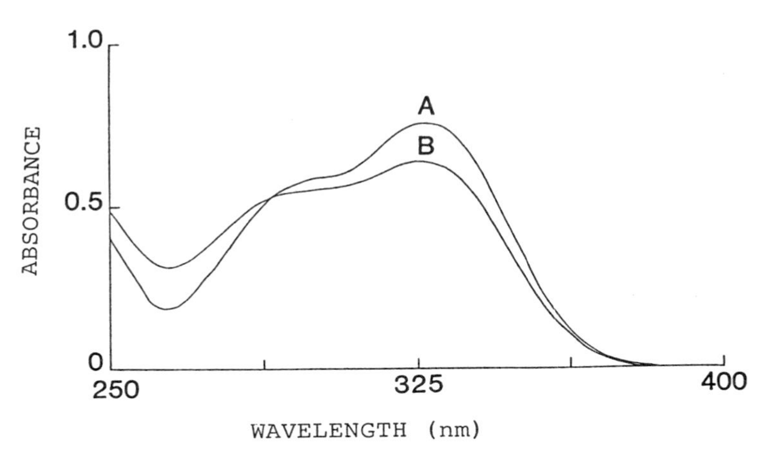

Figure 5. UV spectra of **A**. chlorogenic acid; and **B**. an ethanol extract of potato roots (18).

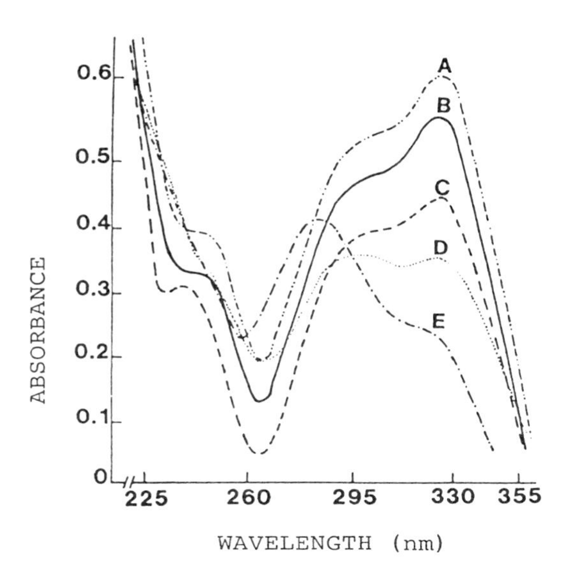

Figure. 6. Effect of baking on the stability of chlorogenic acid measured by UV spectroscopy in a muffin prepared from a mixture of chlorogenic acid containing morning glory seed flour and wheat flour. **A.** chlorogenic acid; **B.** ethanolic extract of morning glory seeds; **C.** ethanolic extract of crumb fraction of oven-baked muffin; **D.** extract of microwave-baked muffin; and **E.** extract of crust fraction of oven-baked muffin (17).

Table VI. Millimolar Concentration of Inhibitors Required to Reduce 400 U PPO/mL by 50% (I_{50}) at 25°C (the lower the value the more potent the inhibitor) (80)

Inhibitor	I_{50}
L-cysteinylglycine	0.43
L-cysteine	0.35
N-acetyl-L-cysteine	0.27
homocysteine	0.23
sodium bisulfite	0.21
2-mercaptoethanesulfonic acid	0.14
L-cysteine ethyl ester	0.12
reduced glutathione	0.12
L-cysteine methyl ester	0.1
2-mercaptopyridine	0.10
N-(2-mercaptopropionyl)glycine	0.06

Temperature and organic content of the soil seems to affect the chlorogenic acid content of Russet Burbank potatoes grown in Canada (27).

Host-Plant Resistance. In reviewing resistance factors in potatoes, Lyon (28) makes the following points about the contribution of phenolics: (a) phenolic compounds and derivatives are an important part of the general defense mechanism, especially against soft rot bacteria. They may function by inhibiting bacterial growth, by inhibiting cell wall-degrading enzymes, and/or as precursors in the formation of physical barriers; (b) various phenolic compounds exhibit different specificities against *Ewinia spp* in vitro; (c) phenolic compounds may inhibit enzymes involved in pathogenesis such as polygalacturonase from *Rhizobia solani* and glucan synthase from *Beta vulgaris*; (d) most of the chlorogenic acid is found in the outer periderm rather than the cortex of the tuber because as mentioned, chlorogenic biosynthesis has an absolute requirement for oxygen; (e) polyphenols react with protein forming insoluble tannins. The precipitated tannins inhibit pectinase in the cell walls by crosslinking reactions; and (f) both phenolics and phytoalexins appear to be involved in the resistance of potatoes to *Erwinia spp*. To what extent each of the native phenolics such as chlorogenic acid and tyrosine and their oxidation and condensation products including semiquinones, quinones, melanins, and suberins contribute to plant resistance is largely unknown.

Dinkle (29) used a histochemical test based on the reaction of chlorogenic with nitrous acid to demonstrate an association of the polyphenol with physiological internal necrosis of potato tubers.

According to Craft and Audia (30), polyphenols, including flavonols, cinnamic acid derivatives, and coumarins accumulate in the sound tissue adjacent to injured tissue in many types of vegetables and fruits. Compounds accumulating in Irish potato tissues as a result of wounding, exposure to pathogens, and virus infection include chlorogenic acid, scopoletin, scopolin, aesculetin, and caffeic acid. This accumulation of polyphenols in sound tissue adjacent to injured tissue seems to give host-plant resistance toward the pathogen. This effect could result from the increased rate of metabolism induced by the injury and/or alteration of the normal metabolic pathways by the injury, leading to the accumulation of unused polyphenols normally metabolized (oxidized) by plant enzymes.

Craft and Audia (30) describe the formation of such wound-barrier layers in sweet potatoes, Irish potatoes, carrots, beets, parsnips, squash, and turnips. The UV spectra of extracts of wounded sweet potatoes, Irish potatoes, and carrots exhibited enhanced maxima at 325 nm, presumably due to chlorogenic acid from suberized but not from interior tissues. The peaks decreased in intensity following oxidation.

Ghanekar et al. (31) found that the chlorogenic acid in wounded periderm was generally higher in infected tubers than in the uninfected ones. In addition, caffeic, chlorogenic, and ferulic acids were more effective against *Erwinia carotovora* when used as mixtures in the proportions found in the periderm of potatoes than individually.

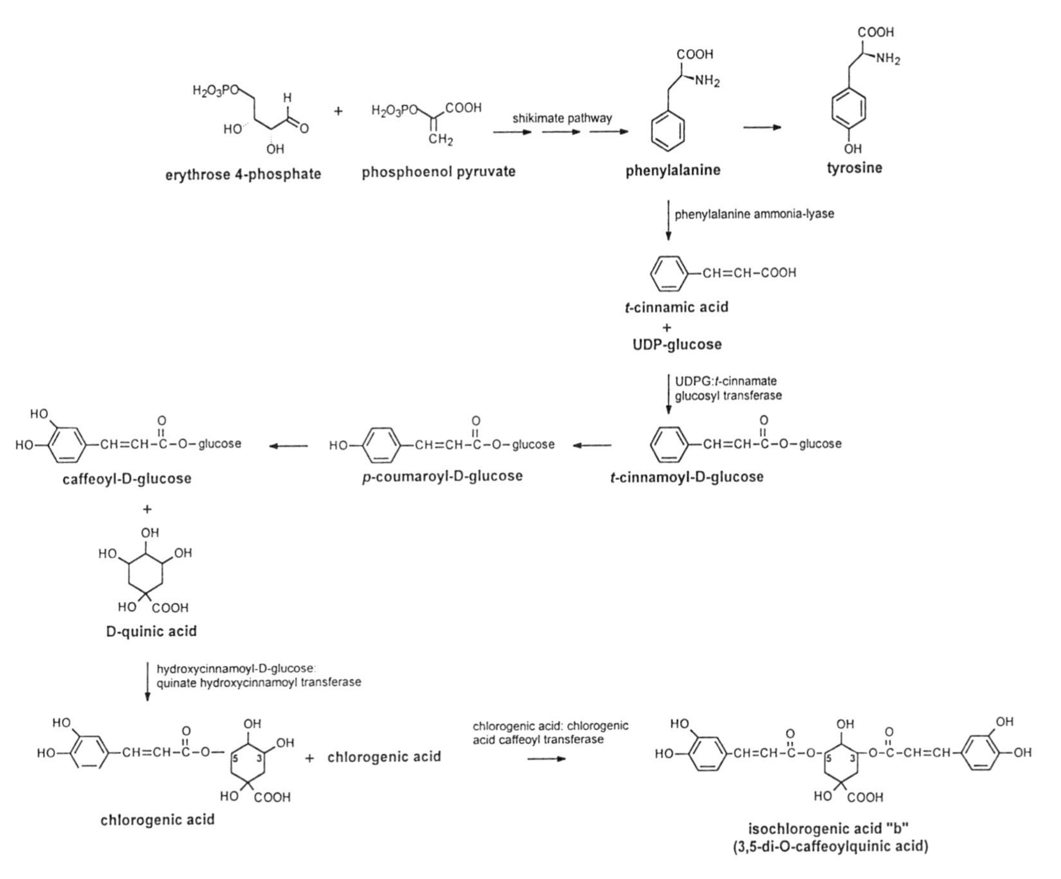

Figure 7. Biosynthetic pathways towards chlorogenic acid (22-26).

These observations raise the possiblity of synergism between structurally related phenolic compounds in imparting resistance to bacteria. Such synergestic action could guide plant breeders to produce improved potato cultivars with the right mixture of structurally different phenolic compounds.

Glandular trichome-bearing leaves of wild potato cultivars such as *Solanum berthaulti* Hawkes are of interest to plant breeders because they confer resistance to several major foliage-feeding insect pests (32). These authors also found that selection of aphid-resistant hybrids can be improved by an improved enzymic browning assay (EBA) based on release of the exudate from tetralobulate (Type A) trichomes of three leaflets in a test tube. The tube contains reagents that change from pink to violet intensity measured at 470 nm. The intensity of the color is related to the polyphenol oxidase and phenolic substrates of the trichomes which interact to form a viscous substance as a result of oxidative transformation to polymeric materials.

Rober et al. (33) found that freshly harvested as well as stored potatoes with soft rot bacteria *Erwinia carotovora* and dry rot fungi *Fusarium spp* produced phenolic compounds and phytoalexins after infection. Uninfected control potatoes synthesized only low levels of these compounds. Infection of potato tubers evidently induces a cascade of metabolic transformations leading to the production of secondary metabolites involved in host-plant resistance.

Kumar et al. (34) found that polyphenolic compounds impart resistance to potato tubers against soft rot caused bu *Erwinia carotovora*, a major cause of losses of stored potatoes. This effect is presumably due to the antibacterial activity of the phenolics or quinones formed by action of polyphenol oxidase on phenolics.

Johnston et al. (35) discuss the possible role of polyphenol oxidation in resistance of potatoes against slugs, which cause considerable damage to main crop potatoes. When potato tissues are damaged, polyphenols are enzymatically oxidized to quinones which then polymerize to dark pigment. The rate of dark pigment formation seems to be important in imparting resistance. As mentioned earlier for other pathogenes, quinones and polyphenolic tannins may inhibit slug digestion of potato tissue by binding to digestive enzymes in the gut and/or binding to proteins lowering protein digestibility (36). Another possibilty is that the polyphenol derivatives bind to proteins in the gut adversely affecting absorption of food (37,38). Spencer et al. (39) discuss the chemistry of such polyphenol complexation reactions.

Jonasson and Olsson (40) found that although differences in glycoalkaloid and sugar content could be related to susceptibility to attack of potatoes to wireworm, *Agrioles obscurus*, this was not the case for chlorogenic acid. Total glycoalkaloid content was the key factor in predicting larval feeding, accounting for 65% of the total variation. Differences in sugar levels of the diet (fructose plus glucose) accounted for 13%. Differences in chlorogenic acid or sucrose levels did not significantly affect the survival of the larvae.

Lyon and Barker (20) found no correlation between chlorogenic acid content in different potato clones and their resistance to infection with potato leafroll virus.

Finally, Belknap et al. (41) report a transient 200-fold increase in phenylalanine ammonia lyase during bruising of Lemhi potatoes about 48 hr after bruise induction and dark color formation. Changes in messenger RNAs encoding stress-inducing proteins and on the main structural potato protein (patatin) genes were also noted. These changes indicate that stress resulting in blackspot bruising elicits a wound response similar to those observed in other plant injuries.

Post-Harvest Events

Browning and Other Discolorations. Potatoes may undergo browning and other discolorations both during growth and after harvest (42). Internal blackspot in potatoes is caused by internal enzymatic browning-type reactions initiated by polyphenol oxidase catalysis with tyrosine as the primary substrate. According to Shetty (43), bruising of potatoes induces these undesirable pigmentations, leading to shrinkage, rotting, and major economic losses. These authors suggest that the sub-surface physiological discoloration resulting from mechanical injury is one of the most important disorders in potato production leading to rejection of shipments. In addition to blackspot formation, bruising also induces the post-harvest biosynthesis of glycoalkaloids which could adversely impact food safety (44,45).

Pavek and colleagues extensively studied the relationship of free- and protein-bound tyrosine to internal blackspot resistance (46,47) and effects on the inheritance of blackspot bruise resistance in potato (48). Enzymatic discoloration was highly correlated with total phenolic ($r = 0.89$) and free tyrosine levels ($r = 0.85$) in potatoes. Mapson et al. (49) found a correlation between tyrosine, the activity of polyphenol oxidase, and the rate of browning in different potato varieties. No such correlation was found for chlorogenic acid and browning. Dean et al. (50) found that the relative rate of tyrosine synthesis in a blackspot-resistant potato cultivar was about 55% less than in the susceptible cultivar Lemhi Russet. Mondy and Munshi (51) report that although free tyrosine was positively correlated with discoloration within a cultivar, it did not appear to be the predominant factor determining blackspot susceptibility of potatoes since the black-spot-susceptible cultivar, Ontario, had higher ascorbic acid and lower tyrosine content than the resistant cultivar, Pontiac. Moreover, ascorbic acid and enzymatic discoloration were not consistently correlated with each other in spite of the fact that the vitamin appears to function as an antioxidant in the process of discoloration.

Evidently, tryrosine-initiated blackspot formation varies in different potato cultivars.

After-Cooking Darkening. The chemical basis for after-cooking-darkening of potatotes, reflected as a steely blue-grey color several hours after cooking has been extensively studied (52-57). It can be measured in two ways: (a) by a test involving a waiting period of 1-12 hr for full color development after steaming; and (b) by a chemical test based on reaction of chlorogenic acid with a mixture of urea, tartaric acid, and sodium nitrite (56). The latter is a rapid, histological staining method based on the formation of a cherry-red color of nitrosylated chlorogenic acid. The histological method for visualizing chlorogenic acid in potato tissues agreed with after-cooking-darkening in a blanching fry-test.

Siciliano et al. (53) found that blackening tendency is directly related to the size of the potato tubers. This trend presumably results from the fact that the stems of large tubers compared to smaller ones have the following compositional characteristic: low citric acid content, high potassium-citric acid ratios, low citric acid-polyphenolic ratios. Citric acid and other chelating agents inhibit potato-blackening by competing with chlorogenic acid for ferrous ion chelating sites.

A process for controlling after-cooking darkening in French fried potatoes is described in a U. S. Patent (57). Treatment of water-blanched, fried potato strips, which often discolor on exposure to air, with a combination of calcium acetate and an oxidation inhibitor minimizes after-cooking darkening during french frying of potato strips. The preventive effect is presumably also due to the complexation of the calcium ions with chlorogenic acid on the surface of the potato strips, preventing formation and oxidation of the chlorogenic acid-ferrous ion complex that causes darkening. As already mentioned, citric acid effectively reduces the color of chlorogenic acid-iron complexes, presumably by competitively chelating ferrous ions (Swain, 1962).

Muneta and Kaisaki (55) found that ascorbic acid plus ferrous iron also forms a purple pigment similar to the after cooking-darkening pigment formed in potatoes from chlorogenic acid and ferrous ions. The ascorbate complex is less stable than the chlorogenic acid complex. Strong chelating agents including citric acid, EDTA, and sodium hydrogen pyrophosphate suppressed color formation resulting from both complexes.

Muneta and Kalbfleish (58) describe a heat-induced-contact discoloration in boiled potatoes, an irregular brown ring near the edge of the contact zone where the unpeeled potato touches the bottom of the pan. Placing unpeeled potatoes in boiling water, peeling potatoes that touch the bottom of the pan, and keeping potatoes away from the bottom of the pan prevents dark ring formation. The authors suggest that the heat-induced-contact discoloration may result from enzymatic-type browning reactions similar to those that may be responsible for blackspot and prepeeling blackening of potatoes. However, Maillard-type browning reactions of amino groups of amino acids and proteins with reducing sugars such as glucose could also explain the heat-induced contact discoloration. Such reactions are described in detail elsewhere (59-63).

Whether the heat-induced discolorations adversely affect the nutritional value of the boiled and darkened potatoes is not known.

Browning Prevention. The sulfhydryl (SH or thiol) compounds such as cysteine, N-acetyl-L-cysteine, and reduced glutathione and ascorbic and citric acids are good inhibitors of the enzyme polyphenol oxidase (PPO) which catalyzes enzymatic browning in fruits and vegetables including potatoes (64-75). We showed that SH-containing amino acids and peptides are good inhibitors of both enzymatic and non-enzymatic browning in freshly prepared and commercial fruit juices, apples, fresh and dehydrated potatoes, heated amino acid-carbohydrate mixtures, and protein-rich foods (76-78; Figure 8).

In some applications, the effectiveness of some of these compounds approached that of sodium sulfite, whose use is being discontinued because many asthmatic individuals are sensitive to it (81). However, in searching for the most potent and safe replacement for sodium sulfite, it became apparent that the potency of different thiols as PPO inhibitors varies widely, depending on both the structure of the thiol and the environment in which the inhibition is carried out. To compare relative effectiveness of structurally different thiols in a food matrix as well as in pure enzyme solution, we evaluated the effectiveness of a series of sulfhydryl compounds in PPO solutions and in dehydrated potato suspensions at 25°C. These inhibitors were less effective in the potato suspensions, and the relative effectiveness of structurally different thiols differed from those observed in PPO solutions. Table VI shows relative inhibitory potencies of several thiols against pure PPO.

Although the reasons for these differences are not immediately apparent, possibilities include slow diffusion of the inhibitors to PPO and substrates in the heterogenous dehydrated potato suspension, the action of PPO on substrates other than tyrosine such or chlorogenic acid (82), and the reaction of the thiols with carbohydrates. The latter reactions may also be beneficial since they may prevent non-enzymatic browning (76).

These considerations suggest that no general conclusions can be made about the PPO inhibition by thiols in specific foods. Each food category has to be evaluated separately with each sulfhydryl compound in order to find optimum conditions to prevent both enzymatic and non-enzymatic browning under specific conditions of storage, transport, and processing.

Our studies show that SH-containing compounds, especially cysteine ethyl ester and reduced glutathione, are potent inhibitors of polyphenol oxidase. They are therefore potential sulfite-substitutes for browning prevention. Since the effectiveness of a specific sulfhydryl compound is influenced by its structure, additional studies are needed to optimize the anti-browning action of a variety of structurally different SH-containing amino acids and peptides with different physicochemical properties (80).

Finally, the practical significance for potato quality and nutrition of browning prevention through suppression of genes involved in the biosynthesis of polyphenol oxidase through antinsense RNA merits further study (83,84).

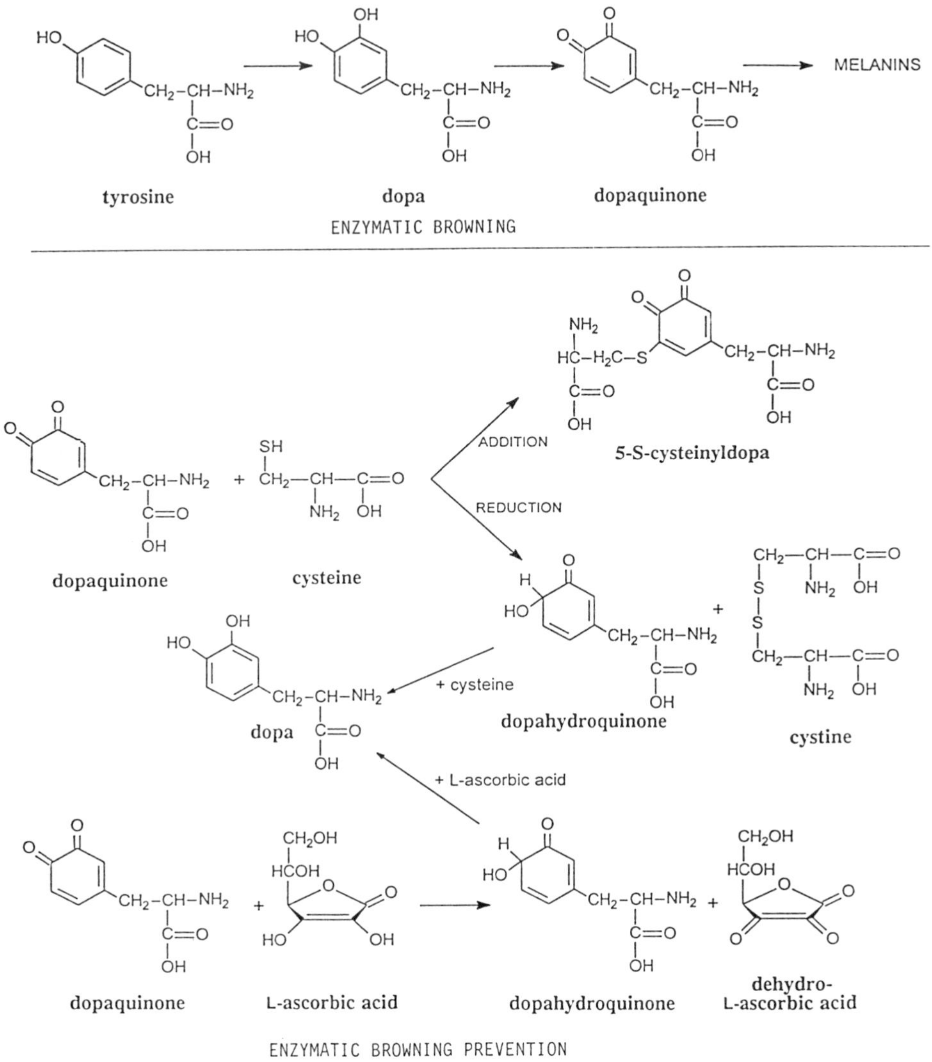

tyrosine → dopa → dopaquinone → MELANINS

ENZYMATIC BROWNING

dopaquinone + cysteine — ADDITION → 5-S-cysteinyldopa

dopaquinone + cysteine — REDUCTION → dopahydroquinone + cystine

dopahydroquinone + cysteine → dopa

dopaquinone + L-ascorbic acid → dopahydroquinone + dehydro-L-ascorbic acid

dopahydroquinone + L-ascorbic acid → dopa

ENZYMATIC BROWNING PREVENTION

Figure 8. Transformation of tyrosine to melanin via DOPA and dopaquinone (enzymatic browning) and postulated inhibition of enzymatic browning by cysteine and ascorbic acid by trapping the dopaquinone intermediate (80).

Effect of Light. Potatoes contain polyphenolic compounds such as chlorogenic acid, glycoalkaloids such as α-chaconine and α-solanine, and proteins, which have the ability to inhibit trypsin, chymotrypsin, and carboxypeptidase (19, 85-87). Light, mechanical injury, temperature extremes, and sprouting induce an increase in the glycoalkaloid content which, depending on conditions used, can range up to five times original levels. Exposure to light after harvest also leads to surface greening and to increases in glycoalkaloid content and of chlorophyll biosynthesis.

However, the increase in glycoalkaloid content may be undesirable since it could impart a bitter taste to potatoes (88-92) and make them less safe. Since chlorogenic acid, glycoalkaloids, and protease inhibitors may act as so-called anti-feeding agents in potatoes, protecting potatoes from attack by phytopathogens and insects, the question arises whether suppression of biosynthesis of alkaloids, one of our current objectives (93), will result in compensatory changes in the biosynthesis of chlorogenic acid and/or protease inhibitors. Potatoes were stored in the dark and under fluorescent light for various time periods and analyzed for chlorophyll, chlorogenic acid, α-chaconine, α-solanine, and inhibitors of the digestive enzymes trypsin, chymotrypsin, and carboxypeptidase A.

Exposure of commercial White Rose potatoes to fluorescent light for 20 days induced a time-dependent greening of potato surfaces; an increase in chlorophyll, chlorogenic acid, and glycoalkaloid content (α-chaconine and α-solanine); and no changes in the content of inhibitors of the digestive enzymes trypsin, chymotrypsin, and carboxypeptidase A (19). The maximum chlorophyll level of the light-stored potatoes was 0.5 mg/100 g of fresh potato weight. Storing potatoes in the dark did not result in greening or chlorophyll formation. Chlorogenic acid and glycoalkaloid levels of dark-stored potatoes did increase but less than in the light-stored potatoes. In the light, chlorogenic acid concentration increased from 7.1 mg/100 g fresh potato weight to a maximum of 15.8 after greening. The corresponding values for α-chaconine are 0.66 and 2.03, and for α-solanine 0.58 and 1.71, respectively, or an approximately 300% increase for each glycoalkaloid. The inhibitors of trypsin, chymotrypsin, and carboxypeptidase A (respectively, about 1000, 375, and 100 units per gram of dehydrated potato powder) were not changed. Experiments on delay of greening by immersion in water suggest that the concentration of chlorophyll is 26 times greater, of chlorogenic acid and glycoalkaloids 7 to 8 times greater, and of protease inhibitors about 2 to 3 times lower in the peel of the green potatoes than in the whole tuber.

Griffiths et al. (94) confirmed that exposure of potatoes to light induces significant increases in chlorogenic acid content. This increase appears to be cultivar-dependent since the magnitude of the increase correlated with the initial value found in unexposed potatoes.

The described compositional changes should help define consequences of potato breeding and greening for plant physiology, food quality, and food safety. Specifically, this information may be useful for breeding potato varieties with an initial low chlorogenic acid content to minimize both content and rate of increase, e. g., to minimize after-cooking darkening and/or enzymatic browning during food processing, if that is the objective. However, since chlorogenic acid is reported to exert beneficial effects on health as described below, another desirable objective might be to breed for high-chlorogenic acid varieties and then significantly enhance the chlorogenic acid content by post-harvest exposure to light. The potential value of potatoes with a very high content of chlorogenic acid to prevent disease and promote health merits study.

Effect of γ-Radiation. Different investigators report different results on the effect of γ-radiation on chlorogenic acid content of potatoes (95-107). Such radiation effectively prevents sprouting and protects potatoes during storage and possibly also against damage by fungi and other phytopathogens (103). This is a desirable objective since when potatoes sprout, they decrease in weight, quality, and market value (97).

Penner and Fromm (101) used a quantitative TLC method for the direct determination of chlorogenic acid in irradiated potatoes. Chlorogenic acid content rose immediately after irradiation and then returned to normal values within several weeks of storage.

Bergers (95) found a time-dependent decrease in chlorogenic acid and a glycoside of scopoletin (7-hydroxy-6-methoxy coumarin) after potato tubers were irradiated up to 3 kGy. Irradiation had no effect on the glycoalkaloid content of the potatoes. The author suggests that a dose as low as 0.1 kGy may be sufficient to inhibit sprouting with minimal compositional changes in the treated potatoes. Infection by fungi also leads to an accumulation of scopolin (96). Mondy and Gosselin (97,98) found that irradiation increased discoloration and phenolic content and decreased lipid and phospholipid content of potatoes. Irradiated potatoes stored at 5°C had a higher total phenol content than those at 20°C. These authors also report that a radiation dose of 10 Krad caused less darkening than a higher dose. The increased blackspot formation at higher doses may have been due to rupture of lipid membranes. The rupture may liberate polyphenol oxidase from mitochondria, allowing it to contact phenolic substrates in the vacuole, causing enzymatic darkening. Darkening could therfore be minimized by controlling the dose.

Generally, γ-radiation can suppress wound periderm formation favoring pathogen access to wounds. A defense mechanism against this effect is the formation of quinones by oxidation of phenolic compounds by polyphenol oxidase released by rupture of the potato tissue. The quinones then polymerize to defensive polymers at the site of the tissue injury, as discussed earlier.

Ramarmurthy et al. (102) found an increased formation of potato phenolics during wound-healing of tubers following irradiation (Table VII). The HPLC analysis revealed that three chlorogenic acid isomers accounted for about 88% of the total phenolics measured by HPLC. The formation and accumulation of *neo* and *crypto* isomers of chlorogenic acid during wound healing is also noteworthy. Irradation of tubers to inhibit sprouting caused 50% reduction in both total phenolic and chlorogenic acid content.

Pendharkar and Nair (100) found that γ-irradiation up to 10 Krad had no effect on the defense mechanism at the site of injury of potatoes, i.e., the formation of quinones from phenolic acids. It did, however, reduce by about 50% the ability of tubers to synthesize chlorogenic and caffeic acids in the wounded tissue. The radiation also impaired the induction of cinnamic acid-4-hydroxylase but not phenylalanine ammonia lyase, both of which are involved in chlorogenic acid biosynthesis.

According to Thomas and Delincee (105), wounding of potato tuber tissues induces metabolic and cytological changes which may lead to wound healing. Such healing is accompanied by the deposition of suberin, a polymer derived from polyphenols and fatty acids. Such suberization prevents weight loss and permanent damage to potatoes. These authors also found that γ-radiation at sprout-inhibiting doses did not prevent suberization.

Cheung and Henderson (106) found an increase in polyphenol oxidase activity after potatoes were irradiated at 2 Krad. In contrast, the activity decreased when the dose increased. Prolonged storage of the potatoes also resulted in lower enzyme activity. These results imply that radiation could also be used to control enzymatic browning.

Ogawa and Uritani (99) found considerable browning in stored and irradiated potatoes, especially in the cortex and along the xylem. Browning was accompanied by a marked increase in polyphenol content and peroxidase activity and a transient decrease in polyphenol oxidase activity. The extent of browning strongly dependend on the storage period from harvest to irradiation. These authors recommend that to minimize browning, potatoes should be stored at ambient temperature for about one month before irradiation.

Thomas (104) found that (a) peeling of potatoes prior to boiling reduces the intensity of γ-radiation-induced after-cooking darkening of the tuber flesh. Presumably, leaching of polyphenols from the flesh into the cooking water minimizes the availability of polyphenols to form dark iron complexes. The amount of leaching into the cooking water was three to five times greater in pre-peeled tubers than in whole tubers. It is also possible that leaching of ferrous compounds into the cooking water also contributes to reduced darkening.

Finally, Leszczynski et al. (107) found that (a) γ-radiation inhibited sprouting of potatoes; (b) the irradiated tubers had a lower starch and higher sucrose content; (c) radiation produced darker potatoes; and (d) radiation had no or minimal effects on the quality of potato chips made from the treated potatoes. Since radiation inhibited sprouting during storage, the production of good quality chips was easier from irradiated than from control potatoes.

Role of Polyphenols in the Diet

Flavor and Taste. Sinden et al. (91) found a correlation between glycoalkaloid content and potato flavor, but no significant correlation between phenolic content and either bitterness or burning sensation. In contrast, Mondy et al. (89) report

Table VII. Changes in Phenolic Compounds (mg/100 g Fresh Weight) in Potatoes during Wound Healing (adapted from 111)

	Wound Healing Period, Days					
Phennolic Compound	0	4	6	8	10	15
3-0-caffeoylquinic acid (neochlorogenic acid)		13.6	24.2	31.4	15.7	11.7
4-0-caffeoylquinic acid (cryptochlorogenic acid)		5.8	10.4	13.4	6.8	4.9
5-0-caffeoylquinic acid (chlorogenic acid)	4.5	34.1	60.8	78.8	38.2	28.5
caffeic acid	1.8	5.7	6.4	7.8	6.8	4.9
p-coumaric acid	1.6	4.5	5.0	5.1	4.4	3.5
ferulic acid	1.3	3.1	4.0	4.4	4.0	3.2
total	9.3	66.7	110.8	140.9	75.8	56.7

Table VIII. Oxidation Potentials (in Volts) of Selected Phenolic Acids at pH 4.7 (adapted from 102)

Phenolic Acid	Potential
3,4-dihydoxycinnamic (caffeic)	0.35
chlorogenic	0.39
3,4-dihydroxyphenylpropionic (dihydrocaffeic)	0.43
3,4,5-trihydroxybenzoic (gallic)	0.46
3,4-dihydroxybenzoic (protochatechuic)	0.52
4-hydroxy-3,5-dimethoxybenzoic (syringic)	0.55
4-hydroxy-3-methoxycinnamic (ferulic)	0.57
4-hydroxy-3-methoxybenzoic (vanillic)	0.72
4-hydroxycinnamic (p-courmaric)	0.73
4-hydroxyphenylacetic	0.77
4-hydroxybenzoic	0.99

a positive correlation between phenolic content and bitterness and astringency of potatoes.

Since potatoes contain both glycoalkaloids and polyphenols in varying amounts depending on variety, the net effect of taste and flavor could be the result of combined, possibly synergistic, effects of both components (88,90).

Antioxidants. Polyphenolic compounds in potatoes show antioxidative activity in several food systems (108-112). For example, Onyeneho and Hettiachchy (108) evaluated the effectiveness of freeze-dried extracts from the peels of six potato varieties for their ability to prevent soybean-oil oxidation. Using an active oxygen method, they found that 20 g of soybean oil treated with 50 mg of the extracts had lower peroxide value (PV, 22 - 28) than did a control oil sample (PV, 109). HPLC and TLC studies suggested that the chlorogenic and protocatechuic acids were the main antioxidants in the extracts. Peels from red potatoes contained greater amounts of polyphenols than those from brown-skinned varieties.

In related studies, Rodriguez de Sotillo et al. (109,110) confirm the strong antioxidant activity of freeze-dried extracts of potato peel waste in sunflower oil. The total polyphenolic content of the peel determined by HPLC consisted of 50.3% chlorogenic acid, 41.7% gallic acid, 7.8% protocatechuic acid, and 0.21% caffeic acids. Al-Saikhan et al. (111) found that the phenolic content and antioxidative activity of four potato cultivars was genotype-dependent and not related to flesh color.

These results suggest the possible value of potato peel in the prevention of oxidative rancidity of food oils.

Chlorogenic acid and other polyphenols also exhibit strong *in vitro* antioxidant activity for lipoproteins (112). *In vivo* oxidations of lipoproteins (LDL) appears to be a major cause of heart disease. It is thus possible that chlorogenic acid and other polyphenols may also lessen heart disease.

As part of an effort to develop an HPLC method for plant phenolics with amperometric detection, Felice et al. (113) measured the anodic oxidation potentials of structurally different compounds. The low values for caffeic and chlorogenic acids (Table VIII) suggest that the presence of the acrylic acid group in these compounds conjugated with the aromatic ring facilitates oxidation to the corresponding quinones. The free electron of peroxy radicals present in oxidized fatty acids or lipoproteins can be abstracted. Generally, phenols are more reactive with the peroxyl radicals than the corresponding quinones. On the other hand, quinones are better metal-chelating antioxidative agents. Both aspects operate in the antioxidative effect. The free electron on the resulting quinone radical is then stabilized by dissipation of the charge through the conjugated system (Figure 9). The latter radical is therefore much less reactive than the peroxy radical (114,118).

Thus, the oxidation potentials of polyphenols, both free and in various food environments, may be useful for predicting their relative potencies as antioxidants. This aspect merits further study.

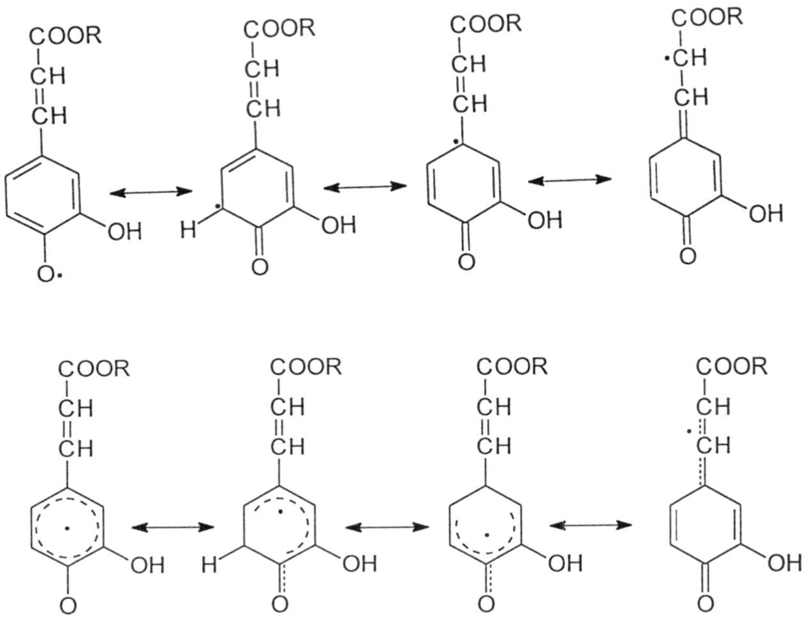

Figure 9. Resonance stabilization of a phenoxyl radical through delocalization of the free electron throughout the conjugated system of chlorogenic acid. The top structures show resonance forms in which the free electron is localized on oxygen or carbon and the bottom ones in which it is delocalized. The actual resonance structure is a hybrid of the depicted forms (143). The depicted quinone radical is more stable (has a lower ground-state energy) than, for example, a linoleic acid peroxy radical. Mixing chlorogenic acid with such a peroxide would therefore result in transfer of the electron from the peroxide to the electron sink of chlorogenic acid (antioxidative effect).

While behaving as antioxidants in foods and *in vivo*, semiquinones and quinones formed on oxidation of polyphenols can also simultaneously react with other molecules including amino acids and structural and functional proteins (2, 61-63, 72, 118). For example, Figure 10 shows some possible protein derivatives that can form from reaction of chlorogenic acid quinone with active-hydrogen-bearing protein functional groups such as NH_2 groups of lysine, SH groups of cysteine, and OH groups of serine. Note that chlorogenic acid quinone can form protein adducts both at the acrylic acid side chain (adduct A) and on the benzene ring (adducts B, C) and that the carbon atom to which the protein becomes attached becomes asymmetric. In each case, the addition reaction can therefore create more than one diastereoisomer. The nutritional and toxicological significance of consuming such protein derivatives is largely unknown.

Elsewhere, we examined in detail the kinetics, mechanisms, and synthetic aspects of related nucleophilic addition reactions of protein functional groups to conjugated double bonds (119-128). The oxidation of a polyphenol to a quinone followed by participation of the quinone in nucleophilic addition reactions is mechanistically analogous to the corresponding oxidation of a prophyrin to a dehydroporphyrin intermediate followed by the addition of nucleophiles to the dehydroporphyrin (118,129,130). These studies should facilitate the design of experiments for the synthesis of chlorogenic acid quinone adducts for biological evaluation.

Antimutagenic and Anticarcinogenic Effects. Nitrites in food can react with secondary amines to form mutagenic and carcinogenic nitrosamines. Chlorogenic acid and other polyphenols are reported to block nitrosamine formation by competitively reacting with the nitrite (131,132).

Camire et al. (133) found that a model system containing cellulose plus chlorogenic acid bound 100% of the carcinogen benzo(a)pyrene compared to 66% by cellulose alone, and 32% by a cellulose-quercetin mixture. Potato peels bound more of the carcinogen than did wheat bran, cellulose, or arabinogalactan. Extrusion of the peel at 110°C reduced the affinity of the carcinogen for the potato peel. Such extrusion may result in destruction of chlorogenic acid.

The authors speculate that chlorogenic acid in the peel may interact with benzopyrene to form an insoluble complex, possibly as does chlorophyll with carcinogenic heterocyclic amines (63).

These observations suggest that additional studies are merited on the health-promoting properties of high-chlorogenic-acid potatoes, especially freeze-dried potato peels.

Generally, inhibition of cancer development by polyphenolic compouds could be due to their ability to scavenge and trap potentially DNA-damaging electrophiles and free radicals, to inhibit enzymes that activate pre-carcinogens to carcinogens, and to induce carcinogen-detoxifying enzymes (134-138). This aspect also merits study.

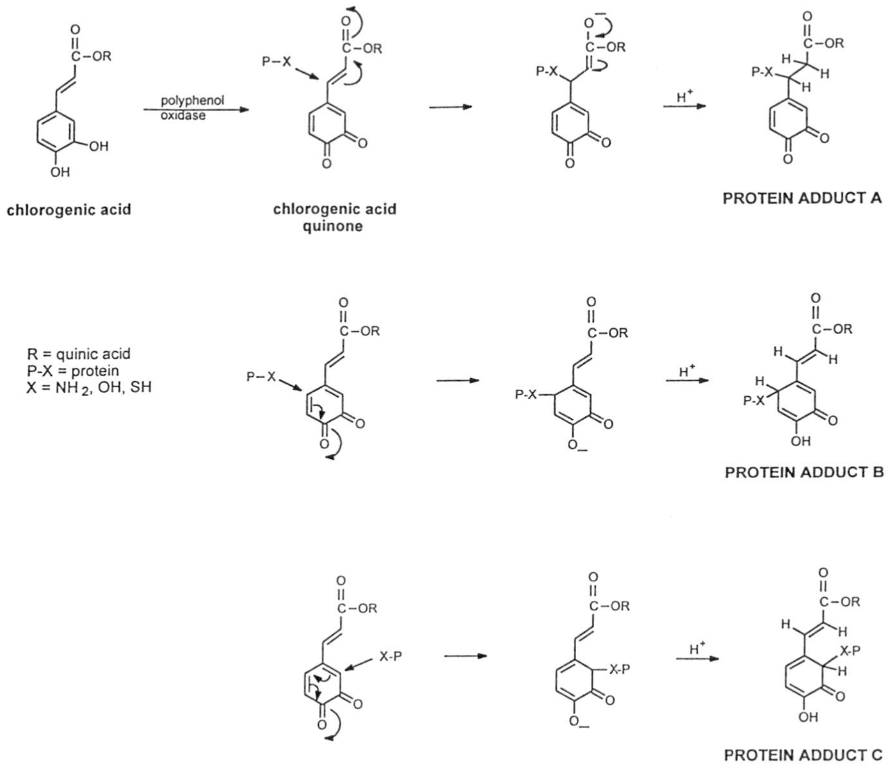

Figure 10. Nucelophilic addition reactions of protein functional groups to the conjugated systems of chlorogenic acid quinone. Addition can take place at the acrylate ester side chain to form product A and at two positions of the quinone ring to produce the derivates B and C. In addition, each of the trigonal carbon atoms with which P-X combines becomes tetrahedral and asymmetric, creating the possibility of two diastereoisomers in each case. Losses of chlorogenic acid during baking shown in Figure 6 may be due to these and related transformations during food processing.

Antiglycemic Properties. Thompson et al. (139) report that the polyphenol content of potatoes, legumes, and cereals correlated negatively with the blood glucose response (glycemic index) of normal and diabetic humans consuming them in a controlled study. The glucose-lowering effect of the polyphenols may arise from their ability to inhibit amylases (which catalyze the hydrolysis of starch to glucose) and proteolytic enzymes (that catalyze the hydrolysis of proteins to free amino acids in the digestive tract) and/or to direct complexation between the polyphenols and starch, preventing digestion.

In conclusion, future studies should be optimizing beneficial effects of polyphenols in the potato and other plants and in the diet. A major concern of food safety research is whether phytochemicals such as chlorgenic acid behave differently, depending on whether they are consumed alone or part of a complex diet which could affect their stability, interaction with other dietary components, and consequently their efficacy.

Research Needs

The preceding analysis of the current knowledge of potato polyphenols in the plant and in the diet shows that these studies constitute an evolving, potentially beneficial area of food science, food safety, and nutrition with many unsolved problems. In addition to recommendations made earlier to facilitate progress to enhance the value of the potato plant as a health-promoting food, we are challenged to respond to additional research needs outlined below.

1. Define relative antioxidative, antimutagenic, anticarcinogenic, and antiglycemic properties of structurally different chlorogenic acid isomers.
2. Determine the content of these isomers in commercial potato varieties and processed potato products.
3. Enhance the content of the most potent antioxidative isomers by exposure of potatoes to light and by plant breeding and plant molecular biology techniques.
4. Since proteins isolated from various leaves have high nutritional value (62), prepare and evaluate protein concentrates from potato leaves which contain 10-16 time greater amounts of chlorogenic acid than potato tubers (Table II). Such concentrates may have both desirable nutritional and antioxidative properties, provided leaf glycoalkaloids are removed during protein isolation (45, 141).
5. Assess whether antioxidative potencies of polyphenols in terms of their abilities to trap damaging hydroxyl (OH•), alkoxyl (RO•), peroxyl (RCOO•), and superoxide anion ($O_2^{\bullet -}$) radicals can be predicted from the *net* electrochemical oxidation-reduction potentials of mixtures of structurally different polyphenols actually present in potatoes and other plant foods.

6. Determine whether two or more polyphenols can act synergistically in reducing oxidative stress.
7. Define effects of food processing on polyphenols in potatoes.
8. Carry out animal and human feeding studies with high-chlorogenic acid potato diets to assess whether beneficial effects of chlorogenic acid in vitro are confirmed in vivo (142).

Acknowledgements

I am grateful to E. N. Frenkel, R. A. Jacob, and S. Schwimmer for reviewing this paper, and to Gary M. McDonald for computer-drawing the structural formulae.

Presented at the Symposium on Antinutrients and Phytochemicals, Division of Agricultural and Food Chemistry, American Chemical Society Meeting, Chicago, IL, August 20-28, 1995.

Literature Cited

1. Deshpande, S. S.; Sathe, S. K.; Salunkhe, D. K. In *Nutritional and Toxicological Aspects of Food Safety*; Friedman, M., Ed.; Plenum: New York, **1984**; pp 457-495.
2. Hurrell, R. F.; Finot, P. A. In *Nutritional and Toxicological Aspects of Food Safety*; Friedman, M., Ed.; Plenum: New York, **1984**; pp 423-435.
3. Lee, C. Y.; Whitaker, J. R. (Eds.), *Enzymatic Browning and its Preventions, ACS Symp. Ser. No. 600*, American Chemical Society; Washington, D. C. **1995.**
4. Wong, D. W. S. *Mechanism and Theory in Food Chemistry*; AVI-Van Nostrand Reinhold: New York, **1989**.
5. Schwimmer, S. *Sourcebook of Food Enzymology*; AVI: Westport, CT, **1981**.
6. Voigt, J.; Noske, R. *Nahrung* **1964**, 8, 19-26.
7. Reeve, R. M.; Hautala, E.; Weaver, M. L. *Am. Potato J.* **1969**, 46, 374-386.
8. Hasegawa, D.; Johnson, R. M.; Gould, W. A. *J. Agric. Food Chem.* **1966**, 14, 165-169.
9. Brandle, W.; Herrmann, K. *Z. Lebensm. Unters. Forsch.* **1984**, 178, 192-194.
10. Molgaard, P.; Ravn, H. *Phytochemistry* **1988, 27, 2411-2421.**
11. Tisza, S.; Molnar-Perl, I.; Friedman, M.; Sass, P. *J. High Resol. Chromatogr.* **1996**, 19, 54-58.

12. Nagels, L.; Van Dongen, W.; De Brucker, J.; De Pooter, H. *J. Chromatogr.* **1980**, 187, 181-187.
13. Sosulski, S.; Krygier, K.; Hoggs, L. *J. Agric. Food Chem.* **1982**, 30, 337-340.
14. Griffiths, D. W.; Bain, H.; Dale, M. F. B. *J. Sci. Food Agric.* **1992,** 68, 105-110.
15. Malmberg, A. G.; Theander, O. *J. Agric. Food Chem.* **1985**, 33, 549-551.
16. Friedman, M; Dao, L.; Gumbmann, M. R. *J. Agric. Food Chem.* **1989**, 37, 708-712.
17. Friedman, J.; Dao, L. *J. Agric. Food Chem.* **1990**, 38, 805-808.
18. Dao, L.; Friedman, M. *J. Agric. Food Chem.*, **1992**, 40, 2152-2156.
19. Dao, L.; Friedman, M. *J. Agric. Food Chem.* **1994**, 42, 633-639.
20. Lyon, G. D.; Barker, H. *Potato Res.* **1984,** 27, 291-295.
21. Leja, M. *Acta Physiol. Plant.* **1989, 11,** 201-206.
22. Herrmann, K. M. *Plant Physiol.* **1995**, 107, 7-12.
23. Schmid, J.; Amrhein, N. *Phytochemistry* **1995**, 39, 737-749.
24. Villegas, R. J. A.; Kojima, M. *J. Biol. Chem.* **1986**, 261, 8729-8733.
25. Moriguchi, T.; Villegas, R. J. A.; Kondo, T.; Kojima, M. *Plant Cell Physiol.* **1988**, 29, 1221-1226.
26. Tanaka, M.; Kojima, M. *Archiv. Biochem. Biophys.* **1991**, 284, 151-157.
27. Kaldy, M. S.; Lynch, D. R. *Am. Potato J.* **1983**, 60, 375-377.
28. Lyon, G. D. *Plant Pathol.* **1989**, 38, 313-339.
29. Dinkle, D. H. *Am. Potato J.* **1964**, 40, 149-153.
30. Craft, C. C.; Audia, W. M. *Botan. Gazz.* **1962**, 123, 211-219.
31. Ghanekar, A. S.; Padwal-Desai, S. F.; Nadkarni, G. B. *Potato Res.* **1984**, 27, 189-199.
32. Ave, D. A.; Eannetta, N. T.; Tingery, W. M. *Am. Potato J.* **1986,** 43, 533-558.
33. Rober, K. C. *Biochem. Physiol. Pflanzen* **1989**, 184, 277-284.
34. Kumar, A.; Pundhir, V. S.; Gupta, K. C. *Potato Res.* **1991**, 34, 9-16.
35. Johnston, K. A.; Kershwa, W. J. S.; Pearce, R. S. In *British Crop Protection Council Monograph. Slugs and Snails in World Agriculture.* **1989**, 41, 281-288.
36. Oste, R. E. In *Nutritional and Toxicological Consequences of Food Processing*; Friedman, M., Ed.; Plenum: New York, **1991;** pp 371-388.
37. Friedman, M. (Ed.). *Absorption and Utilization of Amino Acids*; CRC Press: Boca Raton, Fl, **1989**.
38. Griffiths, D. W. In *Nutritional and Toxicological Aspects of Food Safety*; Friedman, M., Ed.; Plenum: New York, **1984**; pp 509-515.
39. Spencer, C. M.; Cai, Y.; Martin, R.; Gaffney, S. H.; Goulding, P. N.; Magnolato, D.; Lilley, T. H.; Haslam, E. *Phytochemistry* **1988**, 27, 2397-2409.

40. Jonasson, T.; Olsson, K. *Potato Res.* **1994**, 37, 205-216.
41. Belknap, W. R.; Rickey, T. M.; Rockhold, D. R. *Am. Potato J.* **1990,** 67, 253-265.
42. Stark, J. C.; Corsini, D. L.; Hurley, P. J.; Dwelle, R. B. *Am. Potato J.* **1985**, 62, 657-666.
43. Shetty, K. K.; Dwelle, R. B.; Fellman, J. K.; Patterson, M. E. *Potato Res.* **1991**, 34, 253-260.
44. Friedman, M. ACS Symp. Ser., **1992**, No. 484, 429-462.
45. Friedman, M.; McDonald, G. M. *Crit. Rev. Plant Sci.* **1996**, in press.
46. Corsini, D. L.; Pavek, J. J.; Dean, B. *Am. Potato J.* **1992**, 69, 423-435.
47. Pavek, J.; Corsini, D.; Nissley, F. *Am. Potato J.* **1985**, 62, 511-517.
48. Pavek, J. J.; Brown, C. R.; Martin, M. W.; Corsini, D. L. *Am. Potato J.* **1993**, 70, 43-48.
49. Mapson, L. W.; Swain, T.; Tomalin, A. W. *J. Sci. Food Agric.* **1963,** 14, 673-684.
50. Dean, B. D.; Jacklowiack, N.; Munck, S. *Potato Res.* **1992**, 35, 49-53.
51. Mondy, N. I.; Munshi, C. B. *J. Agric. Food Chem.* **1993**, 41, 1868-1871.
52. Hughes, J. C.; Swain, T. *J. Sci. Food Agric.* **1962**, 13, 358-363.
53. Siciliano, J.; Heisler, E. G.; Porter, W. L. *Am. Potato J.* **1969**, 46, 91-97.
54. Heisler, E. G.; Siciliano, J.; Porter, W. L. *Am. Potato J.* **1969**, 46, 98-107.
55. Muneta, P.; Kaisaki, F. *Am. Potato J.* **1985**, 62, 531-536.
56. Mann, J. D.; De Lambert, C. *New Zealand J. Crop Hort. Sci.* **1989**, 17, 207-209.
57. Mann, J. D. U. S. Patent 5 391 384 (**1995**); Food Science and Technology Abstracts (FSTA), **1995**, 7, 7J 199.
58. Muneta, P.; Kalbfleisch, G. *Am. Potato J.* **1987**, 64, 11-15.
59. Felton, G. W.; Donato, K. J.; Broadway, R. M.; Duffey, S. S. *J. Insect Physiol.* **1992**, 38, 277-285.
60. Friedman, M. In *Nutritional and Toxicological Consequences of Food Processing*; Friedman, M., Ed.: Plenum: New York, **1991**; pp 171-215.
61. Friedman, M. *J. Agric. Food Chem.* **1994**, 42, 3-20.
62. Friedman, M. *J. Agric. Food Chem.* **1996**, 44, 6-29.
63. Friedman, M. *J. Agric. Food Chem.* **1996**, 44, 631-653.
64. Muneta, P.; Walradt, J. *J. Food Sci.* **1968**, 33, 606-608.
65. Muneta, P. C. *Am. Potato J.* **1981**, 38, 1202-1204.
66. Dudley, E. D.; Hotchkiss, J. H. *J. Food Biochem.* **1989**, 13, 65-75.
67. Friedman, M.; Grosjean, O. K.; Zahnley, J. C. *Food Chem. Toxicol.* **1986**, 24, 497-502.
68. Golan-Goldhirsh, A.; Whitaker, J. R. ; Kahn, V. In *Nutritional and Toxicological Aspects of Food Safety*; Friedman, M., Ed.; Plenum: New York, **1984**; pp 457-495.

69. Langdon, T. T. *Food Technol.* **1987**, 41, 64-67.
70. Matheis, G. Properties of potato polyphenol oxidase. *Chem. Mikrobiol. Technol. Lebensm.* **1987**, 11, 5-12.
71. Matheis, G. *Chem. Mikrobiol. Technol. Lebensm.* **1987**, 11, 33-41.
72. Matheis, G.; Whitaker, J. R. *J. Food Biochem.* **1984**, 8, 137-162.
73. Sanchez-Ferrer, A.; Laveda, F.; Garcia-Carmona, F. *J. Agric. Food Chem.* **1993**, 41, 1219-1224.
74. Sapers, G. M. *Food Technol.* **1993**, No. 10, 75-84.
75. Sapers, G. M.; Miller, R. L. *J. Food Sci.* **1992**, 57, 1132-1135.
76. Friedman, M.; Molnar-Perl, I. *J. Agric. Food Chem.* **1990**, 38, 1642-1647.
77. Molnar-Perl, I.; Friedman, M. *J. Agric. Food Chem.* **1990,** 38, 1648-1651.
78. Molnar-Perl, I.; Friedman, M. *J. Agric. Food Chem.* **1990**, 38, 1652-1656.
79. Friedman, M.; Molnar-Perl, I.: Knighton, D. *Food Additives Contam.* **1992**, 9, 499-503.
80. Friedman, M.; Bautista, F. F. *J. Agric. Food Chem.* **1995**, 43, 69-76.
81. FDA. *Fed. Reg.*, **1990**, 55, 9826-9833.
82. Friedman, M.; Smith, G. A. *Food Chem. Toxicol.* **1984**, 22, 535-539.
83. Thwaites, T. *New Scientist* **1995**, No. 1, 24.
84. Thygesen, P. W.; Dry, I. B.; Robinson, S. P. In *The Molecular and Cellular Biology of the Potato*; Belknap, W. R.; Vayda, M. E.; Park, W. D., Eds.; CAB International: Wallingford, UK **1994**; pp 151-159.
85. Brown, W. E.; Graham, J. S.; Lee, J. S. and Ryan, C. A. In *Nutritional and Toxicological Significance of Enzyme Inhibitors in Foods,* Friedman, M., Ed.; Plenum: New York, **1986**; pp 281-290.
86. Friedman, M.; Dao, L. *J. Agric. Food Chem.* **1992**, 40, 419-423.
87. Lisinska, G.; Leszczynski, W. In *Potato Science and Technology;* Elsevier Applied Science: London and New York, **1989**; pp 129-164.
88. Kaaber, L. *Norwegian J. Agric. Sci.* **1993**, 7, 221-229.
89. Mondy, N. I.; Metcalf, C.; Plaisted, R. L. *J. Food Sci.* **1971**, 41, 459-461.
90. Johns, T.; Keen, S. *Human Ecology* **1986**, 14, 437-452.
91. Sinden, S. L.; Deahl, K. L.; Aulenbach, B. B. *J. Food Sci.* **1976**, 41, 520-523.
92. Mondy, N. I.; Gosselin, B. *J. Food Sci.* **1988**, 53, 756-759.
93. Stapleton, A.; Allen, P. V.; Friedman, M.; Belknap, W. R. *J. Agric. Food Chem.* **1991**, 39, 1187-1203.
94. Griffiths, D. W.; Bain, H.; Dale, M. F. B. *J. Sci. Food Agric.* **1995**, 68, 105-110.
95. Bergers, W. W. A. *Food Chem.* **1981**, 6, 47-61.
96. Clarke, D. D. *Physiol. Plant Pathol.* **1976**, 9, 199-203.
97. Mondy, N. I.; Gosselin, B. *J. Food Sci.* **1971**, 41, 459-461.

98. Mondy, N. I.; Gosselin, B. *J. Food Sci.* **1989**, 54, 982-984.
99. Ogawa, M.; Uritiani, I. *Agric. Biol. Chem.* (Japan) **1979**, 34, 870-877.
100. Pendharkar, M. B.; Nair, P. M. *Potato Res.* **1987**, 30, 589-601.
101. Penner, H.; Fromm, H. *Z. Lebensm. Unters. Forschung* **1972**, 150, 84-87.
102. Ramamurthy, M. S.; Maiti, B.; Thomas, P.; Nair, P. M. *J. Agric. Food Chem.* **1992**, 40, 569-572.
103. Swallow, A. J. In *Nutritional and Toxicological Consequences of Food Processing*; Friedman, M., Ed.; Plenum: New York, **1991**; pp 11-31.
104. Thomas, P. *J. Food Sci.* **1981**, 46, 1620-1621.
105. Thomas, P.; Delincee, H. *Phytochemistry* **1979**, 18, 917-921.
106. Cheung, K. W. K.; Henderson, H. M. *Phytochemistry* **1972**, 11, 1255-1260.
107. Leszczynski, W.; Golachowski, A.; Lisinska, G.; Peksa, A. *Polish J. Food Nutr. Sci.* **1992,** 1/42, 61-70.
108. Onyeneho, S. N.; Hettiaachchy, N. S. *J. Sci. Food Agric.* **1993**, 62, 345-350.
109. Rodriguez, de Sotillo, D.; Hadley, M.; Holm, E. T. *J. Food Sci.* **1994**, 59, 649-651.
110. Rodriquez, de Sotillo, D.; Hadley, M.; Holm, E. T. *J. Food Sci.* **1994**, 59, 1031-1033.
111. Al-Saikhan, M. S.; Howard, L. R.; Miller, J. C., Jr. *J. Food Sci.* **1995**, 60, 341-343 and 347.
112. Vinson, J. A.; Jan, J.; Dabbagh, Y. A.; Y. A.; Serry, M. M.; Cai, S. *J. Agric. Food Chem.* **1995**, 43, 2687-2689.
113. Felice, L. J.; King, W. P.; Kissinger, P. T. *J. Agric. Food Chem.* **1976**, 24, 380-382.
114. Larson, R. *Phytochemistry* **1988**, 27, 969-978.
115. Newmark, H. L. *Nutr. Cancer* **1984**, 6, 58-70.
116. Newmark, H. L. *Can. J. Pharmacol.* **1987**, 65, 461-466.
117. Halliwell, B.; Aeschbach, R.; Loliger, J.; Aruoma, O. I. *Food Chem. Toxicol.* **1995**, 33, 601-617.
118. Friedman, M. *The Chemistry and Biochemistry of the Sulfhydryl Group in Amino Acids, Peptides, and Proteins*; Pergamon Press: Oxford, England, **1973**.
119. Cavins, J. F.; Friedman, M. *Biochemistry* 6, **1967**, 3766-3770.
120. Cavins, J. F.; Friedman, M. *J. Biol. Chem.*, **1968,** 243, 3357-3360.
121. Friedman, M. *Biochem. Biophys. Res. Comm.* **1966**, 23, 626-632.
122. Friedman, M. *J. Amer. Chem. Soc.* **1967**, 4709-4713.
123. Friedman, M; Romersberger, J. A. *J. Org. Chem.* **1968**, 33, 154-157.
124. Friedman, M.; Wall, J. S. *J. Am. Chem. Soc.* **1964**, 86, 3735-3741.
125. Friedman, M.; Wall, J. S. *J. Org. Chem.* **1966**, 31, 2888-2894.
126. Friedman, M.; Cavins, J. F.; Wall, J. S. *J. Am. Chem. Soc.* **1965**, 87, 3572-3580.

127. Friedman, M.; Diamond, M. J.; Broderick, G. L. *J. Agric. Food Chem.* **1982**, 30-72-77.
128. Weisleder, D.; Friedman, M. *J. Org. Chem.* **1968**, 33, 3542-3543.
129. Friedman, M. *Oxidations and Reductions of Porphyrins and Metalloporphyrins*. Ph.D. Thesis, University of Chicago, 1962.
130. Friedman, M. *J. Org. Chem.* **1965**, 30, 859-863.
131. Kikugawa, K.; Hakamada, T.; Hasunuma, M.; Kurechi, T. *J. Agric. Food Chem.* **1983**, 31, 780-785.
132. Stich, H. F.; Rosin, M. P. In *Nutritional and Toxicological Aspects of Food Safety*; Friedman, M., Ed.; Plenum: New York, **1984**; pp 1-29.
133. Camire, M. E.; Zhao, J.; Dougherty, M. P.; Bushway, R. J. *J. Agric. Food Chem.* **1995**, 43, 970-973.
134. Friedman, M.; Wehr, C. M.; Schade, J. E.; MacGregor, J. T. *Food Chem. Toxicol.* **1982**, 20, 887-892.
135. Stevens, R.; Wilson, R. E.; Friedman, M. *J. Agric. Food Chem.* **1995**, 42, 2424-2427.
136. Tanaka, T. *Oncology Reports* **1994**, 2, 1139-1155.
137. Tanaka, T.; Kawamori, T.; Ohnishi, M.; Okamoto, K.; Mori, H.; Hara, A. *Carcinogenesis* **1993**, 14, 1321-1325.
138. Tanaka, T.; Kawamori, T.; Mori, H. *Cancer* **1995,** 75, 1433-1439.
139. Thompson, L. U.; Yoon, J. H.; Jenkins, D. J. A.; Wolwer, J.; Jenkins, A. L. *Am. J. Clin. Nutr.* **1983**, 39, 745-751.
140. Olsson, K. *Impact Damage, Gangrene and Dry Rot in Potato - Important Biochemical Factors in Screening for Resistance and Quality in Breeding Material.* Ph.D. Thesis, The Swedish University of Agricultural Sciences, Svalov, **1989.**
141. Dao, L; Friedman, M. *J. Agric. Food Chem.* **1996**, 44, accepted.
142. Friedman, M.; Henika, P. R.; Mackey, B. E. *J. Nutr.* **1996**, 126, 989-999.
143. Wheland, G. W. *Resonance in Organic Chemistry*; John Wiley & Sons Inc., New York, **1955**; p 7, 381, 601.

Chapter 6

Potato Glycoalkaloids: Chemical, Analytical, and Biochemical Perspectives

S. J. Jadhav[1], S. E. Lutz[1], G. Mazza[2], and D. K. Salunkhe[3]

[1]Food Processing Development Centre, Leduc, Aberta T9E 7C5, Canada
[2]Food Research Program, Research Centre, Agriculture and Agri-Food Canada, Summerland, British Columbia V0H 1Z0, Canada
[3]Department of Nutrition and Food Sciences, Utah State University, Logan, UT 84322

The most recent findings on the structural characteristics, extraction, separation, analysis, biosynthesis and metabolism of potato glycoalkaloids are discussed. Molecular structures of the predominant steroidal alkaloids are presented, and analytical methods including colorimetric, TLC, GLC, HPLC and ELISA techniques are discussed in detail.

Glycoalkaloids are naturally occurring secondary metabolites of potato. These compounds are comprised of a steroidal-like alkaloid skeleton to which one to four sugars may be attached through a series of glycosidic linkages. Glycoalkaloids are toxic and thus are believed to be involved in pest resistance. These toxicants occur in potato tubers, peels, sprouts, and blossoms and their concentration in tubers depends on cultivar, maturity, environmental factors, and stress conditions (*1,2*). Concentrations may vary as a result of fungal or bacterial infection and usually increase in response to wound, apparently as a defense mechanism against potential disease. Glycoalkaloids, therefore, may function as phytoalexins (*2,3*).

Potato tubers with more than 20 mg of glycoalkaloids per 100 g of fresh weight exceed the upper safety limit for food purposes. Moreover, the glycoalkaoids are not destroyed under conventional heat processing regimes. In fact, the toxicity is highly variable among the different glycoalkaoids. Certain factors can induce variation of the toxic levels of the glycoalkaloids in potatoes. There is great genetic variability of the glycoalkaloid level in tubers, and breeding programs for improved cultivars may accidently introduce extensive glycoalkaloid accumulation and thereby increase the potential for toxicoses. Since the role of the glycoalkaloids in host defense to insects and pathogens is uncertain and the presence of these compounds in potato tubers can have a negative impact on quality, the need for these compounds in potatoes is questionable *(4)*. In view of these questions, understanding of the chemical, analytical, and biochemical aspects of the glycoalkaloids in potatoes has become important.

Structures and Chemistry

The glycoalkaloids in potato possess a steroidal skeleton that incorporates nitrogen in the molecule by cyclization of the steroidal side chain. The chemistry of steroidal alkaloids of the *Solanum* genus has been extensively reviewed (*5-8*). Most alkaloids in potato occur as glycosides; upon removal of the sugars alkamines (aglycones) are formed, which have a close structural resemblance with cholesterol. Molecular structures of the predominant steroidal alkaloids of potato are shown in Figure 1.

Sugars released by partial hydrolysis have indicated that the glycoalkaloids, α-solanine and α-chaconine, have branched sugars. They are named β-solatriose and β-chacotriose, respectively, and assigned the structures of O-α-L-rhamnopyranosyl-(1→2 gal)-O-β-D-glucopyranosyl-(1→3 gal)-β-D-galactopyranose and O-α-L-rhamnopyranosyl-(1→2 glu)-O-α-L-rhamnopyranosyl-(1→4 glu)-β-D-glucopyranose. The disaccharides of β-solanine and β-chaconine are as follows: O-β-D-glucopyranosyl-(1→3)-β-D-galactopyranose (β-solabiose) and O-α-L-rhamnopyranosyl-(1→4)-β-D-glucopyranose (β-chacobiose). Optical data and the partial syntheses of γ-solanine (3β-D-galactosidosolanidine) and γ-chaconine (3β-D-glucosidosolanidine) have shown that the alkamines are bound β-glycosidically with D-sugars (*7*). The sugar portions of α-solamarine and β-solamarine were shown to be identical with β-solatriose and β-chacotriose, respectively. Gas liquid chromatography (GLC), mass spectrometry (MS), and thin layer chromatography (TLC) have indicated that commersonine is closely related to demissine, a terminal glucose unit in the former being replaced by the terminal xylose in the latter. The tetrasaccharide moiety (β-lycotetraose) of demissine, like tomatine, possesses the O-β-D-glucopyranosyl-(1→2 glu)-O-β-D-xylopyranosyl-(1→3 glu)-O-β-D-glucopyranosyl-(1→4 gal)-β-D-galactopyranose linkage. The mass spectral fragmentation patterns of the glycoalkaloids were revealed by Price et al. (*9*).

The potato alkaloids have been visualized as two different steroidal skeletons: solanidane (solanidine type), which contains the indolizidine system exemplified by solanines, chaconines, solanidine, leptines, leptinines, and demissidine, and spirosolane (solasodine type), which possesses oxa-azaspirodecane structure as represented by tomatidenol, α- and β-solamarines. The structure of solanidine has been proved to be solanid-5en-3β-ol (*10*). Demissidine is identical with solanidan-3β-ol. The stereochemistry of solanidanes has been elucidated with the aid of IR spectroscopy and X-ray analysis of demissidine hydroiodide. All natural solanidanes possess 20S:22R:25S:NS-configuration. Hydrolysis of leptines I and II, by esterases or by mild alkaline treatment creates leptinines I and II which are 23-hydroxy α-chaconine and 23-hydroxy α-solanine, respectively. The removal of sugars from leptinines I and II affords the aglycone leptinidine. Leptinidine possesses a double bond and a hydroxyl group similar to Δ^5-3β-ol system. The position of the second hydroxyl in leptinidine was provided by the selenium dehydrogenation study and was further defined by IR and NMR as β-oriented (*11*). Leptinidine is therefore 23β-hydroxysolanidine.

Tomatidenol (tomatid-5-en-3β-ol) is C(5,6) dehydro tomatidine and has been prepared from 16-ketopregnane following an analogous synthetic pathway as outlined by Kessar et al. (*12*). Schreiber (*7*) found a small amount of this spirosolenol in an acid hydrolyzate of a glycoalkaloid mixture extracted from potato sprouts. The hydrolysis products and sugar moieties of the glycoalkaloids are listed in Table I and the aglycones are shown in Figure 2.

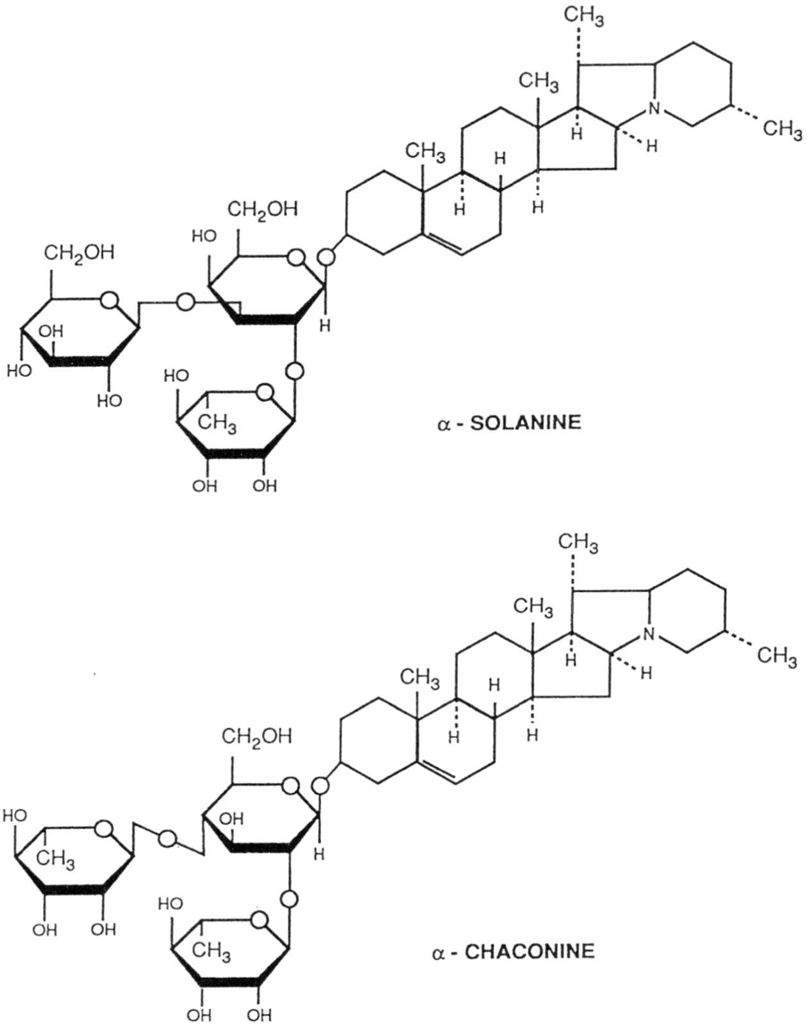

Figure 1. Molecular structures of α-solanine and α-chaconine.

Table I. Glycoalkaloids and Their Hydrolysis Products (Aglycone + Sugar Moiety)

Glycoalkaloid	Aglycone	Sugar Moiety
α-Solanine	Solanidine	D-glu -D-gal< L-rha
β-Solanine	Solanidine	-D-gal-D-glu
γ-Solanine	Solanidine	-D-gal
α-Chaconine	Solanidine	L-rha -D-glu< L-rha
β-Chaconine	Solanidine	-D-glu-L-rha
γ-Chaconine	Solanidine	-D-glu
α-Solamarine	Tomatidenol	D-glu -D-gal< L-rha
β-Solamarine	Tomatidenol	L-rha -D-glu< L-rha
Leptine I	O(23)-Acetylleptinidine	L-rha -D-glu< L-rha
Leptine II	O(23)-Acetylleptinidine	D-glu -D-glu< L-rha
Leptinine I	Leptinidine	L-rha -D-glu< L-rha
Letpinine II	Leptinidine	D-glu -D-glu< L-rha
Commersonine	Demissidine	D-glu -D-gal-D-glu< D-glu
Demissine	Demissidine	D-glu -D-gal-D-glu< D-xyl

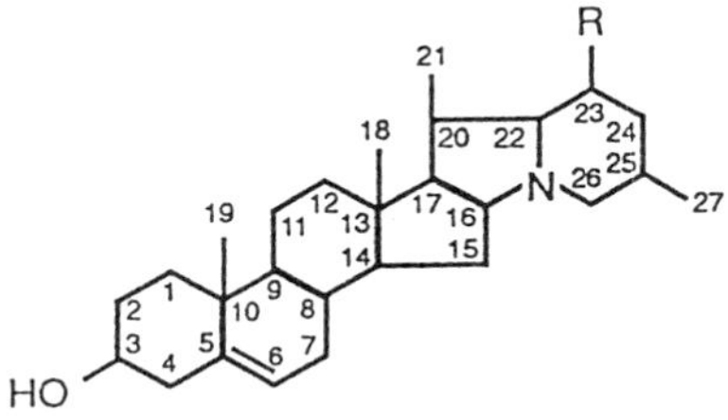

I

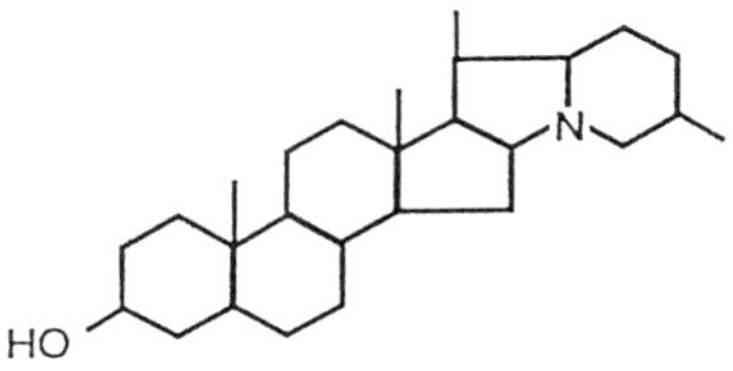

II

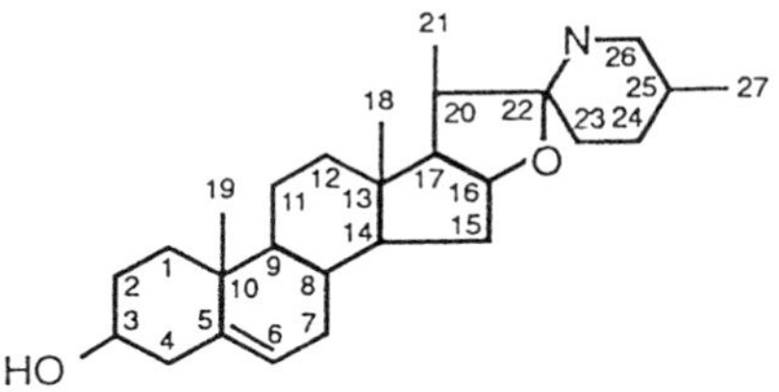

III

Figure 2. Molecular structures of aglycones.
I: Solanidine, R = H; Leptinidine, R = OH; Acetylleptinidine, R = $OCOCH_3$
II: Demissidine (solanidine with C_5 - C_6 saturation)
III: Tomatidenol (tomatidine with C_5 - C_6 unsaturation)

Extraction

The major potato alkaloids occur as salts, and as such are soluble in water. As a result, a crude glycoalkaloid preparation is normally obtained from weak acidic plant extract by precipitation with ammonia at pH above 10 at 70°C (*13,14*). Since toxicological and entomological studies require large amounts of pure crystalline material Achterberg et al. (*13*) developed a procedure for the mass extraction and collection of the glycoalkaloids from freeze-dried potato blossoms. These researchers were able to extract 4 to 6 g of glycoalkaloids from 950-1050 g of freeze-dried blossoms. However, this procedure is laborious, time consuming, uses special equipment, and yields only 10-15% recovery. To overcome these difficulties, Bushway et al. (*14*) developed a new procedure which is relatively simple, rapid, and yields 65-70% recovery (Figure 3). The large amounts of pure glycoalkaloid mixture obtained by this method can be further separated by preparative high performance liquid chromatography (HPLC).

Separation

During the past few decades, agricultural and medical researchers have become increasingly interested in separation and identification of the potato glycoalkaloids because of their profound physiological effects on living cells. Individual glycoalkaloid compounds can be separated by preparative TLC (*15*). HPLC (*16-18*) or by column chromatography (*19-22*). Investigations on the detection of water and ammonia-soluble glycoalkaloids have led to the resolution of as many as 15 substances in potato cultivars on TLC plates (*23*). It appears, therefore, that the glycosidic characteristics seem to dominate the alkaloidal properties since the enzyme activities also seem to suit the glycosidic nature of the glycoalkaloids. Subsequently, Zitnak (*24*) outlined a procedure for the extraction of the glycoalkaloids in order to study quickly and separate effectively the endogenous glycoalkaloids in potato extracts by both qualitative and quantitative TLC. Since solanidine glycosides are not readily available as reference substances, Filadelfi and Zitnak (*25*) developed a simple TLC standard by partial acid hydrolysis of α-solanine and α-chaconine. The resultant hydrolysate contained 8 alkaloids having R_f order = α-solanine (0.26), α-chaconine (0.34), β_2-solanine (0.40), β_1-chaconine (0.44), β_2-chaconine (0.55), γ-solanine (0.67), γ-chaconine (0.71), and solanidine (0.97) when developed on a silica gel G (0.25 mm) plate in a lower organic phase of $CHCl_3$-MeOH-1% aqueous NH_4OH 2:2:1. The tracing of chromatogram of glycoalkaloids extracts from selected genotypes which were banded on silica gel plate and developed with chloroform:methanol:1% NH_4OH (2:2:1) showed the following R_f pattern: commersonine, 0.09; α-solanine, 0.16; demissine, 0.19; α-chaconine, 0.31; and β-chaconine, 0.66 (*20*). Wu and Salunkhe (*26*) described a positive pressure streaker with a disposable pipette for preparative TLC. Jadhav et al. (*2*) have summarized qualitative TLC data on potato glycoalkaloids concerning adsorbent, developer, mode of visualization, and R_f values.

Since 1975, HPTLC material is available in the market. HPTLC layers are slightly thinner than those in conventional TLC material (0.20 mm instead of 0.25 mm), possess a smaller particle size (7 μm instead of 12 μm) and , in particular, a closer grain distribution, giving better separation performance over a shorter separation distance

Freeze-dried blossoms, 220-250 g -(1)

↓

Blend with 3L THF-H_2O-CH_3CN (5:3:2), 18,000 rpm, 10 min -(2)

↓

Filter (Whatman No 1) -(3)

↓

Concentrate to 500 ml on steam bath -(4)

↓

Add 700 ml 0.2 N HCl -(5)

↓

Sonicate 15 min -(6)

↓

Centrifuge (29,000 g, 10 min, 10°C -(7)

↓

Adjust supernatant to 10.5-11.0 pH with NH_4OH,
keep in 70°C water (10 min), cool and centrifuge -(8)

↓

Dry precipitate -(9)

↓

Dissolve in 500 ml 0.2 N HCl, repeat steps 6 to 9 -(10)

↓

Reflux dry product with 600-800 ml MeOH (2-3 h) -(11)

↓

Precipitate with NH_4OH -(12)

↓

Reflux dry product with 95% ethanol (300-500 ml, 2-3 h) -(13)

↓

Vacuum filter (Whatman No. 50), cool 1-2 h, filter -(14)

↓

Dried material 96-98% pure mixture of glycoalkaloids.
Evaporate mother liquor, reflux with 200-500 ml 95% ethanol and crystallize as before (product mixture of 91% purity).
Repeat entire procedure of obtaining crystals one more time to yield product of 85% purity.

Figure 3. The mass extraction of potato glycoalkaloids from blossoms. Adapted from ref. 14.

(50 mm instead of 100-200 mm). Hence separation of glycoalkaoids on HPTLC plate could provide interesting results.

Hunter et al. (*27*) applied HPLC to separate a mixture of steroidal alkaloids and believed that the technique has a high potential for preparative use. Bushway and Storch (*28*) developed a semi-preparative HPLC method to separate crude mixtures of the potato glycoalkaloids.

Analytical Methods

A number of methods have been employed in the determination of glycoalkaloids in potatoes. These are based on colorimetric, TLC, GLC, HPLC, and ELISA techniques. The gravimetric method of direct weighing after precipitation is unreliable and indirect estimation by measuring the reducing power of sugars after hydrolysis is outdated. Similarly, paper chromatographic and polarographic methods are not so common.

Colorimetry. The colorimetric methods require total extraction of the glycoalkaloids from potatoes. In most cases, potato samples are extracted with weak solutions of acetic acid in alcohol or water (*29-32*), followed by precipitation by concentrated liquid ammonia (*29*). A Soxhlet extraction procedure is commonly used although it is time consuming. The Soxhlet extraction with alcohol as used by Dabbs and Hilton (*30*) and Baker et al. (*33*) requires 14 hr or more. Wolf and Duggar (*29*) recommended two successive 18-hr extractions in a Soxhlet apparatus. Sachse and Bachmann (*34*) eliminated the Soxhlet extraction procedure by an exhaustive extraction involving reflux boiling with alcohol for three times, filtering, and washing the residue between the extractions. Smittle (*35*) compared the variation and accuracy of five extraction methods involving trichloroacetic acid (TCA, 5%) in 0 to 75% methanol with Soxhlet extraction of the total glycoalkaloids. The methods using TCA were generally more precise than the Soxhlet method. A bisolvent extraction technique employing a mixture of methanol and chloroform (2:1, v/v) extracted more glycoalkaloids than the Soxhlet extraction with 3% acetic acid in ethanol (*36*). However, Speroni and Pell (*37*) reported better recovery of glycoalkaloids with 5% acetic acid than that with a bisolvent system.

A colorimetric method, incorporating concentrated sulfuric acid and formaldehyde (the Marquis reagent) was used for the analysis of glycoalkaloids (*29-31, 33, 38*). Reproducible results could be obtained with this method when the concentrated sulfuric acid is added over a period of 3 min, followed immediately by the formaldehyde over a period of 2 min. Wu and Salunkhe (*39*), therefore, introduced a simple, inexpensive, and automatic system for delivering these reagents. It is recommended that the intensity of the violet-red color produced in this reaction should be measured after 45 to 55 min at 525 nm. Another reagent that contains a mixture of 85% phosphoric acid and 1% paraformaldehyde and forms a chromogenic product with the glycoalkaloids is known as the Clarke reagent. The suitability of this reagent as well as that of the Marquis reagent for glycoalkaloid analysis has been reviewed by Sachse and Bachmann (*34*). Bretzloff (*40*) developed a rapid colorimetric method for total glycoalkaloid analysis using 5% TCA in 1:1 (v/v) methanol-water mixture as an extractant and utilized antimony trichloride reagent as recommended by Wierzchowski and Wierzchowska (*41*). However, this technique was used to classify large numbers of potato samples into such groups as very low, low, medium, and high in total glycoalkaloids. Bergers (*42*) made modifications with respect to the extraction and work-up procedures to increase the speed and ease of analysis in the colorimetric method of Sachse and Bachmann (*34*). A comparative study of nine colorimetric methods was carried out by Clement and Verbist (*43*) who found the Wang et al. (*36*) method was superior.

Titrimetry. Fitzpatrick and Osman (*44*) described a titrimetric procedure for measuring total glycoalkaloids, including those that are not determined by the colorimetric method because of the absence of a double bond in the aglycone. Demissine is, therefore, measured by this method. This comprehensive method consisted of glycoalkaloid extraction by a methanol-chloroform mixture (2:1, v/v), separation of the methanolic phase in aqueous sodium sulfate, removal of the salt from the dry methanolic extract by filtration followed by the hydrolysis of the glycosides, and extraction of the aglycones with benzene. The basic nitrogen of the aglycone was titrated with a 10^{-3} M solution of bromophenol blue and 10% phenol in absolute methanol until the color changed from blue through the blue-green to a yellow end point. These authors claimed that this analytical method is simple, safe, and rapid compared to the spectrophotometric methods, particularly those requiring extreme caution with the corrosive and toxic nature of the reagents. Later on, it was realized that up to 10 to 20% of the glycoalkaloids may be lost at the stage of the salt removal. Fitzpatrick et al. (*15*), therefore, eliminated this step to improve recovery of the glycoalkaloids and shorten the time of analysis. Mackenzie and Gregory (*45*) pointed out that large and variable amounts of glycoalkaloids would be lost in the discarded chloroform layer and from degradation during hydrolysis. These authors also recommended incorporation of a modified extraction procedure and nonaqueous titration without hydrolysis to improve the Fitzpatrick and Osman (*44*) method. In a modified titration method, Bushway et al. (*46*) eliminated the hydrolysis and partitioning steps. Instead, the authors obtained the glycoalkaloids by ammonium hydroxide precipitation and dissolved in a tetrahydrofuran-water-MeCN (50:30:20) mixture. An aliquot was evaporated and quantitated using the non-aqueous titration procedure. Coxon et al. (*47*) followed direct hydrolysis of the acid extract prior to colorimetric procedure involving bromothymol blue.

Extraction of potato samples with 5% TCA in 75% methanol could result in a partial hydrolysis of glycoalkaloids, which would be indicated by an additional spot on TLC corresponding to solanidine and should be avoided in the quantization of individual glycoalkaloids (*48*). If so, the procedure of Shih and Kuć (*49*) using weak acetic acid may be followed.

TLC. Quantitative assay of individual glycoalkaloids in a relatively short time is possible with a simple and inexpensive system involving TLC and densitometry as adapted by Cadle et al. (*48*). Coxon and Jones (*50*) described a rapid screening method to determine total glycoalkaloid (TGA) content of potato tubers. The TGA concentration in juice expressed from potato periderm and cortex tissue was analyzed by TLC. The authors established a direct relationship between the TGA in expressed juice and the TGA content of the whole tuber. Jellema et al. (*51,52*) reported the separation and analysis of the glycoalkaloids by TLC using an optical brightener as detection reagent and employed a TLC method for determining glycoalkaloids in potato tubers and leaves and in industrial potato protein. With the advent of a fully automated instrumental HPTLC system, operations such as sample application, chromatogram development, post-chromatographic derivatization, photodocumentation, and quantitative chromatogram evaluation have been considerably simplified.

GLC. Gas-liquid chromatography (GLC) has been used to determine the glycoalkaoids as their trimethylsilyl derivatives. Herb et al. (*53*) preferred permethylated over the trimethylsilyl derivatives of the glycoalkaloids for GLC analysis because of the greater volatility and lower molecular weights of the former. Siegfried (*54*) used GLC to separate trimethylated sugars obtained by acid hydrolysis. Osman and Sinden (*55*) found that chromatographic methods could not separate solanidine and demissidine glycoalkaloids containing identical sugars. However, the structural difference resulting from the hydrolysis of glycoalkaloids to solanthrene (Δ^3,Δ^5 diene) and demissidine, respectively, can be useful and aglycones can be measured by GLC in the presence of each other. Indirect estimation of the glycoalkaloids can be made by a sterol binding assay and unbound sterols are quantitated by GLC (*56*). King (*57*) developed a method based on hydrolysis of the glycoalkaloids and GLC separation of the aglycones using N-P detector. Solanidine and demissidine were distinguished by the formation of their respective 3β trifluoroacetates. These and other GLC methods (*47,58*) have the advantage of sensitive detector systems, but problems arise from the low volatility of glycoalkaloids.

HPLC. Bushway et al. (*59*) found HPLC useful for the assay of α-solanine, α-chaconine, and β-chaconine in potato tissues. Separation of these glycoalkaloids was achieved by using three different columns: μBondapak C_{18}, μBondapak NH_2, and a carbohydrate analysis column. These methods were employed to examine the purity of the potato glycoalkaloids isolated by thin-layer and column chromatographic separations. Bushway et al. (*60*) also used HPLC method to compare α-chaconine and α-solanine contents in five varieties of tubers using a carbohydrate column and tetrahydrofuran-water-MeCN (53:17:30) as a mobile phase. Similar studies for potato and potato products were also conducted with radially compressed amino (*14*), C_8 and C_{18} reversed phase (*61*), and a carbohydrate analytical column *(62)*. Carman et al. (*63*) described the extraction of glycoalkaloids from potato products with dilute aqueous acetic acid in the presence of an ion-pairing reagent. The extract was subsequently analyzed on HPLC, thus providing a rapid, simple, and economical method for quantitating the glycoalkaloid content in fresh or processed potato products. These advantages are further supported by several reports (*18, 64-67*). Because of the high accuracy and simplicity of HPLC methods, these techniques are perhaps the methods of choice for quantitative determination and identification of various glycoalkaloids.

ELISA. Morgan et al. (*68*) have described a very promising enzyme-linked immunosorbent assay (ELISA) for total glycoalkaloids in potato tubers. In this method, α-solanine coupled to bovine serum albumin was used to raise rabbit antiglycoalkaloid antiserum which reacted equally with the various glycoalkaloids. Morgan et al. (*17*)

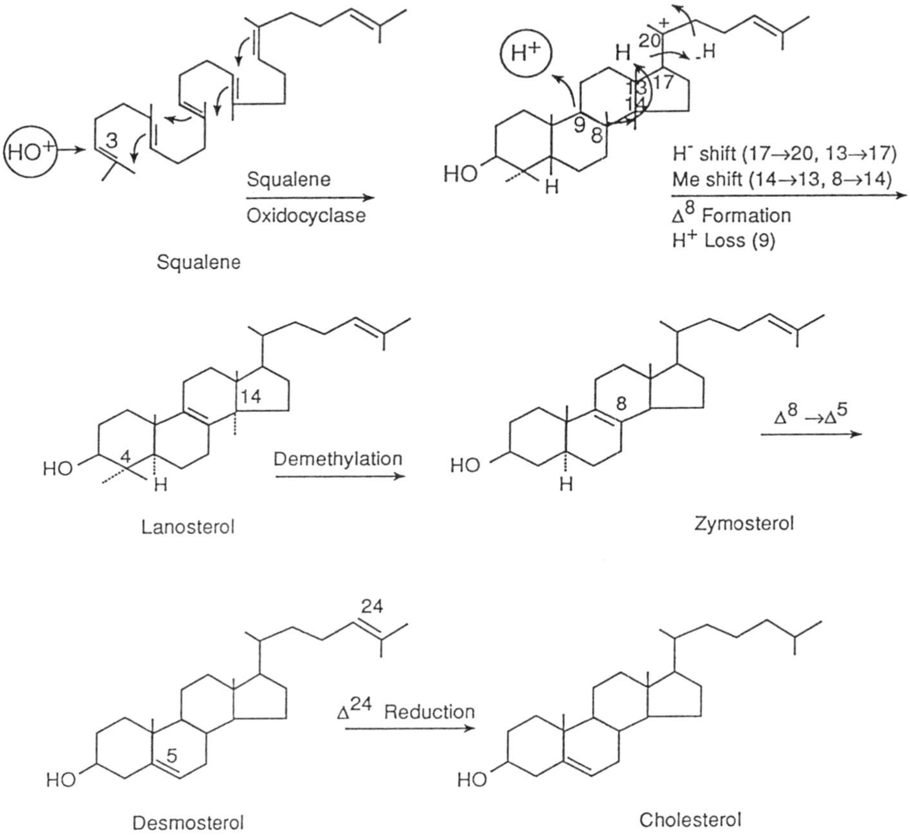

Figure 4. A sequence of some intermediates between squalene and cholesterol in plants. Adapted from ref. 101.

found that the method is sensitive and precise and offers certain advantages over chemical assays. This ELISA method measures demissine and similar Δ^5-glycoalkaloids which are missed by the chemical assays. Recently, Plhak and Sporns (*69*) described new methods for the preparation of solanidine-protein conjugates and improved enzyme immunoassay procedures for solanidine glycoalkaloids in commercial potato cultivars.

Biosynthesis

Biosynthetically, all steroidal compounds, such as sterols, certain sapogenins, terpenes, hormones, and glycoalkaloids are interrelated, and pathways leading to the synthesis of a structurally similar compound could be postulated on the basis of known ones. Thus, the regular pathway starting from acetate via mevalonate, isopentyl pyrophosphate, fernesyl pyrophosphate, squalene, and cholesterol is applicable to steroidal alkaloids. Several authors have reviewed the biochemistry and possible biogenetic relationships of steroids and steroidal alkaloids of the *Solanum* genus (*7, 70-74*).

The first tracer work on the biogenesis of potato alkaloids was initiated by Guseva and Paseshnichenko (*75*), who demonstrated the uptake and utilization of radioactive acetate by potato sprouts. The glycoalkaloids isolated from such sprouts grown under normal illumination had the labelled carbon chiefly in the aglycone, while the labelled carbon of sprouts grown in the dark was in both aglycone and sugar portions of the glycoalkaloids. Maximum radioactivity in the glycoalkaloids occurred when labelled acetate was fed for 2 days. In a later experiment, Guseva et al. *(76)* found that α-chaconine had nearly twice as much specific activity as α-solanine. Mevalonate was more effectively utilized in the biosynthesis of glycoalkaloids by potato seedlings than acetate (*77*). Wu and Salunkhe (*78*) observed that light-exposed tubers incorporated a higher percentage of label from mevalonate into α-chaconine than mechanically injured tubers. The tuber surface had been exposed to light (2152 lux) at 13 °C, and the precursor was applied on a 9-cm^2 skin surface area confined by a thin line of vaseline and then exposed to light for 2 days. The mechanical injury treatment involved half of a normal tuber similarly coated with the precursor and incubated in the dark at 13 °C and 80% relative humidity for two days.

The biosynthesis of cholesterol has been investigated by the isolation of radioactive cholesterol from *Solanum tuberosum* fed with mevalonic acid-2-^{14}C (*79*). In plants, the biosynthetic pathway from squalene is thought to proceed via 2,3-oxidosqualene, cycloartenol, lanosterol, lophenol to cholesterol or a closely related phytosterol. The widespread distribution of cycloartenol in the plant kingdom has indicated that cycloartenol may be the first product of cyclization in higher plants. Cycloartenol and lophenol, as well as cholesterol, have been isolated from the potato *(80-82)*. An electrophilic attack by a hydroxyl cation at C-3 position mediates a concerted nonstop cyclization of squalene and is followed by a series of molecular arrangements or modifications leading to the formation of cholesterol. A sequence of some intermediates between squalene and cholesterol in plants is shown in Figure 4. The manner in which lanosterol loses its methyl groups to yield zymosterol is a characteristic of a specific plant. Ripperger et al. (*83*) showed that cycloartenol and lanosterol are the precursors of *Solanum* alkaloids. Tschesche and Hulpke (*84*) reported

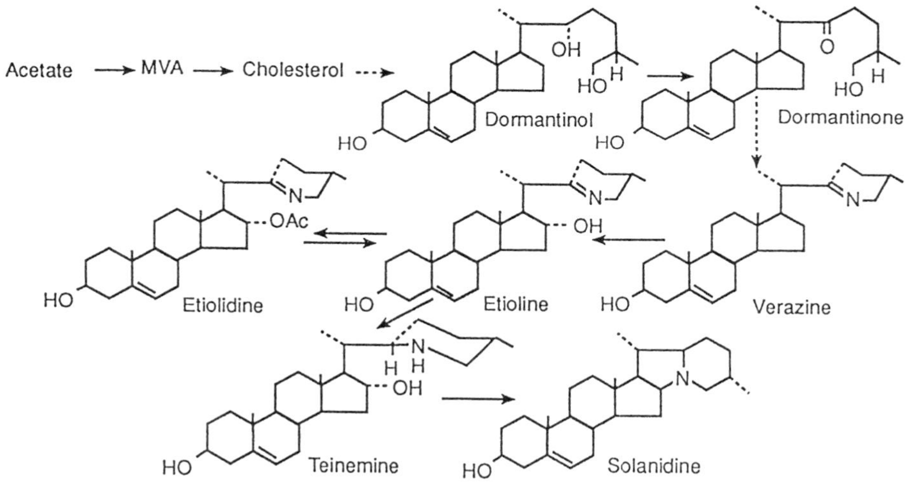

Figure 5. Hypothetical formation of solanidine from sterols and steroidal alkaloids lacking ring E. Adapted from ref. 79.

that cholesterol when applied to leaf surface of potato plants was metabolized to solanidine. During the formation of solanidine from cholesterol, 16β-H is lost from the latter (*85*).

The hypothesis that labelled carbon atoms are distributed in all the steroidal rings of solasodine, synthesized from radioactive acetate, or mevalonate by *S. aviculare*, is consistent with the known biosynthetic and cyclization scheme of squalene (*86*) and may be applied to α-solanine and α-chaconine because of their structural similarity to solasodine. By analogy with solasodine, biologically derived solanidine, demissidine or tomatidenol from radioactive acetate would be expected to possess 15 of 27 carbon atoms formed from methyl carbon and 12 from carboxyl carbon of the acetate (Table II). Also, a label from mevalonate-2-^{14}C would be located at 5 different positions. The origin of the nitrogen atom in potato alkaloids is unknown. However, Heftmann (*72*)

Table II. Hypothetical distribution of labelled carbon atoms in the glycoalkaloid aglycones from acetate-2-^{14}C, acetate-1-^{14}C and mevalonate-2-^{14}C

Precursor	Labelled C atom in aglycones (Figure 2)
Acetate-2-^{14}C	1, 3, 5, 7, 9, 13 15, 17, 18, 19, 21 22, 24, 26, 27
Acetate-1-^{14}C	2, 4, 6, 8, 10, 11, 12, 14, 16, 20, 23, 25
Mevalonate-2-^{14}C	1, 7,15, 22, 27

hypothesized that cholesterol may undergo cyclization in the side chain subsequent to the formation of 26-hydroxycholesterol; the hydroxyl group is then replaced by an amino function. The presence of 26-hydroxycholesterol in growing potato plants has been reported by Heftmann and Weaver (*87*). Tschesche et al. (*88*) found that the C-26 or C-27 hydroxyl group is directly replaced by an amino group during the biosynthesis of C-27 alkaloids of *S. lycopersicum* and *S. laciniatum* administered with 25(RS)-25, 26 $^{3}H_2$-4-^{14}C-cholesterol, a finding consistent with the hypothesis of Heftmann. Kaneko et al. (*89*) suggested that L-arginine is the most likely source of nitrogen for solanidine biosynthesis in *Veratrum grandiflorum* and postulated a biogenetic pathway for solanidine involving sterol intermediates and steroidal alkaloids lacking ring E (Figure 5).

Jadhav et al. (90) studied the incorporation of labelled carbon from β-hydroxy-β-methylglutaric acid (HMG), L-leucine, L-alanine, and D-glucose into the glycosidic steroidal alkaloids of potato sprouts. The higher amount of radioactivity in the glycoside moiety than in the aglycone part of the glycoalkaloids indicated predominant glycosylation when labelled D-glucose was administered to the potato sprouts. Subsequently, glucosylation of solanidine was apparently catalyzed by the crude enzyme preparations from sprouts of potato tubers (*91*). This enzyme system is known to glucosylate a sterically unhindered 3β-hydroxy group of a steroid if it belongs to the 5α-H or Δ^5-series. These results and *in vitro* synthesis of γ, β, and α forms of solanine and chaconine (one, two, and three sugars in the glycosidic part, respectively) by potato tissue (*92, 93*) supported the hypothesis that α-solanine and α-chaconine are synthesized in a stepwise manner from solanidine.

Metabolism

Potatoes contain an enzymatic system capable of hydrolysing the glycoalkaloids. Their role in plant metabolism is not clearly understood. Guseva and Paseshnichenko (*94*) observed that enzyme preparations from the juice of potato sprouts hydrolyse α-solanine to β- and γ-solanines by a stepwise mechanism. However, the same enzyme system removes the rhamnose substituent at the two-position of the glucose residue in α-chaconine, giving rise to β_2-chaconine, which is further hydrolysed to solanidine without forming γ-chaconine as an intermediate product. Swain et al. (*95*) confirmed the presence of rhamnosidase, glucosidase, and galactosidase activities along with the nonstepwise hydrolysis of α-chaconine by the enzyme mixture from sprouts. Also, these authors (*95*) were first to report that the enzyme preparation from dormant tubers is capable of producing β_1-chaconine (2-rhamnosylglucoside of solanidine), β_2-chaconine, γ-chaconine, and solanidine from α-chaconine in a stepwise manner, while producing β-solanine and solanidine from α-solanine by a concerted (anomalous) mechanism. A large accumulation of solanidine accompanying cellular destruction of tuber tissue has been attributed to the hydrolytic enzyme activity on the glycoalkaloids (*96, 97*). However, Holland and Taylor (*98*) found that the blight fungus *Phytophthora infestans* itself can convert α-solanine to solanidine. Rumen microorganisms can initially hydrolyse glycoalkaloids to solanidine and then reduce double bond at position 5 (*99*).

Zitnak (*100*) made several observations concerning heat sensitivity, pH dependency, water extractability, and light sensitivity of the hydrolytic enzyme system associated with the glycoalkaloids in potato tissue homogenate. The potato blossom rhamnosidase initiated a selective hydrolysis of α-chaconine to β-chaconine within 2 to 3 hr of incubation without affecting α-solanine and completed the reaction after 24 hr. The rate of hydrolysis increased at pH 5, leading to a complete degradation of the steroidal system, apparently by a separate enzyme, after 72 hr of incubation. At pH 6 and 7, the degradation of the steroid ring occurred more slowly. α-Solanine was affected by a pH between 5 and 7. A mixture of sprout-blossom homogenate increased the rate of terminal degradation at a faster rate than that in blossom tissue alone. Netted Gem (Russet Burbank cultivar) had rhamnosidase activity only in photostimulated eye tissues and produced solanidine as an end product. Cultivar differences in the enzyme systems, hydrolytic activity in protein precipitates of acetone and ammonium sulfate, and a slow reaction on α-solanine were also noted.

Conclusions

The glycoalkaloids have a negative influence on the quality of potato because of their toxicity to humans and other animals. There appears to be no clear relationship between the constitutive levels of the glycoalkaloids and resistance to disease in spite of their toxicity to many pathogens. Hence the idea of eliminating these compounds from potatoes is being suggested. This approach however would require an alteration of critical stages in the glycoalkaloid biosynthesis through genetic engineering. In addition, there is a need to develop a better understanding of why structurally related glycoalkaloids vary greatly in their pharmacological and toxicological activities. Reducing the content and/or the toxicity of alkaloids in new potato cultivars by direct genetic modification will require considerable efforts in the biosynthesis, chemistry, biochemistry, and toxicity of the variety of glycoalkaloids present in potatoes.

Literature Cited

1. Jadhav, S.J.; Salunkhe, D.K. Formation and control of chlorophyll and glycoalkaloids in tubers of *Solanum tuberosum* L. and evaluation of glycoalkaloid toxicity. *Adv. Food Res.* **1975,** *21*, 307.
2. Jadhav, S.J.; Sharma, R.P.; Salunkhe, D.K. Naturally occurring toxic alkaloids in foods. *CRC Crit. Rev. Toxicol.* **1981,** *9*, 21.
3. Kuć, J.; Tjamos, E.; Bostock, R. Metabolic regulation of terpenoid accumulation and disease resistance in potato. In *Isopentenoids in Plants*; Nes, W.D., Fuller, G., Tsai, L.-S., Eds.; Marcel Dekker, New York, **1984,** p. 103.
4. Hammerschmidt, R.; Zook, M.N. Steroid glycoalkaloid synthesis in potato: Research Project, Michigan State University, East Lansing, MI, **1995.**
5. Prelog, V.; Jeger, O. The chemistry of solanum and veratrum alkaloids. In *The Alkaloids,* Vol. 3; Manske, R.H.F., Holmes, H.L., Eds.; Academic Press, New York, **1953;** p. 247.
6. Prelog, V,; Jeger, O. Steroid alkaloids: The solanum group. In *The Alkaloids,* Vol. 7; Manske R.H.F., Ed.; Academic Press, New York, **1960;** p. 343.
7. Schreiber, K. Steroid alkaloids: *Solanum* group. In *The Alkaloids,* Vol. 10; Manske, R.H.F., Ed.; Academic Press, New York, **1968;** p. 1.
8. Sato, Y. Steroidal alkaloids. In *Chemistry of the Alkaloids;* Pelletier, S.W., Ed.; Van Nostrand Reinhold, New York, **1970;** p. 591.
9. Price, K.R.; Mellon, F.A.; Self, R.; Fenwick, G.R.; Osman, S.F. Fast atom bombardment mass spectrometry of *Solanum* glycoalkaloids and its potential for mixture analysis. *Biomed. Mass Spectrum.* **1985,** *12*, 79.
10. Kessar, S.V.; Rampal, A.L.; Gandhi, S.S.; Mahajan, R.K. Synthetic studies in steroidal sapogenins and alkaloids -IX. Synthesis of solanidine. *Tetrahedron,* **1971,** *27*, 2153.
11. Schreiber, K; Ripperger, H. Notiz zur konfiguration von leptinidin an C-23. *Chem. Ber.* **1967,** *100*, 1381.
12. Kessar, S.V.; Gupta, Y.P.; Singh, M.; Mahajan, R.K. Synthetic studies in steroidal sapogenins and alkaloids -X. Synthesis of tomatid-5-ene-3-ol and solasodine. *Tetrahedron,* **1971,** *27*, 2869.

13. Achterberg, C.L.; Clauson, D.M.; Blease, J.A.; Barden, E.S. New procedure for the mass extraction and collection of potato glycoalkaloids. *Am. Potato J.* **1979,** *56*, 145.
14. Bushway, R.J.; Barden, E.S.; Bushway, A.W.; Bushway, A.A. The mass extraction of potato glycoalkaloids from blossoms. *Am. Potato J.* **1980,** *57*, 175.
15. Fitzpatrick, T.J.; Mackenzie, J.D.; Gregory, P. Modifications of comprehensive method for total glycoalkaloid determination. *Am. Potato J.* **1978,** *55*, 247.
16. Coxon, D.T. Methodology for glycoalkaloid analysis. *Am. Potato J.* **1984,** *61*, 169.
17. Morgan, M.R.A.; Coxon, D.T.; Bramham, S.; Chan, H.W.-S.; Van Gelder, W.M.J.; Allison, M.J. Determination of glycoalkaloid content of potato tubers by three methods including enzyme linked immunosorbent assay. *J. Sci. Food Agric.* **1985,** *36*, 282.
18. Friedman, M.; Levin, C.E. Reversed-phase high performance chromatographic separation of potato glycoalkaloids and hydrolysis products on acidic columns. *J. Agric. Food Chem.* **1992,** *40*, 2157.
19. Chaube, S.; Swinyard, C.A. Teratological and toxicological studies of alkaloidal and phenolic compounds from *Solanum tuberosum* L. *Toxicol. Appl. Pharmacol.* **1976,** *36*, 227.
20. McCollum, G.D.; Sinden, S.L. Inheritance study of tuber glycoalkaloids in a wild potato, *Solanum chacoense* bitter. *Am. Potato J.* **1979,** *56*, 95.
21. Nishie, K.; Norred, W.P.; Swain, A.P. Pharmacology and toxicology of chaconine and tomatine. *Res. Commun. Chem. Pathol. Pharmacol.* **1975,** *12*, 657.
22. Bushway, R.J. Sources of alkaloid and glycoalkaloid standards for potato breeding programs and other research. *Am. Potato J.* **1983,** *60*, 793.
23. Zitnak, A. Isolation of new glycoalkaloids from the potato plant. *Proc. Can. Soc. Hortic. Sci.* **1965,** *4*, 92.
24. Zitnak, A. Separation of glycoalkaloids from *Solanum tuberosum* L. by thin-layer chromatography. *Proc. Can. Soc. Hortic. Sci.* **1968,** *7*, 75.
25. Filadelfi, M.A.; Zitnak, A. A simple TLC standard for identification of potato glycoalkaloids. *Can. Inst. Food Sci. Technol. J.* **1983,** *16*, 151.
26. Wu, M.T.; Salunkhe, D.K. Positive pressure streaker with disposable pipette for thin-layer chromatography. *J. Chromatogr.* **1978,** *147*, 429.
27. Hunter, I.R.; Walden, M.K.; Wagner, J.R.; Heftmann, E. High-pressure liquid-chromatography of steroidal alkaloids. *J. Chromatogr.* **1976,** *119*, 223.
28. Bushway, R.J.; Storch, R.H. Semi-preparative high-performance liquid chromatographic separation of potato glycoalkaloids. *J. Liq. Chromatogr.* **1982,** *5*, 731.
29. Wolf, M.J.; Duggar, B.M. Estimation and physiological role of solanine in the potato. *J. Agric. Res.* **1946,** *73*, 1.
30. Dabbs, D.H.; Hilton, R.J. Methods of analysis for solanine in tubers of *Solanum tuberosum. Can. J. Technol.* **1953,** *31*, 213.
31. Gull, D.D.; Isenberg, F.M. Chlorophyll and solanine content and distribution in four varieties of potato tubers. *Proc. Am. Soc. Hortic. Sci.* **1960,** *75*, 545.

32. Shih, M.; Kuć, J.; Williams, E.B. Suppression of steroid glycoalkaloid accumulation as related to rishitin accumulation in potato tubers. *Phytopathol.* **1973,** *63*, 821.
33. Baker, L.C.; Lampitt, L.H.; Meredith, O.B. Solanine, glycoside of the potato. III. An improved method of extraction and determination. *J. Sci. Food Agric.* **1955,** *6*, 197.
34. Sachse, J.; Bachmann, F. About alkaloid determination in *Solanum tuberosum* L. *Z. Lebensm. Unters. -Forsch.* **1969,** *141*, 262.
35. Smittle, D.A. Comparison and modification of methods of total glycoalkaloid analysis. *Am. Potato J.* **1971,** *48*, 410.
36. Wang, S.L.; Bedford, C.L.; Thompson, N.R. Determination of glycoalkaloids in potatoes (*S. tuberosum*) with a bisolvent extraction method. *Am. Potato J.* **1972,** *49*, 302.
37. Speroni, J.J.; Pell, E.J. Modified method for tuber glycoalkaloid and leaf glycoalkaloid analysis. *Am. Potato J.* **1980,** *57*, 537.
38. Pfankuch, E. Photometric determination of solanine. *Biochem. Z.* **1937,** *295*, 44.
39. Wu, M.T.; Salunkhe, D.K. Devices for improving color development in glycoalkaloid determination. *J. Food Sci.* **1976,** *41*, 220.
40. Bretzloff, C.W. Method for rapid estimation of glycoalkaloids in potato tubers. *Am. Potato J.* **1971,** *48*, 158.
41. Wierzchowski, P.; Wierzchowska, Z. Colorimetric determination of solanine and solanidine with antimony trichloride. *Chem. Anal.(Warsaw),* **1961,** *6*, 579.
42. Bergers, W.W.A. A rapid quantitative assay for solanidine glycoalkaloids in potatoes and industrial potato protein. *Potato Res.* **1980,** *23*, 105.
43. Clement, E.; Verbist, J.F. The determination of solanine in *Solanum tuberosum* L. tubers; a comparative studies of nine colorimetric methods. *Lebensm. Wiss. Technol.* **1980,** *13*, 202.
44. Fitzpatrick, T.J.; Osman, S.F. Comprehensive method for determination of total potato glycoalkaloids. *Am. Potato J.* **1974,** *51*, 318.
45. Mackenzie, J.D.; Gregory, P. Evaluation of a comprehensive method for total glycoalkaloid determination. *Am. Potato J.* **1979,** *56*, 27.
46. Bushway, R.J.; Wilson, A,M.; Bushway, A.A. Determination of total glycoalkaloids in potato tubers using a modified titration method. *Am. Potato J.* **1980,** *57*, 561.
47. Coxon, D.T.; Price, K.R.; Jones, P.G. A simplified method for the determination of total glycoalkaloids in potato tubers. *J. Sci. Food Agric.* **1979,** *30*, 1043.
48. Cadle, L.S.; Stelzig, D.A.; Harper, K.L.; Young, R.J. Thin-layer chromatographic system for identification and quantitation of potato tuber glycoalkaloids. *J. Agric. Food Chem.* **1978,** *26,* 1453.
49. Shih, M.; Kuć, J. α- and β-Solamarine in Kennebec *Solanum tuberosum* leaves and aged tuber slices. *Phytochem.* **1974,** *13*, 997.
50. Coxon, D.T.; Jones P.G. A rapid screening method for the estimation of total glycoalkaloids in potato tubers. *J. Sci. Food Agric.* **1981,** *32*, 366.

51. Jellema, R.; Elema, E.T.; Malingre, T.M. Flurodensitometric determination of potato glycoalkaloids on thin-layer chromatograms. *J. Chromatogr.* **1981,** *210*, 121.
52. Jellema, R.; Elema, E.T.; Malingre, T.M. A rapid quantitative determination of the individual glycoalkaloids in tubers and leaves of *Solanum tuberosum* L. *Potato Res.* **1982,** *25*, 247.
53. Herb, S.F.; Fitzpatrick, T.J.; Osman, S.F. Separation of potato glycoalkaloids by gas-chromatography. *J. Agric. Food Chem.* **1975,** *23*, 520.
54. Siegfried, R. Determination of solanine-chaconine ratio in potato sprouts and tubers. *Z. Lebensm. Unters. Forsch.* **1976,** *162*, 253.
55. Osman, S.F.; Sinden, S.L. Analysis of mixture of solanidine and demissidine glycoalkaloids containing identical carbohydrate units. *J. Agric. Food Chem.* **1977,** *25*, 955.
56. Roddick, J.G. A sterol binding assay for potato glycoalkaloids. *Phytochem.* **1980,** *19*, 245.
57. King, R.R. Analysis of potato glycoalkaloids by gas-liquid chromatography of alkaloid components. *J. Assoc. Off. Anal. Chem.* **1980,** *63*, 1226.
58. Van Gelder, W.M.J. Determination of the total C_{27}-steroidal alkaloid composition of *Solanum* species. *J. Chromatogr.* **1985,** *331*, 285.
59. Bushway, R.J.; Barden, E.S.; Bushway, A.W.; Bushway, A.A. High-performance liquid chromatographic separation of potato glycoalkaloids. *J. Chromatogr.* **1979,** *178*, 533.
60. Bushway, R.J.; Barden, E.S.; Wilson, A.M.; Bushway, A.A. Analysis of potato glycoalkaloids by high-performance liquid chromatography. *J. Food Sci.* **1980,** *45*, 1088.
61. Morris, S.C.; Lee, T.H. Analysis of potato glycoalkaloids with radially compressed high performance liquid chromatographic cartridges and ethanolamine in the mobile phase. *J. Chromatogr.* **1981,** *219*, 403.
62. Bushway, R.J. High performance liquid chromatographic determination of the metabolites of the potato glycoalkaloids, α-chaconine and α-solanine in potato tubers and potato products. *J. Liq. Chromatogra.* **1982,** *5*, 1313.
63. Carman, A.S., Jr.; Kuan, S.S.; Ware, G.M.; Francis, O.J., Jr.; Kirschenheuter, G.P. Rapid high performance liquid chromatographic determination of the potato glycoalkaloids α-solanine and α-chaconine. *J. Agric. Food Chem.* **1986,** *34*, 279.
64. Bushway, R.J.; Bureau, J.L.; King, J. Modification of the rapid high performance liquid chromatographic method for the determination of potato glycoalkaloids. *J. Agric. Food Chem.* **1986,** *34*, 277.
65. Hellenäs, K.E. A simplified procedure for quantification of potato glycoalkaloids in tuber extracts by HPLC. Comparison with ELISA and a colorimetric method. *J. Sci. Food Agric.* **1986,** *37*, 776.
66. Kobayashi, K,; Powell, A.D.; Toyoda, M.; Saito, Y. HPLC method for the simultaneous analysis of α-solanine and α-chaconine in potato plants cultured in vitro. *J. Chromatogr.* **1989,** *462*, 357.
67. Saito, K.; Horie, M.; Hoshino, Y.; Nose, N.; Nakazawa, H. HPLC determination of glycoalkaloids in potato products. *J. Chromatogr.* **1990,** *508*, 141.

68. Morgan, M.R.A.; McNerhey, R.; Matthew, J.K.; Coxon, D.T.; Chan, H.W.-S. An enzyme-linked immunosorbent assay for total glycoalkaloids in potato tubers. *J. Sci. Food Agric.* **1983,** *34*, 593.
69. Plhak, L.C.; Sporns, P. Enzyme immunoassay for potato glycoalkaloids. *J. Agric. Food Chem.* **1992,** *40*, 2533.
70. Clayton, R.B. Biosynthesis of sterol, steroids, and terpenoids. Part II. Phytosterols, terpenes, and the physiologically active steroids. *Q. Rev. Chem. Soc.* **1965,** *19*, 201.
71. Heftmann, E. Biochemistry of plant steroids. *Annu. Rev. Plant Physiol.* **1963,** *14*, 225.
72. Heftmann, E. Biochemistry of steroidal saponins and glycoalkaloids. *Lloydia,* **1967,** *30*, 209.
73. Heftmann, E. Biosynthesis of plant steroids. *Lloydia,* **1968,** *31*, 293.
74. Heftmann, E.; Mosettig, E. *Biochemistry of Steroids,* Van Nostrand-Reinhold, Princeton, N.J., **1960.**
75. Guseva, A.R.; Paseshnichenko, V.A. A study of the biogenesis of potato glycoalkaloids by the method of labeled atoms. *Biokhimiya,* **1958,** *23*, 412.
76. Guseva, A.R.; Borikhina, M.G.; Paseshnichenko, V.A. Utilization of acetate for the biosynthesis of chaconine and solanine in potato sprouts. *Biokhimiya,* **1960,** *25*, 282.
77. Guseva, A.R.; Paseshnichenko, V.A.; Borikhina, M.G. Synthesis of radioactive mevalonic acid and its use in the study of the biosynthesis of steroid glycoalkaloids from *Solanum*. *Biokhimiya,* **1961,** *26*, 723.
78. Wu, M.T.; Salunkhe, D.K. Difference between light-induced and wound-induced biosynthesis of alpha-solanine and alpha-chaconine in potato tubers. *Biol. Plant.* **1978b,** *20*, 149.
79. Johnson, D.F.; Heftmann, E.; Houghland, G.V.C. Biosynthesis of sterols in *Solanum tuberosum*. *Arch. Biochem. Biophys.* **1964,** *104*, 102.
80. Hartmann, M.A.; Benveniste, P. Effect of aging on sterol-metabolism in potato-tuber slices. *Phytochem.* **1974,** *13*, 2667.
81. Johnson, D.F.; Bennett, R.D.; Heftmann, E. Cholesterol in higher plants. *Science,* **1963,** *140*, 198.
82. Schreiber, K.; Osske, G. Sterols and triperpenoids II. The isolation of cycloartenol from leaves of *Solanum tuberosum, Kulturpflanze,* **1962,** *10*, 372.
83. Ripperger, H.; Mortiz, W.; Schreiber, K. Zur biosynthese von solanum-alkaloiden aus cycloartenol oder lanosterin. *Phytochem.* **1971,** *10*, 2699.
84. Tschesche, R.; Hulpke, H. Biosynthesis of sterol derivatives in plants. VII. Biogenesis of solanidine from cholesterol. *Z. Naturforsch.* **1967,** *22b*, 791.
85. Canonica, L.; Ronchetti, F; Russo, G.; Sportotti, G. Fate of 16 β-hydrogen atom of cholesterol in the biosynthesis of tomatidine and solanidine. *J. Chem. Soc. Chem. Commun.* **1977,** *8*, 286.
86. Guseva, A.R.; Paseshnichenko, V.A. A study of solasodine biosynthesis by the method of oxidative breakdown. *Biokhimiya,* **1962,** *27*, 853.
87. Heftmann, E.; Weaver, M.L. 26-Hydroxycholesterol and cholest-4-en-3-one, first metabolites of cholesterol in potato plants. *Phytochem.* **1974,** *13*, 1801.

88. Tschesche, R.; Goossens, B.; Töpfer, A. Biosynthesis of steroid derivatives in plants. 22. Introduction of nitrogen and common occurrence of 25(R) steroid alkaloids and 25(S) steroid alkaloids in solanaceae. *Phytochem.* **1976,** *15*, 1387.
89. Kaneko, K.; Tanaka, M.W.; Mitsuhashi, H. Origin of nitrogen in the biosynthesis of solanidine by *Veratrum grandiflorum*. *Phytochem.* **1976,** *15*, 1391.
90. Jadhav, S.J.; Salunkhe, D.K.; Wyse, R.E.; Dalvi, R.R. Solanum alkaloids: Biosynthesis and inhibition by chemicals. *J. Food Sci.* **1973,** *38*, 453.
91. Jadhav, S.J.; Salunkhe, D.K. Enzymatic glucosylation of solanidine. *J. Food Sci.* **1973,** *38*, 1099.
92. Osman, S.F.; Zacharius, R.M. Biosynthesis of potato glycoalkaloids. *Am. Potato J.* **1979,** *56*, 475.
93. Osman, S.F.; Zacharius, R.M.; Naglak, D. Solanidine metabolism in potato tuber slices and suspension cultures. *Phytochem.* **1980,** *19*, 2599.
94. Guseva, A.R.; Paseshnichenko, V.A. Enzymic spliting of potato glycoalkaloids by the method of labeled atoms. *Biokhimiya,* **1957,** *22*, 843.
95. Swain, A.P.; Fitzpatrick, T.J.; Talley E.A.; Herb, S.F.; Osman, S.F. Enzymatic hydrolysis of alpha-chaconine and alpha-solanine. *Phytochem.* **1978,** *17*, 800.
96. Zitnak, A. The occurrence and distribution of free alkaloid solanidine in Netted Gem potatoes. *Can. J. Biochem. Physiol.* **1961,** *39*, 1257.
97. Zacharius, R.M.; Kalan, E.B.; Osman, S.F.; Herb, S.F. Solanidine in potato tuber tissue disrupted by *Erwinia atroseptica* and by *Phytophthora infestans*. *Physiol. Plant Pathol.* **1975,** *6*, 301.
98. Holland, H.L.; Taylor, G.J. Transformations of steroids and the steroidal alkaloid, solanine, by *Phytophthora infestans*. *Phytochem.* **1979,** *18*, 437.
99. King, R.R.; McQueen, R.E. Transformations of potato glycoalkaloids by rumen microorganisms. *J. Agric. Food Chem.* **1981,** *29*, 1101.
100. Zitnak, A. The significance of glycoalkaloids in the potato plant. *Proc. Can. Soc. Hortic. Sci.* **1964,** *3*, 81.
101. Heftmann, E. Steroids. In *Plant Biochemistry;* Bonner, J.; Varner, J.E., Eds; Academic Press, N.Y. **1965,** p. 693.

Chapter 7

Biological Activities of Potato Glycoalkaloids

Leslie C. Plhak[1] and Peter Sporns[2]

[1]Department of Food Science, Louisiana Agricultural Experiment Station, Louisiana State University, Baton Rouge, LA 70803–4200
[2]Department of Agricultural, Food and Nutritional Science, University of Alberta, Edmonton, Alberta T6G 2P5, Canada

Potatoes (*Solanum* sp.) are a rich source of nutrients and food energy but they are also a source of secondary plant products, including glycoalkaloids (GA). GA have been responsible for human and livestock poisonings, including deaths. Toxic effects are mainly due to membrane disruption and cholinesterase inhibition. Even though potatoes are the world's fourth most important crop, there is surprisingly little known about the metabolism and long term accumulation of GA in humans.

The potato, a New World food, has been used as a food source for more than 13 thousand years and cultivated for 3 to 5 thousand years (*1-4*). The important role of the potato during this early era in what is now Peru and Bolivia, is evidenced by archaeological finds that show that potatoes were worshipped, placed in tombs with the dead, and depicted in the form of pottery (*1, 3, 5*). After its introduction to Europe, approximately 50-100 years passed before the potato was accepted as a food. It was associated with Old World members of the Solanaceae family, including the deadly nightshade, belladonna, henbane and mandrake, plants known primarily by herbalists for their narcotic and hallucinogenic effects. It was believed to cause leprosy, dysentery, syphilis, consumption, rickets, leishmaniasis and have "unwholesome" aphrodisiac effects (*3, 5, 6*).

During the past 100 years, cultivation of potatoes for use as a food has spread to much of the developing world. In 1982, the potato was produced by 130 of the world's 167 independent countries (*3*). Based on yield, the potato ranks 4th in the world, after wheat, corn and rice. It is valued at over $100 billion (US) annually (*3, 6*). On a dry-weight basis, the protein content of the potato approaches 10%, comparable to wheat and higher than most rice and corn varieties. If multiple harvesting is taken into account, the potato yields more protein and energy per land area than all other crops. It has relatively good protein quality, having substantial levels of lysine while only moderately limiting in the sulfur-containing amino acids (*7*), and is a good source of ascorbic acid, thiamin, iron, magnesium and phosphorus (*8*).

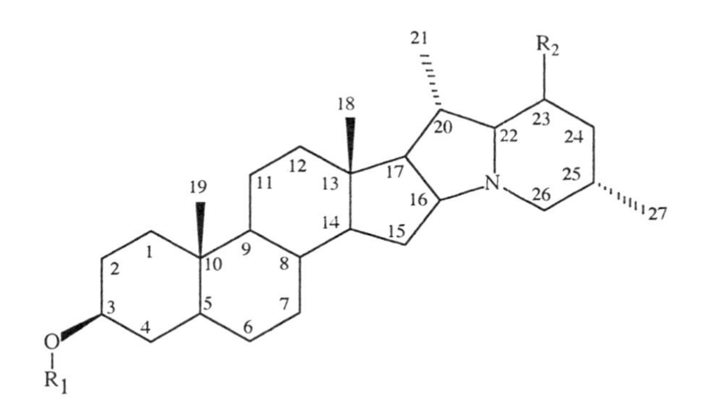

Alkaloid	5-ene	R_1	R_2
solanidine	yes	H	H
demissidine	no	H	H
leptinidine	yes	H	OH
acetylleptinidine	yes	H	OAc

Figure 1. Solanidane Skeleton

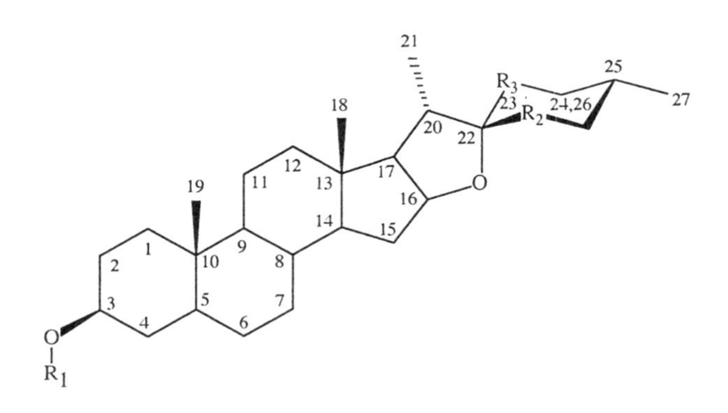

Alkaloid	5-ene	R_1	R_2	R_3
tomatidenol	yes	H	NH	CH_2
tomatidine	no	H	NH	CH_2
solasodine	yes	H	CH_2	NH

Figure 2. Spirosolane Skeleton

While the potato has an extremely important role in the world's food supply (*3, 9*), the potato plant, including its tubers, contains naturally-occurring toxic compounds. One class of toxic constituents found in *Solanum* sp. is the glycoalkaloids (GA). The first GA was discovered and named solanine by Defosses in 1820. When first reported, Baup predicted that this compound would "find a use in medicine" (translated from Baup, *10*). More than a century later, GA have indeed been the focus of much research activity.

GA can also be found in other *Solanum* sp. foods such as tomatoes and eggplants. However, the GA levels in these ripe fruits are very low (*11*). The only documented cases of human GA poisoning are reported for potatoes which always contain significant GA levels. Therefore, due to the importance of potatoes as a food and the ubiquitous presence of GA in potatoes, most of the scientific literature, and this review, focus on potato GA.

Structure and Biosynthesis

Solanum sp. GA contain a C27 steroidal alkaloidal aglycone, and a carbohydrate moiety. Their chemistry, including composition, structural determination, reactions and physical constants has been reviewed (*12-14*).

Aglycones vary in their basic skeletal structure (e.g. solanidane vs. spirosolane systems), in their saturation at C_5, or their isomerization at C_{22}. In 1981, 75 different aglycones having the C_{27} cholestane skeleton have been described in *Solanum* sp. (*14*). Only about 7, however, are of significant importance in the tuber-bearing species (*15*). These are illustrated in Figures 1 and 2.

GA are biosynthesized from acetyl-CoA, via several intermediates including mevalonate, farnesyl pyrophosphate, squalene and cholesterol (*13, 16-18*). It is relatively certain that all steroids, in both animals and plants, are synthesized through the same biochemical pathways to cholesterol (*13*). Conversion of cholesterol to a N-containing steroid, however, is less understood. Heftmann (*19*) proposed that cyclization of the cholesterol side-chain may be occurring prior to the formation of 27-hydroxycholesterol, followed by replacement of the hydroxy group with an amino group. L-arginine has been postulated as the source of the amino group (*20*).

The carbohydrate moieties of GA are mono- or disaccharides, or branched tri- and tetrasaccharides, usually consisting of D-glucose, D-galactose, L-rhamnose, or D-xylose units (Table I). The more common GA sugars include β-solatriose, β-chacotriose, β-commertetraose and β-lycotetraose. β-Solatriose possesses the structure O-α-L-rhamnopyranosyl-(1→2gal)-O-β-D-glucopyranosyl-(1→3gal)-β-D-galactopyranose. β-Chacotriose has the structure O-α-L-rhamnopyranosyl-(1→2glu)-O-α-L-rhamnopyranosyl-(1→4glu)-β-D-glucopyranose. Glycosylation has been demonstrated to occur enzymatically, in a stepwise manner to the aglycone (*21, 22*). Solanidine glycosyltransferase has been isolated and characterized by Stapleton *et al.* (*23*).

Different aglycone-carbohydrate combinations give rise to a diverse group of GA. Table I illustrates the most common GA and their composition. Two of the more common GA in commercially cultivated potatoes (*S. tuberosum*) are α-chaconine and α-solanine. These both have the same aglycone, solanidine, but differ by their sugar substitutions and usually exist in a 60:40 ratio, respectively (*24*). Hydrolytic enzymes are present in potato tissue that can remove sugars by a stepwise mechanism (*25*).

GA synthesis or accumulation seems to be associated with high metabolic activity within the plant. GA levels are highest in the unripe fruits, flowers, sprouts, and leaves and lowest in the stems, roots and tubers (*15*). Table II summarizes the distribution of GA that generally occurs within the potato plant and various tuber

Table I. Common GA and Their Composition

Glycoalkaloid	*Alkaloid*	*Glycoside (Sugars)*
α-chaconine	solanidine	β-chacotriose (D-glu, 2 L-rha)
α-solanine	solanidine	β-solatriose (D-gal, D-glu, L-rha)
dehydrocommersonine	solanidine	β-commertetraose (D-gal, 3 D-glu)
demissine	demissidine	β-lycotetraose (D-gal, 2 D-glu, D-xyl)
commersonine	demissidine	β-commertetraose (D-gal, 3 D-glu)
leptinine I	leptinidine	β-chacotriose (D-glu, 2 L-rha)
leptinine II	leptinidine	β-solatriose (D-gal, D-glu, L-rha)
leptine I	acetylleptinidine	β-chacotriose (D-glu, 2 L-rha)
leptine II	acetylleptinidine	β-solatriose (D-gal, D-glu, L-rha)
β-solamarine	tomatidenol	β-chacotriose (D-glu, 2 L-rha)
α-solamarine	tomatidenol	β-solatriose (D-gal, D-glu, L-rha)
α-solamargine	solasodine	β-chacotriose (D-glu, 2 L-rha)
α-solasonine	solasodine	β-solatriose (D-gal, D-glu, L-rha)
α-tomatine	tomatidine	β-lycotetraose (D-gal, 2 D-glu, D-xyl)

Table II. Concentration of GA in various tissues of the potato plant, *S. tuberosum*. (Adapted from *15, 26, 102*).

Plant Tissue	*[GA], (mg/100g), fwb*
flowers	215 - 500
sprouts	195 - 1770
leaves	40 - 100
fruits	42
roots	18 - 40
stems	3
whole tuber	2 - 15
tuber skin (2-3% of tuber)	30 - 64
tuber peel (10-15% of tuber)	15 - 30
peel and eye (3mm disk)	30 - 50
tuber flesh	1.2 - 6

tissues. Commercial cultivars typically contain 2 to 15 mg GA/100 g, fresh weight basis (fwb), for unpeeled tubers (*26*). The safety question arises when certain conditions allow or even stimulate GA synthesis and accumulation.

Factors Affecting Glycoalkaloid Accumulation

Several factors are related to GA accumulation in tubers: genetic factors, environmental conditions during growth and development, storage temperature and time, the presence of light, disease or insect damage, or mechanical stress (*27-29*).

Genetic Variation

GA have been demonstrated in over 300 species of the botanical families Solanaceae, Apocynaceae, Buxaceae, and Liliaceae (*30-32*). Within the Solanaceae, the *Solanum* (potato) and *Lycopersicon* (tomato) genera are of greatest importance. *Solanum* includes more than 200 species, of which about 160 are tuber-bearing. In Peru, eight species are cultivated. Only one, *S. tuberosum*, is grown in Canada and the U.S.A. (*3*). There is some evidence (*33*) that during the domestication of *S. tuberosum*, selection for reduced toxicity occurred.

Wild or so-called "bitter" varieties of *Solanum* are often employed in breeding programs to gain better yield, hardiness, disease resistance, solids content or chipping quality (*28, 34-38*). Wild types generally contain much higher GA levels, up to 100 mg/100 g , fwb, for unpeeled tubers (*38*). In the past, reduction of GA during breeding has not been a high priority since GA synthesis in the plant is under polygenic control (*28*) and GA are important in pest and disease resistance in potatoes (*34-36*). In addition, some wild species contain different types of GA that are not quantitatively detected by the commonly used methods of analysis (*39-41*) and therefore routine screening is not always included in breeding programs.

The problem of GA inheritability can be illustrated by the cultivar Lenape, developed in the late 1960's (*42*). *S. tuberosum* was bred with the wild variety *S. chacoense*, a genetic source of higher yields (due to its resistance to the Colorado potato beetle), higher solids content and chipping quality (*28*). In 1970, Lenape was rapidly withdrawn, just prior to its release, when it was discovered to have 27 to 65 mg /100 g GA (*28, 43*). The resistance of *S. chacoense* to the Colorado potato beetle is believed to be directly related to its high content of leptine GA (*44*) and/or the presence of glandular trichomes (*37, 45*).

S. vernei is another wild species used in breeding programs for a source of resistance to potato cyst nematodes and its high solids content. Van Gelder and Scheffer (*46*) demonstrated that solasodine GA were inherited in hybrids from this wild species crossed with *S. tuberosum*.

GA have also been implicated in potato leafhopper resistance (*47*). In a study by Sanford *et al.* (*34*) selection for leafhopper resistance led to an increase in foliar GA. Leafhopper resistance is also associated with the presence of hairs and/or glandular trichomes on leaves (*37*).

There is a great potential to use wild *Solanum* species as genetic sources for higher yields, hardiness, resistance or quality. In order to do so, however, it must be certain that offspring do not inherit the tendency to produce high levels of GA. In order to ensure a safe potato, therefore, a rapid, simple, reliable, inexpensive and comprehensive method of GA analysis is required for routine screening of breeding lines (*41*).

Growth, Development and Storage of Tubers

Location, climate, altitude, air pollution, soil type, soil moisture, the presence of fertilizer, infestation by pests (microbial or nonmicrobial), the application of pesticides, length of growing season, maturity of the tubers and tuber damage during growth, harvest or transportation all are believed to affect GA synthesis and accumulation (*17, 24, 28*).

Physiological stress to the harvested tuber may cause significant increases in the GA content. Extremes in the storage temperature, exposure to light, mechanical damage and sprouting can all stimulate GA synthesis (*17*). Methods of GA control are therefore aimed at reducing exposure to the above conditions and slowing down tuber metabolism. Treatments that have been found to slow down GA formation during storage include certain types of packaging, the application of some sprout inhibitors (e.g. isopropyl-N-(3-chlorophenyl)-carbamate), exposure to γ-irradiation and the application of wax, detergents, surfactants or oils (*24, 25, 48*). Some of these treatments, however, have practical limitations because of safety considerations and marketing trends.

Effect of Processing

The oldest known potato processing method is the production of chuño, believed to be about 2000 years old (*5*). In this process, still practiced in Peru, potatoes are left out and exposed to the high mountain elements for 2-3 weeks. During this time, the cool/dry/low pressure conditions allow freeze-drying to take place. Potatoes are then stepped on to remove their skins and the remainder is placed in a slow moving riverbed, with straw, and left for another 2 weeks to 2 months and then redried. This process reduces the bitterness and the resulting dry powder, which can be stored for 2 to 3 years, is used in soups, stews and sweet desserts (*3*). While this method of processing causes a reduction in the nutrient content, it also reduces the GA content from about 30 mg/100 g in the fresh bitter potatoes to about 4 to 16 mg/100 g in chuño (*7*).

Large-scale processing of potatoes has become increasingly important in industrialized nations, especially since the appearance and increased consumption of fast foods and snack foods. The percentage of potatoes consumed in the US that were processed by frying, dehydration or canning increased from 2% in 1940 to 51% in 1970 (*7*). In 1980, Americans alone consumed 2 billion kg French fries and 0.4 billion kg potato chips (*3*).

The production of pre-peeled and/or pre-cut potatoes is a growing industry for supply to the food service sector (*7*). Peeling may remove a significant amount of GA, by removal of the tissue where GA are most concentrated. At the same time, however, tissue damage may cause an increase in GA synthesis in the peeled product through a wound-response mechanism (*17*). The question of GA accumulation also arises in potato-processing plants if pre-cut potatoes are left in conditions of high light intensity and temperature for a period of time before cooking (*25, 46*). The use of anoxia water treatment, by inhibiting respiration, has been shown to be effective for inhibition of light-induced and wound-induced GA formation in potato tissue (*49*).

GA are not easily removed or destroyed by cooking due to their heat stability and poor solubility characteristics. Boiling, baking, microwaving, frying or dehydration of potatoes does not cause a reduction in the GA concentration (*50, 51*). Frying, due to the loss of water, can result in a three- to four-fold increase in the GA concentration (*52*).

Potato tissue containing GA contents in excess of 14 to 22 mg/100 g have been rated as bitter by taste panelists (*53*). GA contents over 22 mg/100 g were found to produce a mild to severe burning sensation in the mouth and throat (*53*). These same sensations have also been correlated with the phenolic content (*54*). There are differences in the bitterness of different GA and in the perception of bitterness between individuals (*55*). In addition, it has been suggested by some authors that bitterness in processed products can be masked by a high oil content or the addition of flavorings and salt. Taste, therefore, may not be a good indicator of GA content.

Biological Activities

Toxicology

Since GA levels increase in young growing tissue, it is obvious that they serve some useful function for the plant. GA are toxic to insects *(56, 57)* and mammals so they likely play a protective role in these cases. It has also been suggested that GA bitterness acts as an anti-herbivore function and that GA possess anti-fungal activity *(58-60)*.

Potatoes have been responsible for a number of documented episodes of human GA poisoning - up to 30 deaths and over 2000 cases of non-fatal poisoning. It has been suggested, however, that this is an underestimation of the actual number of cases. The British Medical Journal *(61)* reported that many cases, both mild and severe, would be diagnosed as gastroenteritis (a term based on symptoms, rather than cause) and not be recorded as poisoning. If the facilities or expertise are not available to analyze for GA, or if there are no leftovers from a suspected meal, or no tubers of the same lot available for analysis, then confirmation is impossible.

In fact, even though GA are found in relatively large amounts in potatoes, one of the major drawbacks in understanding GA effects has been the difficulty in accurate analysis of these compounds and their metabolites. Recent reviews describe the analytical problems and variety of approaches that have been used *(46, 62, 63)*.

In confirmed cases, where the GA content of the potatoes was reported and the amount of potatoes consumed was estimated, toxic or lethal doses could be calculated. Toxic dose refers to the mg of GA per kg body weight consumed by those who suffered toxic, not necessarily lethal, effects. The symptoms are both gastrointestinal and neurological in nature. They include vomiting, diarrhea (sometimes including blood), severe abdominal pain, drowsiness, apathy (in some cases this alternates with restlessness or shaking attacks), confusion, weakness, depression and in some cases unconsciousness *(64-67)*.

In those cases where people died, and the information is available, a lethal dose can be calculated. This is the mg of GA per kg body weight consumed by those who suffered lethal effects. In these cases, the toxic symptoms, as mentioned before, were severe and death was attributed to strangulation of the bowel *(68)*, respiratory failure *(69)*, and/or cardiac arrest *(64)*.

Morris and Lee *(70)* calculated toxic and lethal doses based on available and estimated information involving approximately 1400 cases during 15 episodes. These ranged from 2 to 6 mg/kg body weight. From this, the authors concluded that the safety factor between the average GA level occurring in potatoes and a toxic dose is only about four-fold for a 500 g serving. Therefore, it should not be surprising that episodes of GA poisoning, some fatal, have occurred. It has been suggested that potato GA are one of the most serious toxic components in the human diet *(71, 72)*.

GA levels in potato tubers below 20 mg/100 g, fresh-weight basis (fwb) have often been quoted as "safe" or "acceptable" for human consumption *(28, 41, 73)*. The suggestion for 20 mg/100 g as the tolerance limit came from findings, as early as 1924, that potatoes involved in poisoning showed GA contents exceeding 25 mg/100 g *(46)*. It is probably more correct, therefore, to state that a GA content over 20 mg/100 g is potentially unsafe or even hazardous for human consumption.

There is little information available on the subacute or possible chronic toxicity of GA *(74)*. It is remarkable that there has not been more research into human metabolism of GA. What is known is that GA can be found in the bloodstream of anyone consuming potatoes and that this level is dependent on dietary intake for both males and females (*75*). It has been suggested that the pH of the food consumed with GA may have an effect, with acidic foods rendering the GA less toxic (*70*). GA have fairly long human biological half lives (11 and 19 h, for α-solanine and α-chaconine, respectively, *76*) which could be perceived as even longer for complete clearance (*75*), since the glycoalkaloids are partially converted to their more slowly eliminated aglycone. Of concern is the potential for long term liver deposition (*77*) and the production of unknown GA metabolites (*76, 78*).

Membrane Destabilization

As plant sterols, GA are considered to be important in regulation of the plant's membrane fluidity and integrity. GA can also be considered as saponins. Saponins are triterpenoid or steroidal glycosides of plant origin that have the ability to form a soapy lather when agitated with water. Saponins can have hemolytic activity, cholesterol-binding properties, bitterness and toxicity (*79*). The gastrointestinal symptoms that GA can give rise to are believed to be due to their saponin-like properties.

Membrane destabilization, measured as cell lysis in various cell systems, was shown to occur synergistically when certain mixtures of GA, including α-solanine and α-chaconine, were present (*80-85*). The aglycone portion of the GA interacts with membrane sterols, while the corresponding GA carbohydrates bind with each other to form strong detergent-resistant complexes. The formation of these complexes then leads to the observed membrane disruption (*85*). The order of GA potency was found to be α-tomatine > α-chaconine > α-solanine (*84*).

Cholinesterase Inhibition

GA also act as cholinesterase-inhibitors, a likely cause of their neurological effects. Susceptibility of cholinesterase to inhibition by GA is related to the isozyme(s) present and the type of GA (*17, 86*).

Recently there has been renewed interest in the interactions between GA and the two human cholinesterases, acetylcholinesterase and butyrylcholinesterase. The human cholinesterases hydrolyze or scavenge a variety of drugs including cocaine and related narcotics, and muscle relaxants such as succinylcholine. Cholinesterase inhibitors (e.g. tacrine, atropine) are use to treat neurodegenerative diseases like Alzheimer's disease or organophosphorus, carbamate and nerve gas poisoning. Since GA also inhibit human cholinesterases, they can interfere with all of the above agents, an inhibition further complicated by the different enzyme affinities of the genetically diverse human cholinesterases (*87*).

Mutagenic and Teratogenic Effects

α-Solanine and α-chaconine have not been found to be mutagenic both in the Ames test using *Salmonella* strains TA98 and TA100 and also in micronucleus assays (*88*). Both fish (*89*) and frog (*90*) embryo model systems have demonstrated GA teratogenicity. The latter researchers however add the caution that their data does not mean that GA would be teratogenic in a normal diet for pregnant mammals.

There have also been numerous studies into the mutagenic and teratogenic effects of GA in mammals and attempts to extrapolate these results into humans. Crawford and Myhr (*91*) noted greater liver mutational frequencies in transgenic mice but no teratogenicity. A complication in much of this research is the variety of both GA administration methods and the species examined. It is clear that the route of administration of GA has a distinct effect on toxicity (*92*). GA are considerably less toxic when given orally, likely because of their poor adsorption and rapid excretion. Also, there are large species differences. For example while oral doses of only 2 mg/kg can cause toxicity in man, doses a hundred times greater cause little effect in sheep (*70*).

While teratogenic effects such as neural tube disorders have been reported in experimental animals (*93, 94*), another more recent study found no teratogenic effects (*95*). Moreover, earlier teratogenic studies have been criticized for not identifying the teratogenic agent or administration of such high GA doses that toxic responses occurred in the pregnant mothers (*96*).

Hepatotoxicity (*97*), cardiotonic activity (*98*) , immunomodulation (*99*) , and even the use of GA as treatment for skin cancers (*100, 101*) have been noted. However, it is likely that many of these studies do not indicate unique GA properties, but demonstrate an indirect manifestation of cell toxicity due to membrane destabilization.

Overall, as others have noted (*74, 77*), there is a need for further research into the bioactivity of GA, especially the human metabolic fate of these important dietary components.

Literature Cited

1. Ugent, D. *Science* **1970**, *170*, 1161-1166.
2. Hawkes, J. G.; Lester, R. N.; Skelding, A. D. (Eds). *The Biology and Taxonomy of the Solanaceae.* Academic Press: London, U.K. 1979.
3. Rhoades, R. E.; Rogers, M. *National Geographic.* **1982**, *161*, 668-694.
4. Fenwick, G. R. *Proc. Nutr. Soc. Aust.* **1986**, *11*, 11-23.
5. Salaman, R. N. *The History and Social Influence of the Potato.* Cambridge University Press: Cambridge, U.K., 1949
6. Niederhauser, J. S. *Food Technol.* **1992**, *46*, 91-95.
7. Woolfe, J. A. *The Potato in the Human Diet.* Cambridge University Press: Cambridge, U.K. 1987.
8. Jadhav, S. J.; Salunkhe, D. K. *Adv. Food Research.* **1975**, *21*, 307-354.
9. Rhoades, L. A.; Johnson, L. *National Geographic.* **1991**, *179*, 74-105.
10. Baup, M. *Ann. Chim. (Paris)* **1826**, *31*, 108-109.
11. Maga, J. A. *CRC Crit. Rev. Food Sci. Nutr.* **1980**, *12*, 371-405.
12. Prelog, V.; Jeger, O. In *The Alkaloids: Chemistry and Physiology, Vol III.* R. H. F. Manske and H. L. Holmes, eds.; Academic Press: New York, NY. 1953; pp. 247-312.
13. Schreiber, K. In *The Alkaloids: Chemistry and Physiology, Vol X,* R. H. F Manske and H. L. Holmes, eds.; Academic Press: New York, NY. 1968; pp. 1-192.
14. Ripperger, H.; Schreiber, K. *In The Alkaloids: Chemistry and Physiology, Vol XIX,* R. H. F Manske and R. G. A. Rodrigo, eds.; Academic Press: New York, NY. 1981; pp. 81-192.
15. Van Gelder, W. M. J. In *Handbook of Natural Toxins. Vol 6. Toxicology of Plant and Fungal Compounds.* R. F. Keeler and A. T. Tu, eds., Marcel Dekker, Inc., NY. 1991; pp. 101-134.
16. Goodwin, T. W. In *The Biochemistry of Plants. A Comprehensive Treatise Vol. 4. Lipids: Structure and Function.* P. K. Stumph, ed.; Academic Press: New York, NY. 1980; pp 485-507.
17. Sharma, R. P.; Salunkhe, D. K. In *Toxicants of Plant Origin. Vol. 1, Alkaloids.* P. R. Cheeke, ed.; CRC Press, Boca Raton, FL. 1989; pp. 179-236.
18. Heftmann, E. *Phytochem.* **1983**, *22*, 1843-1860.
19. Heftmann, E. *Lloydia.* **1967**, *30*, 209-230.

20. Kaneko, K.; Tanaka, M. W.; Mitsuhashi, H. *Phytochem.* **1976**, *15*, 1391-1393.
21. Jadhav, S. J.; Salunkhe, D. K. *J Food Sci.* **1973**, *38*, 1099-1100.
22. Lavintman, N.; Tandecarz, J.; Cardini, C. *Plant Sci. Lett.* **1977**, *8*, 65-70.
23. Stapleton, A.; Allen, P.V.; Friedman, M.; Belknap, W. R. *J. Agric. Food Chem.* **1991**, *39*, 1187-1193.
24. Maga, J. A. *CRC Crit. Rev. Food Sci. Nutr.* **1980**, *12*, 371-405.
25. Jadhav, S. J.; Sharma, R. P.; Salunkhe, D. K. *CRC Crit. Rev. Toxicol.* **1981**, *9*, 21-104.
26. Slanina, P. *Fd. Chem. Toxicol.* **1990**, *28*, 759-761.
27. Ross, H.; Pasemann, P.; Nitzche, W. *Z. Pflanzenzüchtg*. **1978**, *80*, 64-79.
28. Sinden, S. L.; Sanford, L. L.; Webb, R. E. *Am. Potato J.* **1984**, *61*, 141-156.
29. Olsson, K. *Potato Research*. **1986**, *29*, 1-12.
30. Schreiber, K. In *The Biology and Taxonomy of the Solanaceae*, J. G. Hawkes, R. N. Lester and A. D. Skelding, eds.; Academic Press: London, U.K. 1979; pp. 193-202.
31. Hardman, R. *Planta Medica.* **1987**, *53*, 233-238.
32. Gaffield, W. Keeler, R. F.; Baker, D. C. In *Handbood of Natural Toxins. Vol. 6, Toxicology of Plant and Fungal Compounds*. R. F. Keeler and A. T. Tu, eds.; Marcel Dekker, Inc.: New York, NY. 1991; pp. 135-158.
33. Johns, T.; Alonso, J. G. Petota. Euphytica. **1990**, *50*, 203-210.
34. Sanford, L. L.; Deahl, K. L.; Sinden, S. L.; Ladd, T. L. Jr. *Am. Potato J.* **1990**, *67*, 461-466.
35. Deahl, K. L.; Cantelo, W. W.; Sinden, S. L.; Sanford, L. L *Am. Potato J.* **1991**, *68*, 659-666.
36. Osman, S. F.; Sinden, S. L.; Irwin, P.; Deahl, K; Tingey, W. M. *Phytochem.* **1991**, *30*, 3161-3163.
37. Flanders, K. L.; Hawkes, J. G.; Radcliffe, E. B.; Lauer, F. I. *Euphitica*. **1992**, *61*, 83-111.
38. Osman, S. F.; Herb, S. F.; Fitzpatrick, T. J.; Schmiediche, P. *J. Agric. Food Chem.* **1978**, *26*, 1246-1248.
39. Gregory, P.; Sinden, S. L.; Osman, S. F.; Tingey, W. M.; Chessin, D. A. *J. Agric. Food Chem.* **1981**, *29*, 1212-1215.
40. Coxon, D.T. *Am. Potato J.* **1984**, *61*, 169-183.
41. Gregory, P. *Am. Potato J.* **1984**, *61*, 115-122.
42. Akeley, R. V.; Mills, W. R.; Cunningham, C. E.; Watts, J. *Am. Potato J.* **1968**, *45*, 142-145.
43. Zitnak, A.; Johnston, G. R. *Am. Potato J.* **1970**, *47*, 256-260.
44. Sinden, S. L.; Sanford, L. L.; Cantelo, W. W.; Deahl, K. L. *Environ. Entomol.* **1986**, *15*, 1057-1062.
45. Collins, E. *Sci. Amer.* **1987**, *257*, 31.
46. Van Gelder, W. J. M.; Scheffer, J. J. C. *Phytochem.* **1991**, *30*, 165-168.
47. Tingey, W. M. *Am. Potato J.* **1984**, *61*, 157-167.
48. Mondy, N. I.; Seetharman, K.; Munshi, C. B. *J. Food Sci.* **1992**, *57*, 1357-1358.
49. Salunkhe, D. K.; Wu, M. T. *J. Food Protect.* **1979**, *42*, 519-525.
50. Bushway, A. A.; Bushway, A. W.; Belyea, P. R.; Bushway, R. J. *Am. Potato J.* **1980**, *57*, 167-171.
51. Bushway, R. J.; Ponnampalam, R. *J. Agric. Food Chem.* **1981**, *29*, 814-817.
52. Sizer, C. E.; Maga, J. A.; Craven, C. J. *J. Agric. Food Chem.* **1980**, *28*, 578-579.
53. Sinden, S. L.; Deahl, K. L.; Aulenbach, B. B. *J. Food Sci.* **1976**, *41*, 520-523.
54. Mondy, N. I.; Metcalf, C.; Plaisted, R. L. *J. Food Sci.* **1971**, *36*, 459-461.
55. Zitnak, A. In *Nightshades and Health.* N. F. Childers and G. M. Russo, eds., Somerset Press: Somerville, NJ. 1977; pp. 41-91.
56. Jonasson, T.; Olsson, K. *Potato Res.* **1994**, *37*, 205-216.

57. Sanford, L. L.; Domek, J. M..; Cantelo, W. W..; Kobayashi, R. S.; Sinden, S. L. *Am. Potato J.* **1996**, *73*, 79-88.
58. Arneson, P. A.; Durbin, R. D. *Phytophathology* **1968**, *58*, 536-537.
59. Fewell, A. M.; Roddick, J. G. *Phytochem.* **1993**, *33*, 323-328.
60. Fewell, A. M.; Roddick, J. G.; Weissengberg, M. *Phytochem.* **1994**, *37*, 1007-1001.
61. Anon. *Br. Med. J.* **1979**, Dec. 8, 1458-1459.
62. Maga, J. A. *Food Rev. Int'l.* **1994**, 10, 385-418.
63. Sporns, P.; Abell, D. C.; Kwok, A. S. K.; Plhak, L. C.; Thomson, C. A. In *Immunoassays for Residue Analysis*. R. C. Beier and L. H. Stanker, eds.; ACS Symposium Series 621, 1996; pp. 256-272.
64. Willimott, S. G. *Analyst.* **1933**, *58*, 431-439.
65. Reepyah, L. A.; Keem, A. *Sov. Med.* **1958**, 129-131.
66. McMillian, M.; Thompson, J. C. *Quarterly J. Med.* **1979**, *48*, 227-243.
67. Health and Welfare Canada. *Foodborne and Waterborne Disease in Canada. Annual Summaries 1983, 1984*. Polyscience Publications Inc. and Supply and Services Canada: Ottawa, ON. 1988.
68. Harris, F. W.; Cockburn, T. *Analyst.* **1918**, *43*, 133-137.
69. Hansen, A. A. Science. **1925**, *61*, 340-341.
70. Morris, S. C.; Lee, T. H. *Food Technol. Australia.* **1984**, *36*, 118-124.
71. International Food Biotechnology Council. *Reg. Toxicol. Pharmacol.* **1990**, *12* (Part 2), S11-S78.
72. Hall, R. L. *Food Technol.* **1992**, *46*, 109-112.
73. Osman, S. F. *Food Chem.* **1983**, *11*, 235-247.
74. Smith, D. B.; Roddick, J. G.; Jones, J. L. *Trends Food Sc. Tech.* **1996**, *7*, 126-131.
75. Harvey, M. H.; McMillan, M.; Morgan, M. R. A.; Chan, H. W. S. *Human Toxicol.* **1985**, *4*, 187-194.
76. Hellenäs, K-E.; Nyman, A., Slanina, P.; Lööf, L.; Gabrielsson, J. *J. Chromatogr.* **1992**, *573*, 69-78.
77. Creeke, P. I.; Lee, H. A.; Morgan, M. R. A.; Price, K. R.; Rhodes, M. J. C.; Wilkinson, A. P. In *Immunoassays for Residue Analysis* R. C. Beier and L. H. Stanker, eds.; ACS Symposium Series 621, 1996, pp. 202-218.
78. Claringbold, W. D. B.; Few, J. D.; Brace, C. J.; Renwick, J. H. *J. Steroid Biochem.* **1980**, *13*, 889-895.
79. Price, K. R.; Johnson, I. T.; Fenwick, G. R. *CRC Crit. Rev. Food Sci. Nutr.* **1987**, *26*, 27-135.
80. Roddick, J. G.; Rijnenberg, A. L. *Phytochem.* **1987**, *26*, 1325-1328.
81. Roddick, J. G.; Rijnenberg, A. L.; Osman, S. F. *J Chem. Ecol.* **1988**, *14*, 889-902.
82. Roddick, J. G.; Rijnenberg, A. L.; Weissenberg, M. *Phytochem.* **1990**, *29*, 1513-1518.
83. Roddick, J. G.; Rijnenberg, A. L.; Weissenberg, M. *Phytochem.* **1992**, *31*, 1951-1954.
84. Keukens, E. A. J.; Vrije, T.; Fabri, C. H. J. P.; Demel, R. A.; Jongen, W. M. F.; Kruijff, B. *Biochim. Biophys. Acta.* **1992**, *1110*, 127-136.
85. Keukens, E.A.J.; Vrije, T.; van den Boom, C.; de Waard, P.; Plasman, H. H.; Thiel, F.; Chupin, V.; Jongen, W. M. F.; de Kruijff, B. *Biochim. Biophys. Acta*, **1995**, *1240*, 216-228.
86. Alozie, S. O.; Sharma, R. P.; Salunkhe, D. K. *J. Food Biochem.* **1978**, *2*, 259-276.
87. Schwarz, M.; Glick, D.; Loewenstein, Y.; Soreq, H. *Pharmac. Ther.* **1995**, *67*, 283-322.
88. Friedman, M.; Henika, P. R. *Food Chem. Toxic.* **1992**, *30*, 689-694.

89. Crawford, L.; Kocan, R. M. *Toxicology Let.* **1993**, *66*, 175-181.
90. Friedman, M.; Rayburn, J. R.; Bantle, J. A. *J. Agric. Food Chem.* **1992**, *40*, 1617-1624.
91. Crawford, L. and Myhr, B. *Food Chem. Toxic.* **1995**, *33*, 191-194.
92. Dalvi, R. R.; Bowie, W. C. *Vet. Human Toxicol.* **1983**, *25*, 13-15.
93. Keeler, R. F.; Young, S.; Brown, D.; Stallknecht, G. F.; Douglas, D. *Teratol.* **1978**, *17*, 327-334.
94. Renwick, J. H.; Claringbold, W. D. B.; Earhty, M. E., Few, J. D.; McLean, A. C. S. *Teratol.* **1984**, *30*, 371-381.
95. Hellenäs, K-E.; Cekan, E.; Slanina, P. and Bergman, K. *Pharmacol. and Toxicol.* **1992**, *70*, 381-383.
96. Groen, K.; Pereboon-de Fauw, D. P. K. H.; Besamusca; P., Beekhof, P. K. Speijers, G. J. A.; Derks, H. J. G. M. *Xenobiotica*, **1993**, *23*, 995-1005.
97. Caldwell, K. A.; Grosjean, O. K.; Henika, P. R.; Friedman, M. *Fd. Chem. Toxic.* **1991**, *29*, 531-535.
98. Nishie, K.; Fitzpatrick, T. J.; Swain, A. P. and Keyl, A. C. *Res. Comm. Chem. Pathol. Pharmacol.* **1976**, *15*, 601-607.
99. Bähr, V.; Hänsel, R. *Planta Medica.* **1982**, *44*, 32-33.
100. Cham, B. E.; Gilliver, M.; Wilson, L. *Planta Med.* **1987**, *53*, 43-36.
101. Cham, B. E.; Meares, H. M. *Cancer Let.* **1987**, *36*, 111-118.
102. Wood, F. A.; Young, D. A. In *TGA in Potatoes, Publication # 1533.* Agriculture Canada, Ottawa, ON. 1974.

Chapter 8

α-Galactosides of Sucrose in Foods: Composition, Flatulence-Causing Effects, and Removal

Marian Naczk[1], Ryszard Amarowicz[2], and Fereidoon Shahidi[3]

[1]Department of Human Nutrition, St. Francis Xavier University, P.O. Box 5000, Antigonish, Nova Scotia B2G 2W5, Canada
[2]Centre for Agrotechnology and Veterinary Sciences, Division of Food Science, Department of Food Chemistry, Polish Academy of Sciences, 10–718 Olsztyn, Poland
[3]Department of Biochemistry, Memorial University of Newfoundland, St. John's, Newfoundland A1B 3X9, Canada

Alpha-galactosides of sucrose, namely raffinose, stachyose and verbascose, are widely distributed in higher plants, especially leguminous seeds. In addition the presence of galactopinitol, galactinol, mannitriose and melibiose in *Brassica campestris*, soybean and some legumes has been reported. Due to the absence of α-galactosidase activity in human and animal intestine mucosa, these oligosaccharides escape digestion and they are metabolized by bacteria to hydrogen, carbon dioxide and methane. Thus, they are considered to be the principal flatulence-causing factors present in food of plant origin. Furthermore, starch and hemicellulose may contribute to flatulence. A significant positive correlation, between the hydrogen production by rat and the content of α-galactosides of sucrose in legumes consumed, has been reported. Various methods for the removal of flatulence causing oligosaccharides including dehulling, soaking and/or cooking in water and in buffer solutions, irradiation, enzymatic treatment, germination and solvent extraction have been investigated. The pros and cons of these procedures are discussed.

The oligosaccharides raffinose, stachyose and verbascose are broadly distributed in higher plants (*1*). They accumulate in leaves during photosynthesis (*2*) and in seeds during maturation (*3*). It is believed that the function of oligosaccharides of the raffinose family in higher plant is to store or transport carbohydrates (*4, 5*). Moreover, commonly observed accumulation of sucrose and its α-galactosides in leaves and seeds may contribute to cold acclimation and protection of membrane proteins and other biologically active substances from denaturation from the desiccation process toward dormancy (*6, 7, 8*). Galactinol is another α-galactoside sugar found in higher plants. This disaccharide acts as a transfer-intermediate for galactose in the biosynthesis of sugars of the raffinose family (*2*). In foods and feeds sugars of the raffinose family are flatulence causing due to the lack of α-galactosidase activity in human and animal intestinal mucosa. Thus, these carbohydrates escape digestion and are metabolized by the existing bacteria in the lower intestinal tract to hydrogen, carbon dioxide and methane (*9*).

Flatulogenic Factors in Legumes

Consumption of a legume-containing diet is commonly associated with the formation of excessive amounts of intestinal gases. The data presented in Table I shows the amounts of intestinal gas formed following the ingestion of selected legumes.

Table I. Intestinal Gas Formation Following Ingestion of Legumes

Legume [Ref]	*Intake [g]*	*Flatus [mL/h]*
Soybeans [*10*]	100	36
Soybeans [*10*]	200	24
California white beans [*11*]	100	120-137
California white beans [*12*]	100	37
California white beans [*10*]	450	36
Lima beans [*10*]	100	42
Mung beans [*10*]	100	25
Lentils [*10*]	200	34-41

Flatulence. Excessive accumulation of air and/or other gases such as carbon dioxide, hydrogen and methane in the stomach and/or gastrointestinal (GI) tract is known as flatulence. The accumulation of gases may be due to air swallowing, colonic fermentation and passive diffusion from blood stream to bowel. The composition and volume of gases produced is controlled by a number of factors including age, heredity, stress, antibiotics and diet (*14*). Normal individuals eating a typical diet produce from 400 to 1600 mL of gas per day (*15*). Furthermore, it has been demonstrated (16) that GI tract of fasting individuals contains approximately 200 mL of gases. Hydrogen and carbon dioxide are the main gaseous products of microbial intestinal fermentation of nonabsorbed carbohydrates (*17*). On the other hand, skatole, hydrogen sulfide, indole, volatile amines and short-chain fatty acids are linked to the flatus odor. These compounds are detected by smell at concentrations of one part per 100 millions (*18*). Tomomatsu (*19*) calculated (*20*) that 300 g wet feces contained 186 mg of ammonia, 1.4 mg of phenol, 12.2 mg of p-cresol, 8.5 mg of indole, and 3.3 mg of skatole, using the data provided by Kato et al.

Intestinal gases are eliminated by convection via eructation and passage from rectum, metabolism by the intestinal microflora, and diffusion into the blood (*16*). In addition to its social inconvenience, flatulence may cause abdominal discomfort and bloating, abdominal rumbling, repetitive belching, cramps, and constipation, diarrhea, floating stools and anxiety. (*9, 16, 21, 22*).

Role of Diet in Flatulence. There is a common belief about the relationship between the diet, especially those containing legumes, and formation of intestinal gases. The existence of such relationships was confirmed by a number of investigators. Alvarez (*23*), based on the results of survey of 500 people, arranged the flatulogenic foods in decreasing order of potency as onions, cooked cabbage, raw apples, radishes, dry beans, cucumbers, milk, melon, cauliflower, chocolate, coffee, lettuce, peanuts, eggs, oranges, tomatoes and strawberries. According to Van Ness and Cattau (*14*) there is a considerable variability from one individual to another in the list of foods bringing about the symptoms of flatulence. These authors classified foods into normoflatulogenic (meat, fowl, fish, lettuce, cucumber, broccoli, peppers, avocado, cauliflower, tomato, asparagus, zucchini, okra, olives, all nuts, eggs, nonmilk chocolate,), moderately flatulogenic (pastries, potatoes, eggplant, citrus fruit, apples, bread) and extremely flatulogenic (milk and milk products, onions, beans, celery, carrots, raisins, bananas, apricots, prune juice, pretzels, bagels, wheat germs, and Brussels sprouts). More recently a similar list of foods associated with flatulence was published by Price et al. (*9*). These authors, in addition to foods listed above, also included products such as bacon, aubergine and peanuts, but did not attempt to classify them according to their flatulogenic potency. The published data clearly indicate that dietary modification may have a significant influence on the formation of intestinal gases.

Role of Microorganisms in Flatulence. Almost 99% of microorganisms present in the large intestine of humans are anaerobic bacteria (*24*). The colonic and faecal materials contain 10^{11} to 10^{12} organisms per gram (*25*). At least six major bacterial groups have been identified in human intestinal microflora including *Enterbacteriaceae, Bacteroides, Enterococci, Lactobacilli*, Gram positive nonsporing anaerobes (i.e. *Bifidobacterium* species and *Eubacterium* species) and *Clostridia* (*26*). Over 100 species of bacteria reside in the human intestine (*27, 28, 29*). Intestinal bacteria, based on their influence on human health, can be classified into harmful, beneficial, or neutral. *Bifidobacteria* and *Lactobacilli* are examples of beneficial bacteria as they promote immunity and inhibit the growth of pathogenic bacteria. On the other hand, *Escherichia coli*, *Clostridium, Proteus* and some *Bacteroides* species are examples of harmful bacteria as they produce various undesirable compounds from food and may cause some intestinal problems (*29*).

Carbohydrates and their derivatives are principal sources of energy for most intestinal microflora (*30*). Western diet provides at least 20 g/day of carbohydrates (starch that escapes digestion, dietary fibre, oligosaccharides, and mucus carbohydrates) that are subjected to fermentation by intestinal microflora (*31*). The metabolic products of microbial fermentation of carbohydrates include short-chain fatty acids, carbon dioxide, methane, hydrogen and water (Figure 1) (*25*). This fermentation influences salt and water absorption from the colon, bowel habit, the excretion of toxic compounds as well as nitrogen and sterol metabolism (*19, 25*).

Steggerda and Dimmick (*32*) have suggested that the quantity and quality of intestinal gas produced depends on the consumption of specific foods as well as the presence of specific types of bacteria. Richards et al. (*33*) later demonstrated, both *in vivo* and *in vitro* tests, that some anaerobic spore-forming bacteria existing in the small and large intestine of dog were implicated in gas production in the presence of navy bean and soybean products. Based on *in vitro* studies, these authors linked the production of gas to *Clostridia*-type bacteria. Kurtzman and Halbrook (*34*) found a correlation between the human flatus and gas production by *Clostridium perfringens* for green and dry lima beans and navy beans. Strains of *Cl. perfringens* type A, generally found in human guts (*35*), had the ability to ferment raffinose, stachyose and melibiose to different extents. For example, raffinose and stachyose were rapidly utilized by *Cl. perfringens* NCTC strains 8238 and 10240, while strain FD-1 only slightly utilized raffinose, and stachyose was not utilized at all (*36*). However, in humans *Cl. perfringens* which has a unique capacity for unusually rapid growth (*37*) did not rank amongst the top 100 microorganisms present (*27, 28*). Therefore, Sacks and Olson (*36*) suggested that after the ingestion of foods containing α-galactosides some strains of *Cl. perfringens* may grow more rapidly than competing microorganisms. On the other hand, they suggested that under sustained supply of α-galactosides the replacement of *Cl. perfringens* by the competing microorganisms is favored. Furthermore, *Cl. perfringens* have the ability to sporulate in the intestine (*37*) and this process is enhanced when raffinose and melibiose are present (*38*). Moreover some strains of *Cl. perfringens* have the ability to release enterotoxin during sporulation (*37*). Thus, according to Sacks and Olson (*37*) the contribution of small amounts of enterotoxin to some flatulence symptoms such as headache, dizziness, and reduced ability to concentrate should not be ruled out.

Presence of spices significantly reduced the ability of *Clostridia* species to produce gas due to their antimicrobial properties. Garg et al. (*39*) demonstrated in their *in vitro* studies that the amount of gas produced in 24 h by *Cl. perfringens, Cl. sporogenes* and *Cl. butyricum* was decreased by up to 85% after addition of 1% garlic or ginger powder to cooked chickpea or dry pea substrate. Savitri et al. (*40*) also showed that common food spices inhibited *in vitro* production of gas by *Cl. perfringens*. Rackis et al. (*41*) in their *in vitro* experiments demonstrated that phenolics such as syringic and ferulic acids which are commonly found in soybean products were effective inhibitors of gas production.

Myhara et al. (*42*) examined the ability of four organisms isolated from human feces: *Peptostreptococcus productus, P. anaerobius, Bacteroides fragilis ss. distasonis,* and *Cl. perfringens*, to grow and produce gas from α-galactosides and fractions of white beans. These authors found that *Cl. perfringens* produced the most gas from substrate with raffinose. On the other hand, presence of melibiose did not promote gas production, thus suggested that presence of fructose moiety in raffinose may be responsible for gas production.

Masai et al. (*43*) evaluated the effect of diets containing 6 x 10^9 cells of freeze dried *Bifidobacteriun longum* and 10 g soybean oligosaccharides on human intestinal flora. Diets were given to six healthy males twice a day for three weeks. After 4 weeks of suspension of diet ingestion, another diet containing only 10 g soybean oligosaccharides was administered twice a day for three weeks. There was a significant (over 2 times) increase in the average number of *Bifidobacteria* and a decrease in the number of *Cl. perfringens* and other *Clostridia* organisms. In addition, the activity of the two detrimental enzymes found in feces, namely glucuronidase and azoreductase was significantly reduced.

Role of Carbohydrates in Flatulence. Steggerda et al. (*44*) tested a series of soybean fractions for their gas-producing potential in human subjects. It was demonstrated that the soybean fraction containing low-molecular weight carbohydrates was the predominant flatulence-causing factor in soybeans (*44*). Later, Rackis et al. (*41*) evaluated the gas production potential of mono- and oligosaccharides when incubated in thioglycollate media with *Clostridia* type organism. Sucrose, raffinose, and stachyose produced gas at a slower rate than glucose, but the ratio of carbon dioxide to hydrogen in gases produced remained unaffected by the type of carbohydrate used. Therefore, the oligosaccharides first break down to monosaccharides by bacteria in the GI tract themselves and then are effectively used to produce gas. Calloway et al. (*10*) have demonstrated that dry lima beans were as high in flatulence-inducing factors as small white beans from California. These authors also found that mung beans and soybeans produced similar excretion of hydrogen in breath as did white beans but only 2/3 as much flatus. However, soybean produced similar volume of flatus when the dosage of carbohydrates from soybean and white beans were equalized. In addition, Calloway et al (*10*) reported that ethanolic extraction of white beans reduced, but did not eliminate the gas-forming properties of white beans. Murphy et al. (*11*) showed, using human subjects fed with small cooked small white beans from California, that the maximum volume of flatus occurred between 4 and 7 h after consumption. The fraction of California small white beans containing fructose, sucrose, raffinose, stachyose and polypeptides showed the maximum ability to increase the level of carbon dioxide in flatus. On the other hand, these authors found that raffinose and stachyose when fed alone at the levels found in California white beans did not increase the content of carbon dioxide in flatus. Wagner et al. (*45*) compared the hydrogen production in rats following ingestion of raffinose, stachyose, California small white (CSW) beans and oligosaccharide-free CSW beans; both stachyose and raffinose produced less hydrogen than CSW beans. On the other hand, oligosaccharide -free CSW was only 0.4 to 0.5 times as active in hydrogen production as CSW beans. Thus, it was concluded that α-galactosides of sucrose are only one of flatulogenic factors present in CSW (*45*). Later, Wagner et al. (*12*) demonstrated that the quantity of hydrogen excreted by laboratory rats was correlated with the quantity of flatus produced by humans. Fleming (*46*) also investigated the relationship between flatulence potential and the content of carbohydrates in leguminous seeds and found a significant positive correlation between hydrogen production and the contents of α-galactosides of sucrose and acid-hydrolysable pentosans in legume seeds. In addition, hydrogen production was found to closely correlate with the quantity of oligosaccharides in the seed meal; cell wall

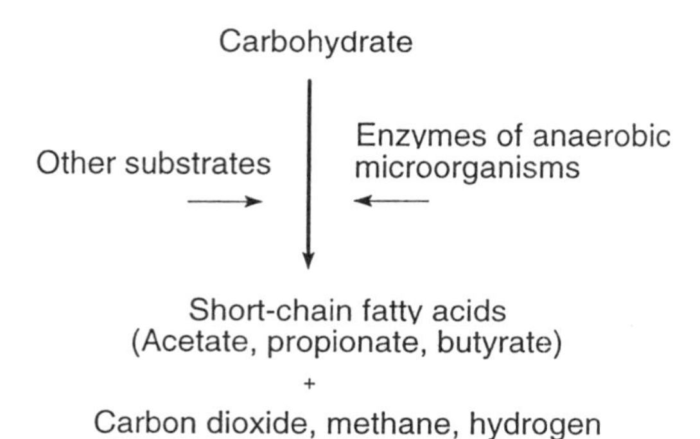

Figure 1. Simple pathway for fermentaiton of selected carbohydrates in human colon and flatus formation.

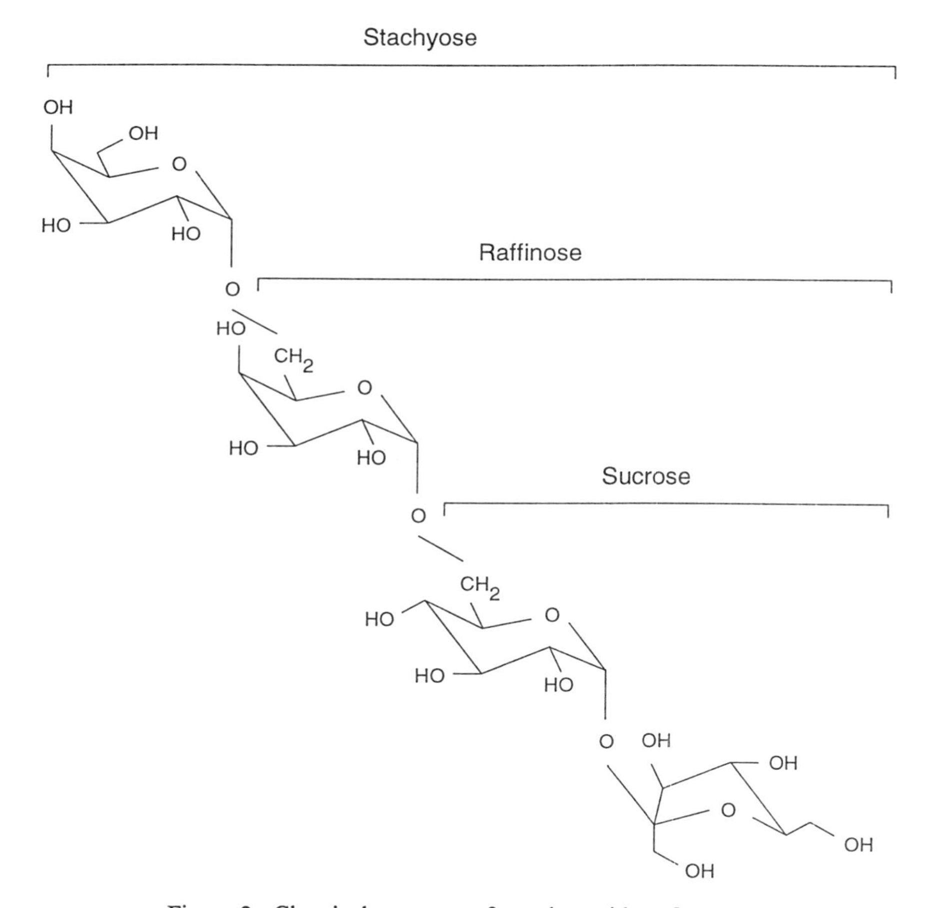

Figure 2. Chemical structure of α-galactosides of sucrose.

fibre constituents also contributed to the flatulence activity of sooth-seeded field pea.(*47*) Moreover, different carbohydrate fractions isolated from chick pea (*Cicer arietinum*), cow pea (*Vigna sinensis*) and horse gram (*Dolichos biflorus*) were tested (*48*). It was demonstrated that not only oligosaccharides but also starch and hemicelluloses contribute to the total flatulence effect of these legumes; starches were the highest gas producer in *in vitro* studies. Although, in *in vivo* studies, oligosaccharide fractions produced the most amount of gas, contribution of starches and hemicellulose was still substantial. The formation of gas by leguminous starches was attributed to their low digestibility. Thus, published results indicate that α-galactosides of sucrose are only one of the flatulence-causing factors present in legumes and their removal would reduce but not eliminate the formation of flatus following ingestion of legumes.

Composition of α-Galactosides in Foods of Plant Origin

The composition of α-galactosides in foods of plant origin has been extensively investigated, as these sugars are considered as a major cause of flatulence.

Chemical Structure and Occurrence. The chemical structure of α-galactosides found in food products are shown in Figure 2. A number of sugars in the raffinose family have been identified in foods of plant origin. Raffinose and stachyose were found in *Brassica* meals (*49, 50, 51*) and some vegetables (*52*); presence of raffinose, stachyose and verbascose was reported in cottonseed, peanut and leguminous seeds (*52*). Presence of ajugose and higher molecular weight sugars of the raffinose family in ethanolic extracts of lupine, wrinkled pea, smooth pea and horsebean (*53*) as well as presence of galactinol, melibiose and manninotriose in *B.campestris* meals (*54*) has also been reported. Melibiose and manninotriose were thought to have originated from the breakdown of stachyose and raffinose during fractionation and isolation of sugars. Schweizer et al. (*55*) reported the presence of galactopinitol, a disaccharide, in 70 % ethanolic extracts of soybean, chick pea, lentil, and bean (*P. vulgaris* var. Borlotti). Sosulski et al. (*56*) have also reported the presence of galactinol, galactopinitol and manninotriose in a number of legume flours and their corresponding protein concentrates. However, only manninotriose was identified in ethanolic extracts of yellow and black varieties of chickpeas (*57*).

Quantification of Oligosaccharides. Tables III to VII summarize the content of sucrose, raffinose, stachyose, and verbascose in foods of plant origin, using various methods for their isolation and fractionation. Factors influencing these values are due to varietal and the maturation stage as well as the extraction conditions and methods of analyses employed. α-Galactosides are usually extracted from plant materials by water/alcohol solvent systems (*46, 58, 59, 60*) which are time-consuming and may not always extract verbascose completely (*61*). Saini (*62*) compared four methods of extraction of low molecular sugars from plant materials (Table II). The author found that treatment of meal with 60% methanol/water solution (*60*) resulted in incomplete extraction of the α-galactosides. On the other hand, extraction of plant material in boiling aqueous ethanol (*63*) or in aqueous ethanol at 65 °C (*64*) or in water at 30 °C (*65*) afforded comparable results for some seeds. Muzquiz et al. (*66*) evaluated two procedures for extraction of α-galactosides from lupin species. The first procedure involved boiling of a finely grounded material in 60% methanol as described by Macrae and Zand-Moghaddam (*60*). The second used ground material which was homogenized three times with 70% methanol for 1 min. Both extraction procedures showed satisfactory recoveries of sucrose, raffinose and stachyose. However, somewhat higher recoveries were achieved when the extraction procedure involved homogenization of the material with aqueous methanol.

Table II. Levels of Sucrose and α-Galactosides of Sucrose in Leguminous Seeds Extracted by Four Methods [g/100g Dry Matters]

Legume [Ref]	*Sucrose*	*Raffinose*	*Stachyose*	*Verbascose*
Soybeans				
[*63*]	7.98	0.93	4.56	—
[*64*]	8.35	1.18	4.83	—
[*60*]	5.66	0.57	3.63	—
[*65*]	8.09	1.29	4.85	—
Chickpeas				
[*63*]	1.98	0.84	3.05	0.42
[*64*]	2.30	1.02	3.58	0.29
[*60*]	2.87	0.94	2.87	0.33
[*65*]	2.15	0.66	3.52	0.00

Adapted from ref. [*62*]

Various chromatographic techniques have been employed for quantification of individual sugars. Paper chromatography (*58*) and thin-layer chromatography (*67*) followed by extraction and colorimetry have been used extensively. Results so obtained are difficult to quantify as these methods required the use of some tedious operations when working with small quantities of sugars. The HPLC methodology has also been used for carbohydrate analysis (*68, 69*), but peak resolution and detector sensitivity may affect the quantification results (*70*). Quantification of oligosaccharide silyl ethers by packed-column GC (*56*) or by capillary column GC (*70*) may also suffer from several shortcomings. These are not always related to quantitative formation of oligosaccharide silyl ethers and their possible thermal degradation at high temperatures of the injection port of the GC (*71*).

Content of α-Galactosides in Legumes. The content of sucrose, raffinose, stachyose, and verbascose in selected legume seeds is summarized in Table III. Black beans (*90*) and pink and black eye beans (*75*) contained the least amount of flatulence-causing sugars. Verbascose was the predominant flatulence-causing sugar in mung bean, faba bean, horse gram, bengal gram, green gram, red gram and black

Table III. Content of Oligosaccharides in Legume Seeds [g/100g Dry Matters]

Legume [Ref]	*Sucrose*	*Raffinose*	*Stachyose*	*Verbascose*
Adzuki bean [*62*]	1.40	0.38	5.04	—
Adzuki bean [*67*]	0.74	0.35	4.06	—
Bengal gram [*72*]	2.30	1.10	2.10	4.50
Bengal gram [*74*]	0.80-2.00	0.80-3.00	0.80-1.80*	
Bengal gram [*83*]	1.30	1.09	1.18	3.41
Black bean [*90*]	1.60	0.07	0.40	1.80
Black eye bean [*75*]	3.00	0.95	0.93	—
Black gram [*74*]	0.20-1.50	0.20-0.80	0.60-1.80*	
Black gram [*78*]	1.46	trace	0.89	3.44
Black gram [*83*]	1.40	0.78	0.80	3.36
Brazilian cultivars [*87*] of *P. vulgaris*	3.10-3.70	1.00-1.40	3.20-4.70	—
Broad bean [*52*]	2.07	0.23	1.07	1.14
Chickpea [*62*]	1.98	0.84	3.05	0.42
Chickpea [*68*]	2.19	3.40	1.60	—
Chickpea, black [*57*]	—	0.09-0.34	0.48-1.64	—
Chickpea, yellow [*57*]	—	trace-0.26	0.35-0.68	—
Common bean [*92*]	1.25	0.45	1.80	0.25
Cowpea [*52*]	2.59	0.37	4.64	0.36
Cowpea [*62*]	4.32	1.63	6.00	—
Cowpea [*76*]	1.80-3.00	1.10-1.30	2.90-4.10	0.60-1.00
Cowpea [*80*]	0.36-1.41	1.26-4.19	1.21-4.94	—
Cowpea [*84*]	1.61	0.45	3.30	—
Cowpea [*86*]	—	0.90-2.60	2.60-3.50	—
Cowpea [*88*]	1.48	1.95	3.56	4.03
Cowpea [*92*]	1.51	0.77	3.00	0.30
Faba bean [*73*]	1.55	0.24	0.80	1.94
Faba bean [*79*]	—	0.30	1.10	1.70
Faba bean [*92*]	1.37	0.52	1.41	1.85
Field pea [*79*]	—	1.10	2.60	1.60
Field pea [*89*]	1.31	0.80	1.28	2.35
Field pea, smooth [*46*]	—	0.34	2.91	2.19
Field pea, wrinkled [*46*]	—	1.47	4.50	2.70
Garbanzo bean [*46*]	—	0.67	2.16	0.43
Garden pea [*52*]	6.23	1.16	3.23	1.91
Garden pea [*68*]	2.47	1.70	1.86	—
Great Northern bean [*81*]	5.11	0.37	5.62	—
Green bean [*52*]	1.94-2.90	2.20-2.5	2.81-3.43	trace
Green gram	0.50-2.10	0.20-0.80	1.80-2.60*	

continued on next page

Table III. Continued ....

Legume [Ref]	*Sucrose*	*Raffinose*	*Stachyose*	*Verbascose*
Green gram [*72*]	1.90	1.10	1.90	3.50
Horse gram [*72*]	2.60	0.780	1.90	3.40
Horse bean [*83*]	1.20	0.96	0.75	2.60
Lentil [*72*]	2.50	1.00	2.60	3.10
Lentil [*73*]	1.81	0.39	1.85	1.20
Lentil [*79*]	—	0.50	2.70	1.00
Lentil [*91*]	1.26	0.22	1.96	—
Lentil [*92*]	1.14	0.45	1.65	0.62
Lentil, green [*46*]	—	0.60	1.70	0.70
Lima bean [*52*]	3.60	0.69	3.03	trace
Lupine [*79*]	—	0.90	5.80	0.70
Mung bean [*46*]	—	0.37	1.67	1.73
Mung bean [*52*]	1.39	0.39	1.67	2.66
Mung bean [*67*]	0.93	0.44	1.96	—
Mung bean [*73*]	1.28	0.32	1.65	2.77
Mung bean [*89*]	1.17	0.40	1.24	2.02
Navy bean [*46*]	—	0.67	3.53	0.50
Navy bean [*73*]	2.23	0.41	2.59	0.13
Navy bean [*85*]	—	0.40	2.20	—
Pink bean [*75*]	1.75	0.20	0.12	—
Pinto bean [*73*]	2.82	0.43	2.97	0.15
Pinto bean [*85*]	—	0.50	2.50	—
Pole bean [*52*]	2.67	0.43	2.62	trace
Rajmah (red bean) [*83*]	1.58	0.89	1.62	3.26
Red gram [*72*]	1.14	1.05	3.00	4.00
Red gram [*83*]	1.25	1.00	2.45	2.96
Red kidney bean [*46*]	—	0.37	4.00	0.40
Red kidney bean [*52*]	2.15	0.31	3.16	trace
Red kidney bean [*68*]	2.87	1.50	2.69	—
Soybean [*52*]	6.42-7.27	1.16-1.26	4.10-4.34	trace
Red gram [*74*]	0.60-1.60	0.30-1.80	1.20-3.60*	
Soybean [*62*]	7.98	0.93	4.56	—
Soybean [*67*]	4.01	1.25	3.80	—
Soybean [*68*]	5.62	2.40	3.40	—
Soybean [*77*]	6.81	1.01	3.53	—
Soybean [*79*]	—	0.80	4.20	0.30
Soybean [*90*]	3.90	0.80	2.80	—
Soybean, black [*82*]	3.66-4.23	0.76-0.81	2.75-4.02	—
Soybean, yellow [*82*]	3.44-3.81	0.81-0.87	2.85-3.59	—
White bean [*67*]	2.61	0.66	3.39	—

* total stachyose & verbascose

gram seeds as it constituted 46 to 79% of the total flatulence-causing sugars. However, stachyose was the principal flatulence-causing sugar in legume seeds such as adzuki bean (3.40-4.56%), white bean (3.39%), navy bean (2.59-3.53%), red kidney bean (2.69-4.00%), lupine (5.8%), great northern bean (5.62%), wrinkled field pea (4.50%), soybeans (3.40-4.56%) and lima bean (3.03%) and predominated in cowpeas (1.21-6.00%) and lentils (1.85-2.70%). Schweizer et al. (*55*) reported that the content of galactopinitol in chickpeas, lentils and beans ranged from 0.03 to 0.80% , on a dry weight basis. According to Rossi et al. (57) the content of manninotriose in chickpea cultivars ranged from 1.24 to 3.10%.

Table IV. Content of Oligosaccharides in Hull-Free Flours of Legumes [g/100g Dry Matters]

Legume [Ref]	*Sucrose*	*Raffinose*	*Stachyose*	*Verbascose*
Field Pea [*59*]	2.20	0.80	2.00	2.30
Faba bean	2.90	0.20	0.20	2.00
Soybean [*93*]	6.17-9.39	0.54-1.18	3.95-4.96	—
Field pea [*78*]	—	1.10	2.60	1.60
Faba bean	—	0.60	1.00	1.50
Soybean [*56*]	6.35	1.15	2.85	—
Lupine	2.63	0.82	4.11	0.48
Cowpea	2.64	0.41	4.44	0.48
Chickpea	2.69	0.45	1.72	0.10
Lentil	3.36	0.31	1.47	0.47
Lima bean	1.85	0.46	2.76	0.31
Navy bean	2.62	0.37	2.36	0.05
Field pea	1.85	0.60	1.71	2.30
Mung bean	0.96	0.23	0.95	1.83
Faba bean	2.00	0.22	0.67	1.45
Cowpea [*80*]	0.21-1.10	0.89-3.39	0.89-3.75	—
Cowpea [*94*]	—	0.25	0.52	—
Cowpea [*88*]	0.85	0.85	2.96	0.95
Cowpea [*86*]	—	0.60-1.80	2.10-2.90	—

The content of sucrose, raffinose, stachyose, and verbascose in hull-free flours of selected legumes is summarized in Table IV. Verbascose was the predominant α-galactoside in hull-free flours of field pea (1.60-2.30%), faba bean (1.45-2.0%) and mung bean (1.83%). On the other hand, stachyose was the principal flatulence-causing sugar in hull-free flours produced from soybean (2.85-3.95%), cowpea (0.52-4.44%), lima bean (2.76%), navy bean (2.36%), chickpea (1.72%) and lupine (4.11%). In most legume flours, except cowpea, the content of raffinose did not exceed 1%.

Content of α-Galactosides in Oilseeds. The content of raffinose, stachyose and sucrose in selected oilseeds is summarized in Table V. Shahidi et al. (*51*) reported that rapeseed and mustard meals contained 3.93-5.73% sucrose, 0.27-0.62% raffinose, and 0.83-1.61% stachyose. They also found that canola meals contained over 50% more raffinose than rapeseed and mustard meals. On the other hand, the European varieties of *B. napus* contained lower levels of sucrose and stachyose, but up to 1.7% raffinose (*49*). Theander et al. (*50*) have reported higher amounts of sucrose (6.51-8.26%) and stachyose (1.43-3.04%) in winter and summer types of *B. napus* and *B. campestris*. Raffinose was the principal flatulence-causing sugar of cottonseed and sunflower seeds (*52, 89*). On the other hand, the level of flatulence-causing sugars in peanuts and safflower seeds (*52*) was comparable to that found in *Brassica* oilseeds.

Content of α-Galactosides in Cereal Grains. Table VI summarizes the content of sucrose, raffinose, stachyose, and verbascose in some cereal grains and brans. Only traces of verbascose were detected in sorghum. A number of cereal grains contained trace amounts of stachyose. Raffinose was the principal α-galactoside of sucrose in cereal grains but its content did not exceed 1%. On the other hand, much higher levels of raffinose were present in cereal brans.

Content of α-Galactosides in Vegetables. The content of sucrose, raffinose and stachyose in some vegetables is shown in Table VII. Stachyose was the principal flatulence-causing sugar in pumpkin, cucumber and squash, but its content did not exceed 2%. On the other hand, spinach and beets contained a maximum of 0.5% raffinose (*52*).

Effect of Processing on α-Galactosides of Sucrose

Legumes are important dietary components of the majority of people living in the developing countries. They serve as a good source of protein, calories and B-vitamins (*84, 97*). However, individuals may still limit their consumption of legumes due to the formation of flatus (*98*). The results of both human and animal studies have clearly demonstrated that carbohydrates of the raffinose-family are principal causes of flatulence in legumes and other plant protein products (*10, 46, 83*). Various methods for the removal of these oligosaccharides such as dehulling, soaking and/or cooking in water and buffer solutions, irradiation, enzymatic treatment, germination and solvent extraction have been evaluated.

Dehulling and Protein Concentration. Legume flours are usually obtained by dry dehulling techniques such as abrasion or air classification followed by milling of the dehulled seeds. Dehulling by abrasion reduced the total content of raffinose, stachyose and verbascose in faba bean of Ackerperle variety from 3.5 to 3.0% (*79*). Meanwhile, dehulling of cowpeas lowered the content of stachyose from 3.1% to 2.5% (*86*) while the content of raffinose remained unchanged. Dehulling of legume seeds, using wet techniques, removed even more oligosaccharides. Dehulling of cowpeas (*Vigna unguiculata*) by 10 min soaking in water, followed by rubbing seeds, separation of testa by flotation in water and washing the cotyledons in water, led to

Table V. Content of Oligosaccharides in Oilseeds [g/100g Dry Matters]

Oilseed [Ref]	*Sucrose*	*Raffinose*	*Stachyose*	*Verbascose*
Cottonseed [*52*]	1.64	6.91	2.36	trace
Sunflower	6.50	3.09	0.14	—
Safflower	1.86	0.52	—	—
Peanut	8.10	0.33	0.99	trace
Brassica napus [*49*]	1.6-1.7	0.2-1.7	0.3-0.4	—
Brassica napus [*50*]	6.5-8.3	0.30	1.4-3.0	—
Brassica campestris	6.8-7.5	0.30	2.40	—
Triton [*51*]	3.90	0.60	1.50	—
Midas	5.10	0.40	1.60	—
Mustard	4.90	0.20	0.80	—
Glandless cottonseed [*89*]	0.77	7.27	0.30	—
Glanded cottonseed	1.15	5.73	1.21	—

Table VI. Content of Oligosaccharides in Cereal Grains and Brans [g/100g Dry Matters]

Grain [Ref]	*Sucrose*	*Raffinose*	*Stachyose*	*Verbascose*
Triticale bran [*95*]	1.78-2.39	1.22-1.56	—	—
Rye bran	2.13	0.28	—	—
Wheat bran	2.18	2.16	—	—
Barley [*52*]	1.18-1.42	0.63-0.79	trace	—
Triticale	0.84	0.72	trace	—
Rye	1.15	0.71	trace	—
Wheat	1.38	0.70	trace	—
Corn	1.42-1.50	0.21-0.31	—	—
Oat	0.88	0.26	trace	—
Rice	0.56	—	—	—
Sorghum	0.84	trace	trace	trace
Sorghum [*96*]	0.80-2.20	0.10-0.13	—	—
Long-Grain Rice [*78*]	trace	—	—	—

Table VII. Content of Oligosaccharides in Vegetables [g/100g Dry Matters]

Vegetable	*Sucrose*	*Raffinose*	*Stachyose*
Pumpkin	2.88	0.65	1.63
Cucumber	1.62	0.92	1.08
Squash	3.44	0.78	1.18
Spinach	0.67	0.48	—
Beet	0.34	0.37	—

Adapted from ref. [*52*]

a 56.0% decrease in raffinose, 16.9 % in stachyose and 76.4% in verbascose. On the other hand, 4 h soaking of cowpeas in water removed only 1.0% of raffinose, 29.8% of stachyose and 49.4% of verbascose.

Protein concentrates are obtained from starchy legumes by dry or wet processing of the corresponding flours. Protein concentrates prepared by air classification of field pea and faba bean flours contained about 60% more flatulence-causing oligosaccharides than the original flours (59, 79). Sosulski et al. (*56*) have reported that protein rich fractions obtained by air classification of ten legume flours contained 40-90% higher amounts of α-galactosides than those present in the original samples.

Production of protein concentrates by wet processing involves extraction of the flour with organic solvents or aqueous solutions. Eldridge et al. (*93*) have reported that concentration of proteins in soybean flours, on commercial scale, by extraction with alcohol or dilute acid solutions markedly reduced the content of raffinose and stachyose in the resultant product. The alcohol was less effective in removal of α-galactosides than dilute acid solution (Table VIII).

A two-phase solvent extraction system consisting of alcohol with or without 10% (w/w) ammonia, possibly containing 5% water and hexane was effective in removing of flatulence-causing sugars from both oilseeds and legume seeds (*51, 89*).

Table VIII. Content of Flatulence-Causing Sugars in Some Soybean Protein Products [g/100g Dry Matters]

Product	*Raffinose*	*Stachyose*
Flours	0.58-1.18	3.95-4.92
Concentrates (dilute acid leach process)	0.00	0.50
Concentrates (aqueous alcohol leach process)	0.11	1.80

Adapted from ref. [*93*]

Absolute methanol was quite effective in removing raffinose, while stachyose was extracted to a lesser extent from both oilseeds and legume seeds. The addition of 5% (v/v) water to methanol improved the extractability of raffinose up to 23 and 70% and stachyose up to 6 and 74% in soybean and cottonseed meals, respectively. Ninety-five percent ethanol and isopropanol were less effective than 95% methanol (Table IX). Presence of ammonia in alkanol greatly enhanced the extraction of flatulence-causing sugars. The extraction of soybean, cottonseed and legume seeds with methanol-ammonia/hexane solvent system removed from 5 to 69% of raffinose, from 6 to 55% of stachyose and up to 8% of verbascose originally present in the seeds (89). The latter solvent system was even more effective in removing flatulence-causing sugars from rapeseed as the content of raffinose and stachyose in the meals was reduced by 78-89% and 25-45%, respectively (*51*).

Soaking and Cooking. Dry beans are usually soaked before cooking in order to facilitate their cooking, especially at higher soaking temperatures. Thus, soaking time

Table IX. Content of Individual Flatulence-Causing Sugars in Glandless Cottonseed as Affected by processing [g/100g Dry Matters]

Solvent extraction system	*Raffinose*	*Stachyose*
Hexane	7.27	0.30
Methanol-hexane	2.65	0.22
(Methanol/water)-hexane	2.15	0.08
(Ethanol/water)-hexane	4.50	0.29
(Isopropanol/water)-hexane	6.32	0.29

Adapted from ref. [*89*]

of legumes may be reduced from 16 h at room temperature to less than an hour at 90°C. Nitrogenous compounds, carbohydrates, phosphorus compounds, Ca, Mg and water-soluble vitamins are extracted from beans into the water (*99*). There is a great diversity in the reported values on the removal of α-galactosides by soaking, cooking or their combination. Selected results of studies on the effect of the length of soaking period on the extraction of oligosaccharides are given in Table X. The amount of extract recovered from beans depended on the cultivar, extraction medium, ratio of the volume of extraction medium to the weight of bean, pH, soaking temperature and the nature of material to be extracted. The extraction of oligosaccharides from beans into water is a simple diffusion process (*99, 83, 101*) and the amount of diffused material depends on the permeability of cell membranes (*102*). However, at elevated soaking temperatures, near the activation temperature of bean α-galactosidase, the amount of oligosaccharides found in the soaking liquid and soaked beans slightly exceeds that found in the raw material. Thus, removal of flatulence-causing sugars at elevated temperatures is due to both diffusion and enzymatic hydrolysis (*99*). However, the content of stachyose and raffinose found in soaked lentils seeds and soaking liquid did not account for the initial level of these sugars in the raw material (Table XI). It was suggested that during soaking a metabolic process resembling germination was taking place. Thus, more research is needed to establish the mechanism(s) involved in the removal of α-galactosides from legumes.

Kon (*99*) has investigated the effect of soaking temperature on cooking and nutritional quality of California small white beans. Soaking of beans at temperatures between 60 and 80°C was found to adversely affect the cooking time required to obtain soft beans for edible purposes. On the other hand, at soaking temperatures of 50°C and below, little solids, including oligosaccharides, were extracted from seeds. When beans were soaked at room temperature or at 50°C only 5% of raffinose and 15% of stachyose were extracted. Increasing the soaking temperature to 60°C and above removed 50 to 57% of stachyose and raffinose originally present in the seeds. On the other hand, Abdel-Gawad (*92*) has reported that soaking of common beans, lentil, faba beans and cowpeas in water (1:4, w/v) at room temperature lead to a 37-46.6% decrease in raffinose, 22-37.7% in stachyose and 17.8-28.6% in verbascose. Similarly, Ogun et al. (*86*) have reported that 12 h of soaking of cowpeas in water at room temperature did not affect the content of α-galactosides in the seeds.

However, hot-soaking carried out by boiling of seeds in water for 2 min and subsequent temperature equilibration of the seed-water mixture to room temperature (1 h) resulted in 16% reduction in the stachyose content. In contrast, much higher losses of α-galactosides of sucrose were reported by Iyenger and Kulkarni (*72*) who soaked black gram seeds in water at room temperature for 12 h. Under these conditions, 75.6% of raffinose, 52.2% of stachyose and 60.0% of verbascose originally present in black gram seeds were removed. Moreover, losses of stachyose and verbascose were more pronounced during the first 3-6 h of soaking, whereas the content of raffinose decreased gradually over a 12 h soaking period. Soaking in a sodium bicarbonate solution only slightly enhanced the extraction of α-galactosides from leguminous seeds (*83, 91*).

Cooking is more effective in the removal of flatulence-causing sugars than soaking. The extent of removal of α-galactosides of sucrose depended on cooking parameters as well as combined effect of cooking with germination or fermentation. Cooking of unsoaked red gram (*Cajanus cajan*), Bengal gram *(Cicer arietinum),* green gram *(Phaseolus radiatus),* lentil *(Lens esculentus*) and horse gram (*Dolichos biflorus*) seeds in water at a bean to water ratio of 2:1 (w/v) for 30 min brought about 46.6-80% decrease in raffinose, 61-80.7% in stachyose and 57-72% in verbascose contents. The maximum removal of raffinose and verbascose was achieved for lentil and Bengal gram, while the content of stachyose was reduced most for red and green gram seeds (*72*). On the other hand, boiling of faba beans, lentils, common beans and cowpeas in water at a 1:5 (w/v) ratio resulted in the removal of flatulence-causing sugars (total) ranging from 41.5% for faba beans to 47.2% for common beans, whereas the loss of raffinose ranged from 56 to 71% (*92*). One hour of boiling of dry whole soybean in tap water, at a bean to water ratio of 1:10 (w/v), removed 74% of raffinose originally present in the seeds (*103*).

Autoclaving of four commonly used Indian pulses (red gram, black gram, green gram and Bengal gram) in water at a 1:2 (w/v) ratio under 15 lbs of pressure for 15 min increased the content of raffinose, stachyose, and verbascose (*74*). This increase was thought to be due to the release of bound oligosaccharides to other seed components and partial hydrolysis of high molecular weight α-galactosides. However, the results of the above study are in contrast to those of others who reported a significant decrease in the content of flatulence-causing sugar (*78, 88, 92*). Abdel-Gawad (*92*) reported that autoclaving of both unsoaked and 12 h-soaked faba beans, lentils, cowpeas, and common beans at 10 psi with double the amount of water was more efficient in removing oligosaccharides than ordinary cooking. Similarly, cooking of black gram cotyledons in water at a 1:4 ratio (w/v) at 116°C (10 psi) for 40 min decreased the content of stachyose and verbascose by 81 and 73%, respectively (*78*).

Combination of soaking at room temperature and cooking may improve the removal of α-galactosides in leguminous seeds. The extent of removal of oligo-sacharides is affected by the ratio of water to bean and the length of soaking period (*100*). Somiari and Balogh (*94*) have reported that soaking of cowpeas for 16 h in water results in the removal of 26.2% of stachyose and 28% of raffinose originally present in the seeds. Cooking of soaked cowpeas for 50 min reduced the content of raffinose by 44% and stachyose by 28.6%. On the other hand, (*87*) 16 h of soaking of *P. vulgaris* (variety CNF 0178) followed by 90 min of boiling in water increased

Table X. Effect of Soaking in Water on the Content of Individual Flatulence-Causing Sugars in Legumes [g/100g Dry Matters]

Legume [Ref]	*Soaking time [h]*	*Raffinose*	*Stachyose*	*Verbascose*
Faba bean [*92*]	0	0.52	1.41	1.85
	4	0.49	1.29	1.71
	8	0.37	1.12	1.50
	12	0.32	1.10	1.32
Faba bean [*83*]	0	0.96	0.75	2.60
	6	0.83	0.69	2.15
	12	0.70	0.48	1.60
Lentil [*92*]	0	0.45	1.65	0.62
	4	0.50	1.73	0.59
	8	0.37	1.32	0.60
	12	0.28	1.04	0.51
Brazilian dry [*100*] bean	0	0.71	3.77	—
	3	0.53	3.55	—
	6	0.79	3.70	—
	12	0.53	2.81	—
	24	0.48	3.26	—
Cowpea [*94*]	0	2.5	4.2	—
	4	2.2	3.4	—
	8	2.1	3.2	—
	12	1.9	3.2	—
	16	1.8	3.1	—
Cowpea [*88*]	0	1.95	3.56	4.03
	4	1.93	2.50	2.04
Red gram [*83*] (pigeon pea)	0	1.00	2.45	2.96
	6	0.80	2.26	2.80
	12	0.60	1.90	2.00

Table XI. Effect of Soaking in Water on Soluble α-Galactoside of Sucrose in Lentils [mg/100g Dry Matters]

	Raffinose	*Mannotriose*	*Stachyose*
Raw Lentils	223	1011	1957
Soaking liquid	25	19	22
Soaked Lentils	ND	539	681

Adapted from ref. [*91*]

only 10% loss of α-galactosides from the seeds (*87*). According to Abdel-Gawad (*92*) boiling of bicarbonate-soaked legume seeds in tap water for 1 h led to greater decrease in the content of individual α-galactoside of sucrose than that observed after cooking of tap water-soaked seeds. Cooking and soaking of beans (*104*) and lentils (*105*) in a 0.2-0.5% sodium bicarbonate solution had a tenderizing effect on both the seed coats and cotyledons. Percival (*106*) reported that alkanity of NaOH brings about swelling of cellulose and dissolving of hemicellulose. Thus, this tenderizing effect may account for the enhanced loss of oligosaccharides reported for sodium bicarbonate-soaked seeds. Although boiling of legume seeds in a sodium bicarbonate solutions was more effective than water in the removal of oligosaccharides, the process may cause an excessive protein loss (*103*).

Enzymatic Treatments. Legumes are traditionally processed to fermented products in parts of Africa and in Far and Near Eastern countries. As an example, Wari, an Indian fermented food, is a hollow brittle cake of 2 to 30 cm spread and 1 to 30 g weight, made of legume dough subjected to microbial fermentation. Wari from black beans and soybeans fermented under controlled laboratory conditions with *Lactobacillus bulgaricus* and *Streptococcus thermophilus* contained 1.38 and 0.70% less flatulence-causing sugars following fermentation and these were reduced by a further 0.4 and 0.3% during drying, respectively (*90*). Different biochemical compositions or the type of endogenous microflora in each seed might be responsible for different responses of these beans to oligosaccharides removal.

Reddy and Salunkhe (*78*) have determined the fate of oligosaccharides during preparation of idli, a bean product prepared by steaming of a blend of fermented black gram and rice. Fermentation of black gram/rice blend for 20 h or 45 h reduced the total content of oligosaccharides by 16.1 and 71.5%., respectively. However, 20 h fermentation of black gram/rice blend was considered to offer the optimum condition for preparation of acceptable Idli products which contained approximately 0.2% stachyose and 1% verbascose. The microorganisms responsible for the fermentation process were not identified (*79*).

Peanut press cake is used in Indonesia in order to prepare a traditional fermented product called ontjom. Worthington and Beuchat (*107*) have evaluated ten fungi for their ability to utilize sucrose, raffinose and stachyose in peanuts. Six fungi including *Neurospora sitophila*, a mold most often used to ferment peanut press cake in production of ontjom, showed a definite α-galactosidase activity. The authors demonstrated that a 21 h fermentation of peanut press cake with *N. sitophila* was sufficient to eliminate raffinose and stachyose.

The possibility of treatment of legumes with α-galactosidase preparations or with microorganisms capable of breaking down raffinose oligosaccharides has been evaluated by a number of investigators. Commercially, α-galactosidase is currently being used to increase the production of sucrose from high-raffinose sugar beets (*108*). Application of α-galactosidase in production of soymilk to eliminate raffinose and other oligosaccharides has shown a great commercial potential. Addition of α-galactosidase preparations to soybean milk results in the hydrolysis of galactooligosaccharides but search for more viable processes is being continued. The use of α-galactosidase from *Mortierella vinacea* in three different forms (undisrupted, disrupted by Sonifier Cell Disruptor, or disrupted and entrapped within 7.5% polyacrylamide gel) were used to hydrolyze the α-oligosaccharides in soybean milk in order to remove raffinose and stachyose (*109*). The disrupted mycelium had the highest efficiency toward the hydrolysis of oligosaccharides in soybean milk. The entrapment of α-galactosidase within polyacrylamide gel slightly changed the optimum temperature and pH and reduced the activity of enzyme by 35%. On the other hand, addition of fungi *Aspergillus oryzae*, to soybean carbohydrates produced a good yield of α-galactosidase and invertase (*110*). The potential of mixtures of these enzymes for elimination of galactooligosaccharides in soybean milk was also tested (*111*). In addition, Cruz et al. (*112*) evaluated several microorganisms isolated from soil enriched in soybean meal for their ability to produce α-galactosidase. Among the microorganisms studied, *Cladosporium cladosporioides* had the highest α-galactosidase activity. α-Galactosidase had an optimum pH of 5.0 with melibiose, raffinose and stachyose as substrates. At the natural pH of soybean milk (pH 6.3) the relative activity of the enzyme was around 90% for all substrates. Thus, it was suggested that α-galactosidase from *Cladosporium cladosporioides* in soymilk could be used only after a considerable research on mycotoxin production by this fungi is done (*112*). Recently, Khare et al. (*113*) entrapped *Aspergillus oryzae* cells within agarose beds. The immobilization process decreased the activity of α-galactosidase from 1.58 units/ml to 0.72 units/ml. Incubation of soybean milk with these entrapped cells for 8 h at 50°C reduced the content of raffinose and stachyose by 52 and 70%, respectively.

Becker et al. (*114*) investigated autolysis conditions of α-galactosides in California small white beans. The optimum temperature for autolysis of oligosaccharides was between 45 and 65°C. Approximately 30, 50 and 70% of α-oligosaccharides were lost after 9, 24 and 48 h of autolysis. The useful range of pH for hydrolysis of stachyose and raffinose extended from 4.5 to 6.0. The α-galactosidase was inactivated at 75°C.

Bhaskar et al. (*115*) investigated the possibility of using germinating seeds of *Cassia sericea Sw.* as an inexpensive source of α-galactosidase.This plant is a wasteland legume shrub which is propagated for the control of the weed parthenium. The germinating seeds of *Cassia sericea Sw.* contained two enzymes with α-galactosidase activity. Both enzymes had optimum activity at pH 5.0 and temperature of 50°C; they also readily hydrolyzed melibiose and raffinose. However, the hydrolysis of stachyose was not achieved.

Akinyele and Akinlosotu (*88*) reported that a 24 h microbial fermentation of cowpeas carried out at 30°C decreased 79.7% of verbascose and 5.9% of stachyose, but increased raffinose by 12.8%. Treatment of cowpea flours for 2 h at 50°C with a crude fungal enzyme preparation having an α-galactosidase activity of 64 units/μg protein eliminated 82.3% of stachyose and 93.3% of raffinose (*94*). The crude enzyme treatment was more effective in the removal of oligosaccharides from cowpea flour than soaking and cooking.

The possibility of elimination of α-galactosides of sucrose by germination was also investigated; germination improved the nutritive value of legumes (*116*) due to the formation of enzymes, including α-galactosidase, that could eliminate or reduce the level of antinutritional factors in seeds. On the other hand, germination may produce undesirable compositional changes due to the initiation of metabolic processes and rootlet development. Jood et al. (*83*) demonstrated that 24 h germination of legumes decreased raffinose by 13-70%, stachyose by 35-75% and verbascose by 66-91%. Germination beyond 48 h resulted in complete loss of these oligosaccharides. During the dark germination of *Bambarra* groundnut seeds the content of stachyose and raffinose decreased rapidly while there was little change in the amount of verbascose present (*3*). Germination of cowpeas at 30°C for 24 h gave rise to produced most complete disappearance of α-oligosaccharides with a minimal rootlet development (*84*). Sathe et al. (*81*) demonstrated that 3 days of dark germination of 12 h-soaked great northern beans *(Phaseolus vulgaris L.)* brought about a 76.4% decrease in the content of stachyose plus raffinose. A 6-day germination of C-20 Navy and Pindak pinto beans removed 70-80% of the total content of raffinose and stachyose (*85*). However, germination did not affect the yield of soluble and insoluble polysaccharides to any great extent. Muzquiz et al. (*66*) investigated the effect of germination at 29 °C in dark on the content of α-galactosides of sucrose in lupine. The greatest rate loss of flatulence-causing sugars occurred during the first 48 h of germination. After 120 h, only small amounts of stachyose and raffinose were present in the seeds (Table XII). Differences in the loss of oligosaccharides of sucrose upon germination may originate from cultivar differences as well as existing differences in the level of endogenous α-galactosidase activity in different seeds.

Table XII. Composition of α-Galactosides in Lupine Seeds During Germination [g/100g of Dry matters]

Species	*Time [h]*	*Raffinose*	*Stachyose*	*Verbascose*
Lupinus albus	0	1.0	7.1	1.0
	24	0.8	2.9	0.6
	48	0.2	0.6	0.1
	72	0.2	0.6	0.08
	120	0.1	0.3	Trace
Lupinus luteus	0	0.9	6.1	5.6
	24	0.6	3.1	3.5
	48	0.5	2.3	2.8
	72	0.1	0.6	1.0
	120	0.01	0.1	0.05

Adapted from ref. [*66*]

Other Treatments. Utilization of dry beans can be expanded by using nontraditional methods of processing such as extrusion (*117*) or irradiation. The extrusion process may have a desirable effect on the nutritional value of extruded products (*118*). The content of oligosaccharides in defatted soybean flour containing 20-40% moisture was not affected by the extrusion process at 150°C (*119*). Furthermore, the content of raffinose + stachyose in the high starch fraction of Pinto beans was reduced, depending on the extrusion conditions, by 47 to 67% (*120*). Maillard reaction between reducing sugars, formed by hydrolysis of oligosaccharides during extrusion process, and amino groups of proteins may account for the lower extractability of oligosaccharides from extruded products (*118*).

Irradiation of dry field beans (*Phaseolus vulgaris*) significantly improved their nutritional value, using a chick growth assay. The increase in the body weight gain of chickens fed on irradiated field beans may be due to inactivation of heat-stable toxic factors (*121*). Gamma-irradiation of hydrated green gram seeds (*Phaseolus Aureus*) at 2.5 kGy reduced the content of stachyose and raffinose by 50% (*122*). Complete disappearance of these sugars was achieved when seeds, irradiated at 2.5-10 kGy, were subjected to a 48 h germination. Thus, gamma-irradiation alone or in combination with germination may serve as an effective means for elimination of flatulence-causing sugars from legume seeds. Analysis of the radiolytic breakdown products of wheat starch suggested random cleavage of glycosidic bonds leading to the formation of oligosaccharides of the maltose series (*123*). Later Schumann and Von Sonntag, (*124*) reported that radiation-induced scission of glycosidic bond is applicable to di-, oligo- and polysaccharides. In addition, irradiation of carbohydrates induced the formation of smaller, non-sugar aldehydic fragments. Raffi et al. (*125*) measured the amounts of radio-induced products such as glyceraldehyde, dihydroxyacetone and 2-hydroxymalonaldehyde in maize starch and concluded that quantities of carbonyl compounds in treated products were too low to cause any health hazards. Rao and Vakil (*122*) studied the effect of gamma-irradiation on raffinose in a model system and reported cleavage of the glycosidic linkage in raffinose leading to the formation of simpler sugars as well as smaller aldehydic fragments.

Legumes are a very important source of dietary proteins for much of the world's population. However, the consumption of legumes is commonly associated with formation of flatus. The α-galactosides of sucrose are still considered as the principal flatulence-causing factor in leguminous seeds. Therefore, most of the research efforts was focused on the role of these sugars in formation of flatus and on the removal of the causative oligosaccharides. However, there are evidences that α-galactosides of sucrose are only one of the flatulence-causing factors present in legume seeds. More research is required to characterize these factors, to establish their contribution to flatulence, as well as to find the procedure for their removal. There are also some evidences that α-galactosides of sucrose may have some beneficial health effects, but further research is needed to document the health benefits resulting from ingestion of these oligosaccharides. Various procedures have been proposed for the removal of α-galactosides from legumes. The data indicate the need to screen the effectiveness of the removal of flatulence-causing factors for each legume seed, using both chemical and physiological assays.

Literature Cited

1. French, D. *Adv. Carbohydrate Chem.* **1954**, *9*, 149-184.
2. Senser, M.; Kandler, O. *Phytochemistry* **1967**, *6*, 1533-1540.
3. Amuti, K.S.; Pollard, C.J. *Phytochemistry* **1977**, *16*, 533-537.
4. Kandler, O.; Hopf, H. In *Encyclopedia of Plant Physiology, New Series, 13A* (Plant Carbohydrates I-Intracellular Carbohydrates); Loewus, F.A.; Tanner, W., Eds.; Spring-Verlag: New York, 1982.
5. Dey, P.M. In *Biochemistry of Storage Carbohydrates in Green Plants*; Dey, P.M.; Dixon, R.A., Eds., Academic: London, 1985.
6. Ovcharov, K.G.; Koshelev, Y.P. *Fiziol. Rast.* **1974**, *21*, 969-974.
7. Santarius, K.A. *Planta.* **1973**, *113*, 105-114.

8. Castillo, E.; de Lumen, B.O.; Reyes, P.S.; de Lumen, H.Z. *J. Agric. Food Chem.* **1990**, *38*, 351-355.
9. Price, K.R.; Lewis, J.; Wyatt, G.M.; Fenwick, G.R. *Die Nahrung* **1988**, *32*, 609-626.
10. Calloway, D.H.; Hickey, C.A.; Murphy, E.L. *J. Food Sci.* **1971**, *36*, 251-255.
11. Murphy, E.L.; Horsley, H.; Burr, H.K. *J. Agric. Food Chem.* **1972**, *20*, 813-817.
12. Wagner, J.R.; Carson, J.F.; Becker, R.; Gumbmann, M.R.; Danhof, I.E. *J. Nutr.* **1977**, *107*, 680-689.
13. Hellendoorn, E.W. *Food Technol.* **1969**, *23*, 87-92.
14. Van Ness, M.M.; Cattau, E.L. *Am. Family Physician* **1985**, *31*, 198-208.
15. Levitt, M.D. *New Engl.J. Med.* **1971**, *284*, 1394-1398.
16. Levitt, M.D. Bond, J.H. *Ann. Rev. Med.* **1980**, *31*, 127-137.
17. Levitt, M.D. *J. Am. Diet. Assoc.* **1972**, *60*, 487-490.
18. Levitt, M.D.; Bond, J.H. Jr. *Gastroenterology,* **1970**, *59*, 921-929.
19. Tomomatsu, H. *Food Technol.* **1994**, *48(10)*, 61-65.
20. Kato, Y., Shirayanagi, S., Mizutani, J., Hayakawa, K. Soya Oligo Japan Co. R&D Center, Tokyo, Personal communication; cited in ref. 19.
21. Levitt, M.D.; Duane, W.C. *New Engl. J. Med.* **1972**, *286*, 973-975.
22. Olson, A.C.; Gray, G.M.; Gumbmann, M.R.; Sell, C.R.; Wagner, J.R. In *Antinutrients and Natural Toxicants in Foods*; Ory, R.L., Ed.; Food & Nutrition Press, Inc.: Westport, Connecticut, 1981, pp. 275-294.
23. Alvarez, W.C. *J. Am. Med. Assoc.* **1942**, *120*, 21-25. Cited in ref. *22*.
24. Moore, W.E.C.; Cato, E.P.; Holdeman, L.V. *Am. J. Clin. Nutr.* **1979**, *32*, S33-42.
25. Cummings, J.H. *The Lancet.* **1983**, 1206-1208.
26. Drasar, B.S.; Hill, M.J. *Human Intestinal Flora.* Academic Press: New York, 1974, pp. 16-17.
27. Moore, W.E.C.; Holdeman, L.V. *Appl. Microbiol.* **1974**, *27*, 961-979
28. Holdeman, L.V.; Good, I.J.; Moore, W.E.C. *Appl. Environ Microbiol.* **1976**, *31*, 359-375.
29. Mitsuoka, T. *Intestinal Bacteria and Health.* Harcourt Brace Jovanovitch, Japan Inc.: Tokyo, 1978.
30. Miller, T.L.; Wolin, M.J. *Am. J. Clin. Nutr.* **1979**, *32*, 164-172.
31. Cummings, J.H. *Gut* **1981**, *22*, 763-779.
32. Steggerda, F.R.; Dimmick, J.F. *Am. J. Clin. Nutr.* **1966**, *19*, 120-124.
33. Richards, E.A.; Steggerda, F.R.; Murata, A. *Gastroenterology* **1968**, *55*, 502-509.
34. Kurtzman,Jr., R.H.; Halbrook, W.U. *Applied Microbiol.* **1970**, *20*, 715-719.
35. Collee, J.G. In *The Normal Microbial Flora of Man*; Skinner, A., Carr, J.G., Eds.; Academic Press: London, 1974, p. 205
36. Sacks, L.E.; Olson, A.C. *J. Food Sci.* **1979**, *44*, 1756-1760.
37. Duncan, C.L. In *Food Microbiology: Public Health and Spoilage Aspects.* AVI: Westport, CT, 1976, p. 170.
38. Nakamura, S.; Nishida, S. *J. Med. Microbiol.* **1974**, *7*, 451-457.
39. Garg, S.K.; Banerjea, A.C.; Verma, J.; Abraham, M.J. *J. Food Sci.* **1980**, *45*, 1601-1602, 1613.

40. Savitri, A.; Bhavanishankar, T.N.; Desikachar, H.S.R. *Food Microbiol.* **1986,** *3*, 261-266.
41. Rackis, J.J.; Sessa, D.J.; Steggerda, F.R.; Shimizu, T.; Anderson, J.; Pearl, S.L. *J. Food Sci.* **1970**, *35*, 634-639.
42. Myhara, R.M.; Nilsson, K.; Skura, B.J.; Bowmer, E.J.; Cruickshank, P.K. *Can. Inst. Food Sci. Technol. J.* **1988**, *21*, 245-250.
43. Masai, T.; Wada, K.; Hayakawa, K.; Yoshihara, I. Mitsuoka, T. *Japan J. Bacteriol.* **1987**, *42*, 313-322.
44. Steggerda, F.R.; Richards, E.A.; Rackis, J.J. *Soc. Exptl. Biol. Med.* **1966**, *121*, 1235-1239.
45. Wagner, J.R.; Becker, R.; Gumbmann, M.R.; Olson, A.C. *J. Nutr.* **1976,** *106*, 466-470.
46. Fleming, S.E. *J. Food Sci.* **1981**, *46*, 794-798, 803.
47. Fleming, S.E. *J. Food Sci.* **1982**, *47*, 12-15.
48. El Faki, H.A.; Bhavanishangar, T.N.; Tharanathan, R.N.; Desikachar, H.S.R. *Nutr. Reports Intern.* **1983**, *27*, 921-927.
49. Hrdlicka, J.; Kozlowska, H.; Pokorny, J.; Rutkowski, A. *Die Nahrung* **1965**, *9*, 71-76.
50. Theander, O.; Aman, P.; Miksche, G.E.; Yasuda, S. *Swed. J. Agric. Res.* **1976**, *6*, 81-85.
51. Shahidi, F.; Naczk, M.; Myhara, R.M. *J. Food Sci.* **1990**, *55*, 1470-1471.
52. Kuo, T.M.; Van Middlesworth, J.F.; Wolf, W.J. *J. Agric. Food Chem.* **1988**, *36*, 32-36.
53. Cerning-Beroard, J.; Filiatre, A. *Cereal Chem.* **1976**, *53*, 968-978.
54. Siddiqui, I.R.; Wood, P.J.; Khanzada, G. *J. Sci. Food Agric.* **1973**, *24*, 1427-1435
55. Schweizer, T.F.; Horman, I.; Wursch, P. *J. Sci. Food Agric.* **1978**, *29*, 148-154.
56. Sosulski, F.W.; Elkowicz, L.; Reichert, R.D. *J. Food Sci.* **1982**, *47*, 498-502.
57. Rossi, M.; Germondari, I.; Casini, P. *J. Agric. Food Chem.* **1984**, *32*, 811-814.
58. Lineback, D.R.; Ke, C.H. *Cereal Chem.* **1975**, *52*, 344-346.
59. Vose, J.R.; Basterrchea, M.J.; Gorin, P.A.J.; Finlayson, A.J.; Youngs, C.G. *Cereal Chem.* **1976**, *53*, 928-936.
60. Macrae, R.; Zand-Moghaddam, A. *J. Sci. Food Agric.* **1978**, *29*, 1083-1086.
61. Quemener, B.; Gueguen, J.; Mercier, C. *Can. Inst. Food Sci. Technol. J.* **1982**, *15*, 109-112.
62. Saini, H.S. *Food Chem.* **1988**, *28*, 149-157.
63. Saini, H.S.; Gladstones, J.S. *J. Agric. Res.* **1986**, *37*, 157-166.
64. Allen, D.G.; Greirson, B.N.; Wilson, N.L. *Proceedings, Fourth International Lupin Conference,* August 18-22, 1986, Geraldton, Western Australia, p. 291; cited in ref. *62*.
65. Kennedy, I.R.; Mwandemele, O.D.; McWhirter, K.S. *Food Chem.* **1985**, *17*, 85-93.
66. Muzquiz, M.; Rey, C.; Cuadrado, C.; Fenwick, G.R. *J. Chromat.* **1992**, *607*, 349-352. 67.
67. Tanaka, M.; Thananunkul, D.; Lee, T.-C.; Chichester, C.O. *J. Food Sci.* **1975**, *40*, 1087-1088.
68. Knudsen, I.M. *J. Sci. Food Agric.* **1986**, *37*, 560-566.

69. Quemener, B. *J. Agric. Food Chem.* **1988**, *36*, 754-759.
70. Karoutis, A.I.; Tyler, R.T.; Slater, G.P. *J. Chromat.* **1992**, *623*, 186-190.
71. Black, L.T.; Bagley, E.B. *J. Am. Oil Chem. Soc.* **1978**, *55* , 228-232.
72. Iyengar, A.K.; Kulkarni, P.R. *J. Food Sci. & Technol.* **1977**, *14*, 222-224.
73. Naivikul, O.; D'Appolonia, B.L. *Cereal Chem.* **1978**, *55*, 913-918.
74. Rao, P.U.; Belavady, B. *J. Agric. Food Chem.* **1978**, *26*, 316-319.
75. Silva, H.C.; Luh, B.S. *Can. Inst. Food Sci. Technol. J.* **1979**, *12*, 103-107.
76. Akpapunam, M. A.; Markakis, P. *J. Food Sci.* **1979**, *44*, 1317-1321.
77. Wang, H.L.; Swain, E.W.; Hesseltine, C.W.; Heath, H.D. *J. Food Sci.* **1979**, *44*, 1510-1513.
78. Reddy, N.R.; Salunkhe, D.K. *Cereal Chem.* **1980**, *57*, 356-360.
79. Eskin, N.A.M.; Johnson, S.; Vaisey-Genser, M.; McDonald, B.E. *Can. Inst. Food Sci. Technol. J.* **1980**, *13*, 40-42.
80. Onigbinde, A.O.; Akinyele, I.O. *J. Food Sci.* **1983**, *48*, 1250-1254.
81. Sathe, S.K.; Desphande, S.S.; Reddy, N.R.; Goll, D.E.; Salunkhe, D.K. *J. Food Sci.* **1983**, *48*, 1796-1800.
82. Bianchi, M.L.P.; DeSilva, H.C.; Braga, G.L. *J.Agric. Food Chem.* **1984**, *32*, 355-357.
83. Jood, S.; Mehta, U.; Singh, R.; Bhat, C.M. *J. Agric. Food Chem.* **1985**, *33*, 268-271.
84. Nnanna, I.A.; Phillips, R.D. *J. Food Sci.* **1988**, *53*, 1782-1786.
85. Chang, K.C.; Chang, D.C.; Phatak, L. *J. Food Sci.* **1989**, *54*, 1615-1619.
86. Ogun, P.O.; Markakis, P.; Chenoweth, W. *J. Food Sci.* **1989**, *54*, 1084-1085.
87. Trugo, L.C.; Ramos, L.A.; Trugo, N.M.F.; Souza, M.C.P. *Food Chem.* **1990**, *36*, 53-61.
88. Akinyele, I.O.; Akinlosotu, A. *Food Chem.* **1991**, *41*, 43-53.
89. Naczk, M.; Myhara, R.M.; Shahidi, F. *Food Chem.* **1992**, *45*, 193-197.
90. Tewary, H.K.; Muller, H.G. *Food Chem.* **1992**, *43*, 107-111.
91. Vidal-Valverde, C.; Frias, J.; Valverde, S. *J. Food Prot.* **1992**, *55*, 301-304.
92. Abdel-Gawad, A.S. *Food Chem.* **1993**, *46*, 25-31.
93. Eldridge, A.C.; Black, L.T.; Wolf, W.J. *J. Agric. Food Chem.* **1979**, *27*, 799-802.
94. Somiari, R.I.; Balogh, E. *J. Sci. Food Agric.* **1993**, *61*, 339-343.
95. Saunders, R.M.; Betschart, A.A.; Lorenz, K. *Cereal Chem.* **1975**, *52*, 472-478.
96. Neucere, N. J.; Sumrell, G. *J. Agric. Food Chem.* **1980**, *28*, 19-21.
97. Iyer, V.; Salunkhe, D.H.; Sathe, S.K.; Rockland, L.B. *Qual. Plant Foods Hum. Nutr.* **1980**, *30*, 27-30.
98. Askar, A. *Food Nutr. Bull.* **1986**, *8*, 15-24.
99. Kon, S. *J. Food Sci.* **1979**, *44*, 1329-1340.
100. Silva, H.C. ; Braga, G.L. *J. Food Sci.* **1982**, *47*, 924-925.
101. Kataria, A.; Chauhan, B.M. *Plant Food Hum. Nutr.* **1988**, *38*, 51-59.
102. Schwimmer, S. *J. Food Sci.* **1972**, *37*, 530-535.
103. Ku. S.; Wei, L.S.; Steinberg, M.P.; Nelson, A.I.; Hymowitz, T. *J. Food Sci.* **1976**, *41*, 361-364.
104. Low, B. *Experimental Cookery from Chemical and Physical Standpoint.* 4th ed., J. Wiley & Sons: New York, 1955.

105. Moharram, Y.G.; Abou-Samaha, A.R.; El-Mahady, A.R. *Lebensm. Unters. u. Forsch.* **1986**, *182*, 307-310.
106. Percival, E.J.V. *Structural Carbohydrate Chemistry.* J.Garnet Miller: London, 1962.
107. Worthington, R.E.; Beuchat, L.R. *J. Agric. Food Chem.* **1974**, *22*, 1063-1066.
108. Whitaker, J.R. In *Enzymes: the interface between technology and economics;* Danehy, J.P., Wolnak, B., Eds., Marcell Dekker: New York, 1980, pp. 67-68.
109. Thananunkul, D.; Tanaka, M.; Chichester, C.O.; Lee, T.-C. *J. Food Sci.*, **1976**, *41*, 173-175.
110. Park, Y.K.; DeSanti, M.S.S.; Pastore, G.M. *J. Food Sci.* **1979**, *44*, 100-103.
111. Cruz, R.; Park, Y.K. *J. Food Sci.* **1982**, *47*, 1973-1975.
112. Cruz, R.; Batistela, J.C.; Wosiacki, G. *J. Food Sci.* **1981**, *46*, 1196-1200.
113. Khare, S.K.; Jha, K.; Ghandi, A.P.; Gupta, M.N. *Food Chem.* **1994**, *51*, 29-31.
114. Becker, R.; Olson, A.C.; Frederick, D.P.; Kon, S.; Gumbmann, M.R.; Wagner, J.R. *J. Food Sci.* **1974**, *39*, 766-769.
115. Bhaskar, B.; Ramachandra, G.; Virupaksha, T.K. *J. Food Biochem.* **1990,** *14*, 45-59.
116. Vanderstoep, J. *Food Technol.* **1981**, *35(3)*, 83-85
117. Gujska, E.; Khan, K. *J. Food Sci.* **1991**, *56*, 431-435.
118. Bjorck, I.; Asp, N.G. *J. Food Eng.* **1983**, *2*, 281-308.
119. Cristofaro, E.; Mottu, F.; Wuhrmann, J. In *Sugars in Nutrition*; Sipple, H.L., McNutt, K.W. Eds.; Academic Press: London, 1974, pp. 316-336.
120. Borejszo, Z.; Khan, K. *J. Food Sci.* **1992**, *57*, 771-772,777
121. Reddy, S.J.; Pubols, M.H.; McGinnis, J. *J. Nutr.* **1979**, *109*, 1307-1312.
122. Rao, V.S.; Vakil, U.K. *J. Food Sci.* **1983**, *48*, 1791-1795.
123. Ananthaswamy, H.N.; Vakil, U.K.; Sreenivasan, A. *J. Food Sci.* **1970**, *35*, 792-794.
124. Schuchmann, M.N.; Von Sonntag, C. *Int. J. Radiat. Biol.*, **1978,** *34*, 397-400.
125. Raffi, J.J.; Agnel, J.P.; Frejavaille, C.M.; Saint-Lebe, L.R. *J.Agric. Food Chem.*, **1981**, *29*, 548-550.

Chapter 9

Glucosinolates in *Brassica* Oilseeds: Processing Effects and Extraction

Fereidoon Shahidi[1], James K. Daun[2], and Douglas R. DeClercq[2]

[1]**Department of Biochemistry, Memorial University of Newfoundland, St. John's, Newfoundland A1B 3X9, Canada**
[2]**Grain Research Laboratory, Canadian Grain Commission, Winnipeg, Manitoba R3C 3G8, Canada**

Glucosinolates are found in canola and other *Brassica* oilseeds and are considered as antinutritional factors since some of their degradation products have goitrogenic and toxic effects. Glucosinolates are also known as precursors of various flavors and off-flavors of certain foods. The pungent odor and biting taste of mustard and radishes arise from their glucosinolate degradation products. Due to toxic effects of glucosinolates at high concentrations, their removal from oilseeds is necessary. Extraction of ground seeds with a two-phase solvent system consisting of hexane, for oil extraction, and methonal, possibly containing water and/or ammonia, removed 50-100% of individual glucosinolates present. However, some glucosinolates degraded during the extraction process and produced epithionitriles, nitriles, isothiocyanate, sulfinylnitrile, sulfinyl isothiocyanate, glucose, thioglucose and thioglucose dimer. Most of these degradation products were extracted into the polar methaolic phase, but some minute residues were also detected in the oil and protein meal fractions.

Glucosinolates are compounds found in canola and other species of Brassica oilseeds and are considered to be goitrogenic antinutritional factors. Goiter enlargement and improper functioning of the thyroid gland due to iodine deficiency is consistently observed when Brassica seeds are fed to experimental animals *(1-3)*. Sulfur-containing compounds in canola oil, derived from degradation of glucosinolates, are associated with various flavor, off-flavor and toxic effects in products and are also implicated as hydrogenation catalyst poisons *(4)*. The general chemical structure of glucosinolates is provided in Figure 1.

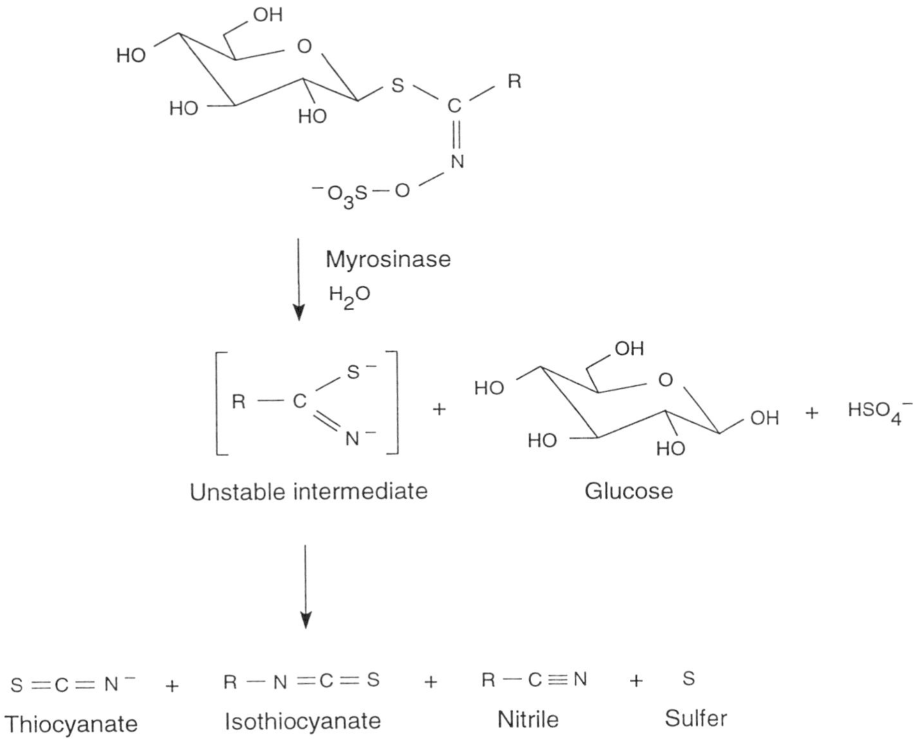

OH
HO
O
S
R
C
HO
HO
N
$^-O_3S-O$

Myrosinase
H_2O

S$^-$
R — C
N$^-$
+
OH
HO
O
OH
+
HSO_4^-
HO
HO

Unstable intermediate
Glucose

$S=C=N^-$ + $R-N=C=S$ + $R-C\equiv N$ + S

Thiocyanate
Isothiocyanate
Nitrile
Sulfer

Figure 1. Chemical structure of glucosinolates and their degradation products.

Table I. Common glucosinolates of Brassica oilseeds

Trivial Name	Side Chain, R	
	Name	Formula
Sinigrin	Allyl	CH_2=CH-CH_2-
Gluconapin	3-Butenyl	CH_2=CH-$(CH_2)_2$-
Progoitrin	2-Hydroxy-3-butenyl	CH_2=CH-CHOH-CH_2-
Glucobrassicanapin	4-Pentenyl	CH_2=CH-$(CH_2)_3$-
Gluconapoleiferin	2-Hydroxy-4-pentenyl	CH_2=CH-CH_2-CHOH-CH_2-
Gluconasturtin	Phenylethyl	C_6H_5-$(CH_2)_2$-
Glucoerucin	4-Methylthiobutyl	CH_3-S-$(CH_2)_4$-
Gluco (Sinalbin)	4-Hydroxybenzyl	*p*-HO-C_6H_4-CH_2-
Glucobrassicin	3-Indolylmethyl	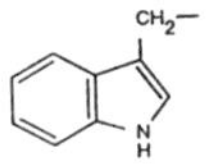 CH_2— N H
4-Hydroxyglucobrassicin	4-Hydroxy-3-indolylmethyl	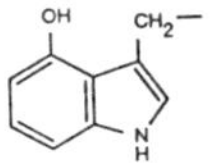 OH CH_2— N H
Neo(Glucobrassicin)	1-Methoxy-3-indoylmethyl	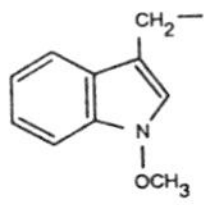 CH_2— N OCH_3

To date, more than one hundred glucosinolates are known to occur in nature *(5)*, and these differ from one another due to their side chain R-groups originating from respective amino acid precursors, with the exception of indolyl glucosinolates. A list of glucosinolates often found in Brassica oilseeds is provided in Table I. Although intact glucosinolates are free from toxicity, their hydrolysis brought about by the action of the endogenous enzyme myrosinase [thioglucose glucohydrolase (E.C. 3.2.3.1)] gives rise to a variety of potentially toxic products. These include isothiocyanates, thiocyanates, nitriles (Figure 1), elemental sulfur and oxazolidinethiones, the latter being formed by cyclization of hydroxyisothioxyanates (Figure 2).

Although glucosinolate degradation products are toxic, they are also responsible for the desirable pungent flavor and biting taste of mustard as well as the characteristic flavors of radish, broccoli, cabbage and cauliflower. Some breakdown products of glucosinolates such as isothiocyanates are fairly reactive and upon reaction with amines or cyclization produce substituted thioamides and oxazolidine-2-thiones, respectively. Isothiocyanates are also responsible for the desirable and pungent flavor of Wassabi which is used extensively in Japanese foods. Meanwhile, the thiocyanate ion and some phenolics may be produced from sinalbin. In acidic solutions, however, nitriles, including cyanoepithioalkanes from progoitrin, are the dominant degradation products.

In the commercial processing of canola/rapeseed, a heat treatment step is included in order to deactivate enzymes such as myrosinase and to facilitate oil release. However, intact glucosinolates left in the meal may undergo decomposition by the action of *Para colobatram* bacteria found in the lower intestinal tract to produce toxic aglucons *(6)*. Therefore, removal of glucosinolates and/or their degradation products from canola/rapeseed is essential for quality improvement of the final protein meal.

This chapter reports on the content of glucosinolates in several Brassica oilseeds and examines the effect of detoxification methods for their removal from seed meals. Of particular interest to this study is the use of a polar solvent mixture consisting of methanol/ammonia/water as an extraction medium and the fate of glucosinolates and/or their degradation products in this process.

Occurrence, Toxicity and Potential Benefits of Glucosinolates

Old varieties of rapeseed are characterized by their high content of glucosinolate (60-200 μmol/g) after oil extraction and high proportion of erucic acid (25-40%) in the oil. Consequently, the defatted meal from these varieties has traditionally been used at reduced levels in animal feed formulations or as a fertilizer. Use of rapeseed, as such, and its meal has also been commonplace in aquaculture farms in China. However, livestocks generally perform poorly in terms of weight gain, even when fed low levels of such meals, and also suffer from breeding problems *(7,8)*. The reduced weight gain of animals has been attributed to the interference of goitrin, a cyclization product of progoitrin (Figure 2), with the uptake of iodine by the thyroid gland. This problem was not circumvented by iodine supplementation in the diet. Therefore, no canola or rapeseed is currently used in the formulation of foods for human

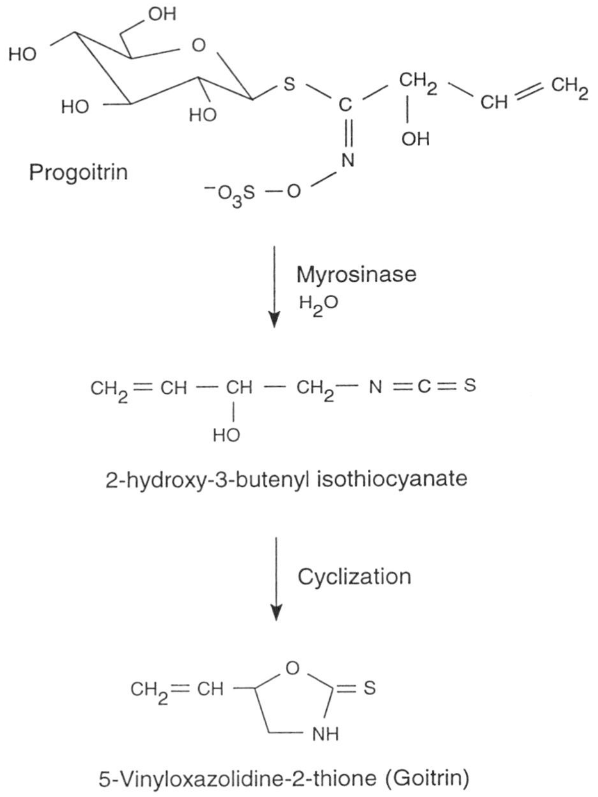

Progoitrin

Myrosinase

H_2O

$CH_2=CH-CH(OH)-CH_2-N=C=S$

2-hydroxy-3-butenyl isothiocyanate

Cyclization

5-Vinyloxazolidine-2-thione (Goitrin)

Figure 2. Enzymatic cyclization of progoitrin and production of goitrin.

consumption, although the meal has a large protein content with a reasonably well-balanced amino acid composition and calculated protein efficiency ratio.

Tookey *et al.* *(9)* have reported the toxic effect of nitriles from decomposition of glucosinolates. As an example, 3-hydroxy-4-pentane nitrile and 3-hydroxy-4,5-epithiopentane nitrile (threo) have LD_{50} values of 170 and 240 mg/kg, respectively. Acute toxicity of these nitriles exceeds that of oxazolidine-2-thiones *(9,10)*. Poor growth as well as liver and kidney lesions were noted in rats fed on 0.1% mixed nitriles for 106 days *(9)*. Despite severe effects of 4,5-epithiopentane nitrile, its polymerized form exhibited no toxic effect when added to rat diets *(11)*. However, it should also be noted that the toxic and goitrogenic effects of glucosinolate breakdown products are generally more pronounced in non-ruminant subjects and that no adverse effect is observed in fish.

Indolyl glucosinolates, in contrast to their aliphatic counterparts, may exert potential benefits by inhibiting chemical and other types of carcinogenesis. For example, glucobrassicin and its myrosinase-catalyzed transformation products have been shown to inhibit carcinogenesis induced by polycyclic aromatic hydrocarbons and other initiators *(12)*. These glucosinolates, found in relatively large amounts in canola, also occur abundantly in cruciferous vegetables. Thus, potential health benefits of indolyl glucosinolates may require design of oilseeds where these glucosinolates are retained or perhaps increased at the expense of harmful glucosinolates.

Conventional Processing of Seeds

The conventional processing of canola/rapeseed is an adaptation of soybean technology but it has adjusted for the small seed size, high oil content and presence of glucosinolates (Figure 3). After cleaning, seeds are crushed in order to fracture the seed coat, to rupture oil cells and to increase surface-to-volume ratio. The crushed seeds are then heated to 90-110°C for 15-20 min in order to inactivate enzymes. After a prepressing step to reduce oil content from 42 to approximately 20%, the resultant cake is flaked and hexane-extracted using percolating bed extractors. Hexane is removed under vacuum with the addition of heat in the form of steam (in a desolventizer-toaster). The resulting meal is further treated with steam to coagulate the protein making it easier for digestion by the ruminants. This causes a darkening of the meal, some loss of protein quality, and brings about a significant reduction in the content of glucosinolates (Table II).

If glucosinolates were degraded during crushing and prior to oil extraction, the sulfur-containing products such as isothiocyanates and oxazolidine-2-thiones might enter the oil. In practice, commercially processed canola oils contain 10-57 ppm of sulfur which may not be easily removed by conventional refining and bleaching operations *(13)*. Therefore, the resultant oils contain 3-5 ppm of sulfur which may interfere with the hydrogenation of the oil and could affect its odor during frying.

Canola versus Rapeseed

The name "canola" refers to the so-called double-low, or double-zero varieties of rapeseed in which the content of glucosinolate in the meal and erucic acid in the oil

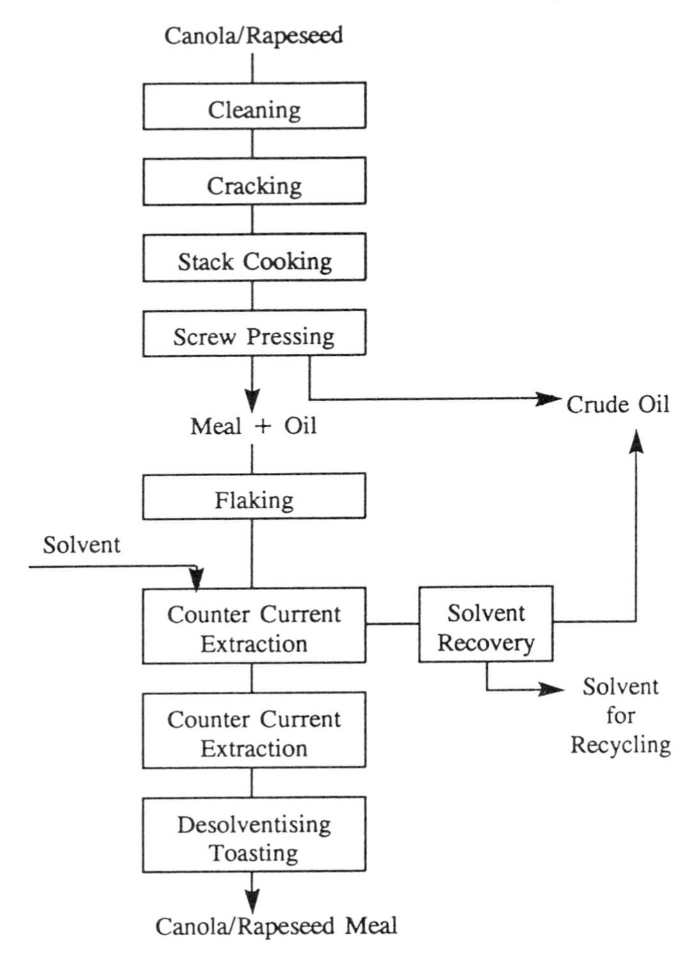

Figure 3. Flow diagram for commercial processing of canola.

has been lowered by breeding techniques. Currently, by definition canola contains less than 30 μmol/g of one or any combination of the four designated glucosinolates in its meal and less than 2% erucic acid in its oil. Effective 1997 this definition will change to 18 μmol/g glucosinolates in the whole seed and less than 1% erucic acid in the oil. The four designated glucosinolates are gluconapin, progoitrin, glucobrassicanapin and gluconapoleiferin. Table III summarizes the total content of glucosinolates in selected Brassica oilseeds, determined by a HPLC *(14)* as well as the thymol *(15)* method. All canola varieties contained less than approximately 30 μmol/g glucosinolates in their deoiled meals as compared with up to approximately 200 μmol/g in the traditional rapeseed and mustard varieties. Recently, new varieties of canola have been developed within the Canadian breeding program, some of which contain as little as 6 μmol (approximately 0.06%) aliphatic glucosinolates per gram of meal.

The content and chemical nature of individual glucosinolates present in canola varieties, as measured by an HPLC methodology, depended on the cultivar being studied. As an example, Triton canola contained 23.7 μmol glucosinolates per gram of deoiled meal, of which 17.9 μg/g consisted of the four designated aliphatic glucosinolates for canola (Tables III and IV). On the other hand, Midas rapeseed contained 144 μmol/g (approximately 1.44%) glucosinolates, mainly progoitrin (101.3 μmol/g) and gluconapin (23.2 μmol/g). Thus, it is evident that the breeding program has had an important effect on reduction of individual glucosinolates, particularly progoitrin. The corresponding slight increase in the content of indolyl glucosinolates may be an added advantage.

Biotechnology of canola/rapeseed has also advanced in another direction. Although canola oil is considered as a unique oil due to the distribution of its fatty acids, having nearly twice as much linoleic as linolenic acid, in some cases a higher stability of oil is desirable. Therefore, in applications where the stability of oil is of greatest most concern, new varieties of canola such as canaplus, which contain negligible amounts of linolenic acid, have been developed.

Antinutrients of Brassica Seeds and their Removal

The antinutrients in canola and rapeseed are dominated by the presence of glucosinolates, phenolics and excess hulls. However, detoxification of these seeds is generally referred to processes where a reduction in the content of glucosinolates is of concern. Detoxification methods applicable to rapeseed/canola are diverse and fragmentary; however, these may be categorized as given below.

1. genetic improvements and breeding techniques;
2. physical extraction of glucosinolates and/or their degradation products;
3. microbiological degradation of glucosinolates by fungi or bacteria and subsequent removal of products;
4. chemical degradation of glucosinolates via oxidation, addition of acid, base or selected salts and subsequent removal of products;
5. enzymatic methods and removal of degradation products by extraction or adsorption;
6. enzyme inactivation procedures;
7. protein isolation;
8. diffusion processes and combination methods.

Although breeding programs have been successful, globally speaking, most of

Table II. Effect of conventional processing on the content of glucosinolates (µmol/g oil-free sample)[1]

Crushing Plant (No. of samples tested)	Glucosinolates		
	Seed	Meal	% Remaining
A (16)	14.5	10.0	69.0
B (5)	24.1	16.6	68.7
C (5)	29.3	12.2	41.5
D (8)	18.6	8.6	47.5
E (13)	23.0	12.4	54.3
F (20)	25.2	11.0	43.8
G (4)	21.8	6.2	28.1

[1]Determined by the GC methodology. Samples were from Canadian crushing plants during 1981-1985.

Table III. The content of total as well as the four specific canola glucosinolates in canola and rapeseed varieties

Canola/Rapeseed	Glucosinolates, µmol/g meal		
	Canola designation	Total (HPLC)[1]	Total (Thymol)[2]
Altex	17.3	23.1	30
Regent	22.4	31.3	36
Tower	20.2	28.6	40
Triton	17.9	23.7	26
Westar	11.7	16.0	20
Midas	137.7	144.0	140
Hu You 9	131.8	134.4	118

[1]From ref. 16.
[2]From ref. 17.

the canola/rapeseed produced in the world still contains levels of glucosinolates which are too high to alleviate concerns in this regard. Furthermore, many of the other processing methods have yet to reach the commercial stage. Disadvantages such as loss of protein, poor functional properties of the resultant meal, or high processing costs may be contemplated. However, despite these drawbacks, an enzyme-assisted process involving aqueous extraction of seeds first reported by Jensen *et al. (16)*, may prove to be beneficial as it has been reported that it produces a degummed oil, a protein concentrate, lipophilic proteins in complex with amphiphilic lipids, hydrophillic and amphiphilic compounds, glucosinolates and hull fibers. The recovery process is comprised of decanting, sieving technology and flash chromatographic techniques *(17)*.

The Methanol/Ammonia/Water-Hexane Extraction Process

Figure 4 depicts the flow-diagram for the alcohol/ammonia/water-hexane extraction of Brassica oilseeds. This process involves exposing crushed seeds to a polar solvent system of methanol/ammonia/water followed by hexane. While the polar phase is responsible for the removal of low-molecular weight polar compounds, the non-polar hexane phase extracts the oil. The use of alcohol/ammonia was first reported by Schllingman and co-workers *(18,19)*. The authors indicated that methanol/ammonia was most effective in removing glucosinolates from rapeseed but failed to provide any supporting data. In addition, ammoniation of other Brassica seeds has been reported *(20-22)*.

The effect of the two-phase solvent extraction system on total glucosinolate content of Brassica seeds using different polar phase constituents, and determined by HPLC as well as thymol methods of analyses, is summarized in Table V. While absolute or 95% methanol had little effect in lowering the concentration of seed glucosinolates, the effect exerted by addition of ammonia to the above systems was quite remarkable. Thus, extraction of canola, rape, and mustard seeds with systems containing 10% (w/w) ammonia in absolute or 95% (vv) methanol, resulted in a removal of some 74 and 80% of their total glucosinolates, respectively.

For individual glucosinolates, the reduction in their content in seeds, on the average, varied between 78 and 88% when methanol/ammonia/water was employed (Table VI). However, no discernible trends were evident when considering the nature of the side chain R groups of glucosinolates. Thus, it might be assumed that the effect exerted by side chain R groups is overwhelmed by the highly polar nature of the thioglucose moiety of glucosinolates.

Although it might be considered that use of methanol in food applications is not safe, the residual methanol left in the seeds was negligible *(23)*. Furthermore, use of higher molecular weight alcohols in place of methanol was inadequate for efficient removal of glucosinolates. It should also be noted that the methanol/ammonia/water system was effective in the removal of other low-molecular-weight polar compounds from seeds as well. These included sugars such as sucrose, raffinose and stachyose *(24)*, phenolic acids *(25)* and tannins *(26)*. Thus, the above solvent system not only

Table IV. Major glucosinolates of selected Brassica oilseed meals (µmol/g deoiled, dried meal)[1]

Glucosinolate	Triton	Midas	Hu You 9	Mustard (Brown)	Mustard[2] (Yellow)
Sinigrin	0.9	0.7	0.1	236.7	0
Gluconapin	5.1	23.2	32.9	0	0.8
Progoitrin	10.9	61.3	92.5	0	5.8
Glucobrassicanapin	0.6	7.3	4.5	0	0
Gluconapoteiferin	1.3	5.9	1.9	0	0
Gluconasturtin	0.1	1.2	1.3	0	0
Glucobrassicin	2.8	2.3	0.2	1.4	ND
4-OH-Glucobrassicin	2.0	2.1	1.8	0.8	0.8
Neoglucobrassicin	0	0	0	0.2	ND

[1]Determined by HPLC method of analysis (Ref. 16). ND, not determined.
[2]Contained mainly *para*-hydroxybenzyl glucosinolate at 101.8 µmol/g.

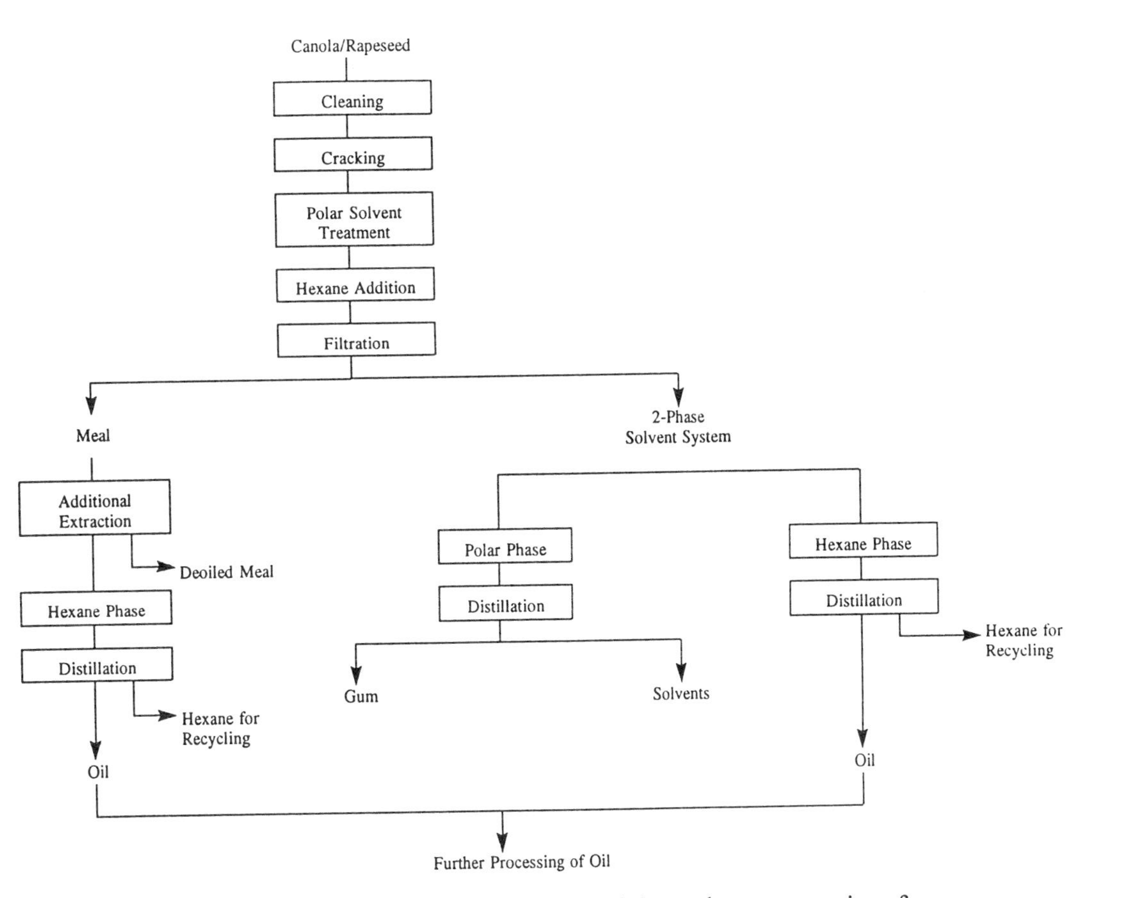

Figure 4. Flow diagram for alkanol/ammonia/water-hexane extraction of canola.

Table V. Total glucosinolate contents (μmol/g) by HPLC and thymol (in parantheses) methods of determination [1]

Solvent	Altex	Reagent	Tower	Triton	Westar	Midas	Hu You 9	Mustard (Brown)
Hexane	23.1 (30)	31.3 (36)	28.8 (40)	23.7 (26)	16.0 (20)	144.0 (140)	134.4 (118)	238.9 (188)
MeOH-Hexane	–	–	–	16.0 (25)	–	–	–	150.0 (145)
MeOH/H_2O-Hexane	–	–	–	12.0 (22)	–	81.0 (99)	–	146.0 (144)
MeOH/NH_3-Hexane	5.0 (40)	6.0 (10)	7.0 (10)	6.0 (12)	4.0 (11)	–	30.0 (32)	63.0 (68)
MeOH/NH_3/H_2O-Hexane	4.0 (70)	5.0 (4)	5.0 (8)	4.0 (8)	3.0 (8)	20.0 (27)	15.0 (220	27.0 (30)

[1]MeOH denotes absolute method, while MeOH/H_2O denotes 95% (v/v) methanol. Ammonia (NH_3), when present, was at 10% (w/v) of the polar phase.

Table VI. Individual glucosinolate contents (µmol/g) as determined by HPLC for hexane-extracted and $MeOH/NH_3/H_2O$-Hexane-extracted (in parentheses) seeds

Glucosinolate	Altex	Reagent	Tower	Triton	Westar	Midas	Hu You 9	Mustard (Brown)
2-OH-3-Butenyl	11.3 (2.1)	14.8 (2.9)	14.4 (2.5)	10.9 (2.2)	7.2 (3.1)	101.3 (10.9)	92.5 (10.9)	0 (0)
2-OH-4-Pentenyl	0.8 (0.1)	1.3 (0.0)	0.6 (0.2)	1.3 (0.2)	0.4 (0.1)	5.9 (0.5	1.9 (0.2)	0 (0)
3-Butenyl	4.7 (1.0)	5.7 (0.8)	4.7 (0.6)	5.1 (0.8)	3.8 (0.9)	23.2 (3.1)	32.9 (3.3)	0 (0)
4-Pentenyl	0.5 (0)	0.6 (0)	0.7 (0.3)	0.6 (0.1)	0.3 (0.1)	7.3 (0.9)	4.5 (0.4)	0 (0)
P-OH-Benzyl	0 (0)	1.9 (0.6)	0.3 (0.1)	0.1 (0.0)	0 (0)	1.2 (0.0	0 (0)	0 (0)
Allyl	0.5 (0.1)	0.5 (0.1)	0.2 (0.1)	0.9 (0.2)	0.3 (0.1)	0.7 (0.2)	0.6 (0.1)	236.7 (26.4)
4-OH-3-Indolylmethyl	2.6 (0.4)	2.7 (0.4)	2.6 (0.3)	2.0 (0.3)	1.8 (0.3)	2.1 (0.2)	1.8 (0.1)	0.8 (0.1)
3-Indolylmethyl	2.7 (0.72)	3.8 (0.8)	5.3 (1.0)	2.8 (0.5)	2.8 (0.5)	2.3 (0.4)	0.2 (0)	1.4 (0.2)

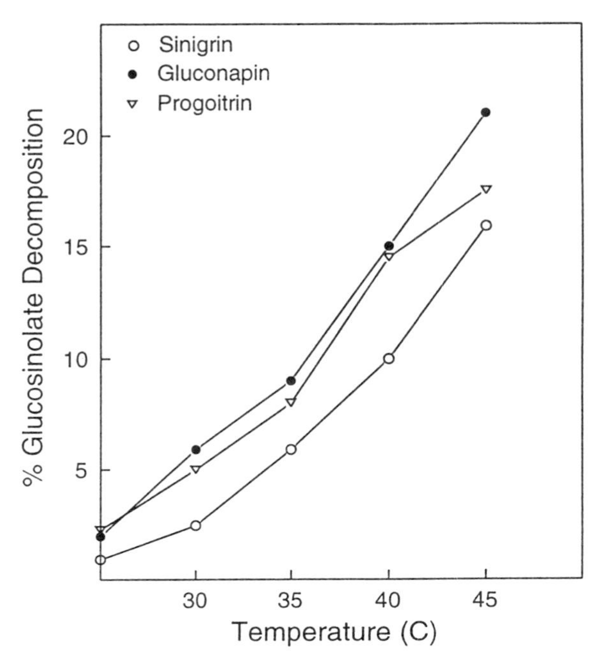

Figure 5. Degradation of sinigrin, gluconapin and progoitrin in a model system over a 24 h upon exposure to methanol/ammonia/water at 25 - 45°C.

Table VII. Distribution of four designated canola glucosinolates in methanol/ammonia/water-hexane extracted canola (μmol/100 g, dry basis) [a]

Glucosinolates & Products	Meal	Oil	Polar Phase
Glucosinolates	41.0	0	308.4
Desulfoglucosinolates	0.4	0	13.6
Isothiocyanates	0.5	0.8	9.7
Oxazolidinethiones	0	0	8.2
Nitrites[b]	1.5	1.3	67.2

[a]Contents are in μmol/100 g seed sample.
[b]Includes hydroxynitriles and epithionitriles.

gave rise to an efficient removal of glucosinolates, but it also helped in the extraction of other undesirable constituents of seeds.

Fate of Glucosinolates

Although the concentration of glucosinolates in all seeds examined was reduced by an order of magnitude, it is not clear whether they were extracted in the intact form or transformed to other products. To address this, a model system study was carried out in order to circumvent problems associated with the complex nature of glucosinolates in the seeds. Furthermore, only the fate of aliphatic glucosinolates was considered since breakdown products of indolyl glucosinolates are apparently harmless.

In model system studies, possible breakdown of pure glucosinolates, namely sinigrin, gluconapin and progoitrin, was examined under simulated experimental conditions of seed extraction with methanol/ammonia/water. Decomposition of glucosinolates was monitored over a 24 h period at temperatures ranging from 25 to 45°C. On the average, the breakdown of glucosinolates tested was in the order of gluconapin>progoitrin>sinigrin and the degree of decomposition of these glucosinolates was directly proportional to the extraction temperature employed (Figure 5). An examination of the chemical nature of these breakdown products indicated that nitriles were the major transformation products. However, minor quantities of isothiocyanates and/or epithionitriles were also produced (5.0-9.4%), but no oxazolidine-2-thiones were detected *(27)*. The corresponding sugars associated with the major aglucone nitrile products were thioglucose and its disulfide dimer, the latter being dominantly present. Thus for sinigrin, gluconapin and progoitrin, dominant aglucones were but-3-enylvnitrile (90.6%) pent-4-enyl nitrile (95.0%) and 1-cyano-2-hydroxybut-3-ene (93.6%), respectively. The minor aglucone decomposition products were associated with the release of glucose and furfuryl alcohol as its dehydration product. Based on the above results, two distinct modes of transformation of glucosinolates may be envisaged as depicted in Figure 6. However, the exact mechanism of decomposition of glucosinolates in methanol/ammonia/water remains speculative.

Application of the two-phase solvent extraction system to canola meal and its effect on transformation of glucosinolates indicated a similar pattern to that observed for pure glucosinolates *(28,29)*. While approximately 80% of the original seed glucosinolates remained in the intact or desulfoglucosinolate form, some 20% were decomposed. Table VII summarizes the distribution of the four designated canola glucosinolates and their transformation products in the meal, oil and polar matters following methanol/ammonia/water extraction. Of the decomposition products, approximately 94.3% (i.e., 16.6% based on the original glucosinolates) were detected in the polar matter, gum phase and only 0.5% were found in each of the oil and meal. The presence of oxazolidinethione in the polar phase may be due to its production during the crushing of seeds since crushed hexane-extracted seeds also contained minute quantities of this compound. Therefore, the polar phase is efficient not only in the removal of any glucosinolate breakdown products formed during the extraction

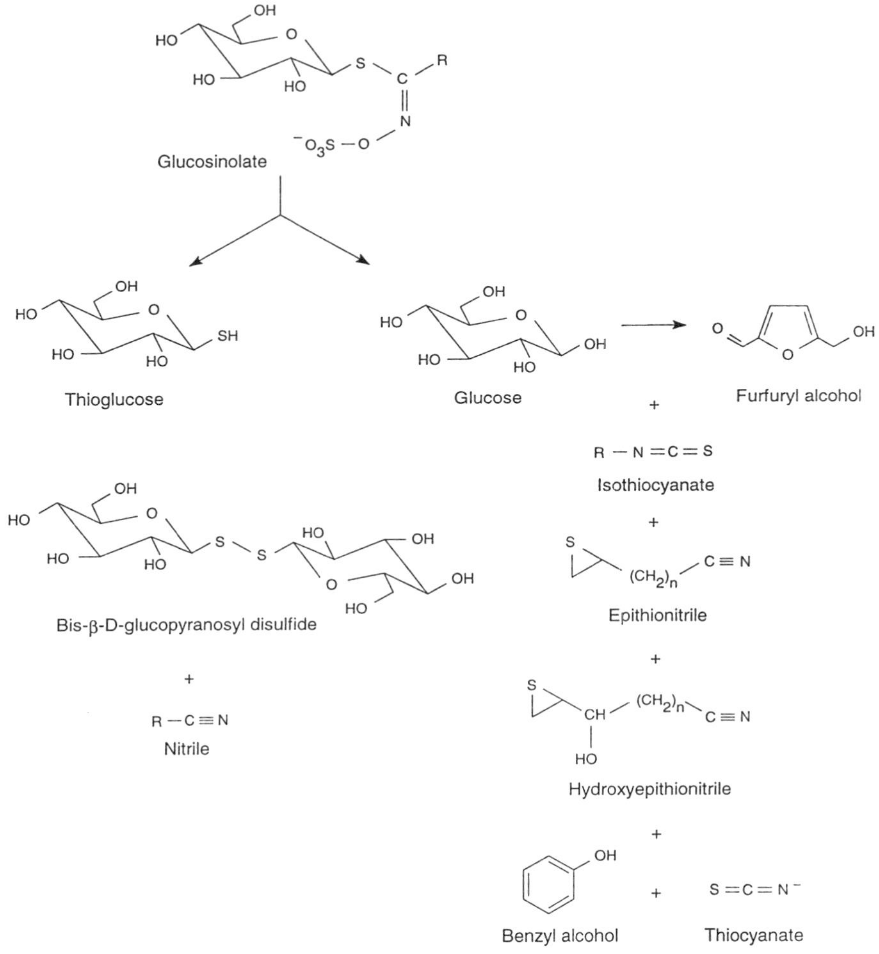

Figure 6. Proposed pathways for degradation of glucosinolates upon extraction of seeds with methanol/ammonia/water.

process, but also in the removal of pre-extraction glucosinolate breakdown products from crushed seeds.

Future trends

While plant breeding programs may successfully remove aliphatic glucosinolates from canola seeds, global transformation of the cultivation of traditional varieties of rapeseed and canola may be overshadowed due to concerns regarding yields and other factors. Therefore, the two-phase solvent extraction process presented here offers an alternative to conventional processing of canola and rapeseed. However, this and other alternate processing methods must address the issue of safety of the resultant oil and meals in adequate animal experimentation prior to the use of products in human food formulations.

Literature cited

1. Maheshwari, P.N.; Stanley, D.W.; Gray, J.I. *J.Food Prot.* **1981**, *44*, 459.
2. Astwood, B.E. J. Pharmacol. *Exp. Ther.* **1943**, *78*, 78.
3. Clandinin, D.R.; Bayley, I.; Cahllero, A. *Poultry Sci.* **1966**, *45*, 833.
4. de Man, J.M.; Pogorzelska, E.; de Man, L. *J. Am. Oil Chem. Soc.* **1983**, *60*, 558.
5. Sørensen, H. In *Canola and Rapeseed: Production, Chemistry, Nutrition and Processing Technology*; Shahidi, F., Ed.; Van Nostrand Reinhold: New York, NY, **1990**, pp. 149-172.
6. Oginsky, E.L.; Stein, A.E.; Greer, M.A. *Proc. Soc. Exptl. Biol. Med.* **1965**, 119, 360.
7. Hill, R. *Brit. Vet. J.* **1979**, *153*, 3.
8. Butler, W.J.; Pearson, A.W.; Fenwick, G.R. *J. Sci. Food Agric.* **1982**, *33*, 866.
9. Tookey, H.L.; Van Etten, C.H.; Daxenbichler, M.E. In *Toxic Constitutents of Plant Foodstuffs*. Second Edition, Liener, I.E., Ed.; Academic Press: New York, NY, **1980**, pp. 103-142.
10. Nishie, K.; Daxenbichler, M.E. *Food Cosmet. Toxicol.* **1980**, *18*, 159.
11. Dietz, H.M.; Panigrahi, S.; Harris, R.V. *J. Agric. Food Chem.* **1991**, *39*, 311.
12. Loft, S.; Otte, J.; Poulsen, H.E.; Sørensen, H. *Food Chem. Toxic.* **1992**, *30*, 927.
13. Daun, J.K.; Hougen, F.W. *J. Am. Oil Chem. Soc.* **1977**, *54*, 351.
14. Shahidi, F.; Gabon, J.E. *J. Food Quilty* **1989**, *11*, 421.
15. De Clercq, D.R.; Daun, J.K. *J. Am. Oil Chem. Soc.* **1986**, *66*, 788.
16. Jensen, S.K.; Sørensen, H. In *Canola and Rapeseeds Production, Chemistry, Nutrition and Processing Techology*; Shahidi, F., Ed.; Van Nostrand Reinhold: New York, NY, **1992**, pp. 331-341.
17. Bjergegaard, C., Li, P.W. Michaelsen, S., Moller, P., O.H.E.J. and Sørensen, H. In *Bioactive Substances in Foods of Plant Origin*; Kozlowka, H., Fornal, J. and Zdunczyk J., Eds. Polish Academy of Sciences, Olsztyn, Poland, **1994**, pp. 1-15.
18. Schlingmann, M.; Praere, P. Fette. Seifen. Anstrichmittel. **1978**, *80*, 283.

19. Schlingmann, M.; von Rymon-Lipinski, G.W. Can. Patent 11201979, **1982**.
20. Kirk, W.D.; Mustakas, G.C.; Griffin, E.L., Jr. *J. Am. Oil Chem. Soc.* **1966**, *43*, 550.
21. Kirk, W.D.; Mustakas, G.C.; Griffin, E.L., Jr.; Booth, A.N. *J. Am. Oil Chem. Soc.* **1971**, *48*, 845.
22. McGregor, D.I.; Blake, J.A.; Pickard, M.O. In *Proceedings of the 6th International Rapeseed Conference*, Paris, France, **1983**, Vol. 2; pp. 1426-1431.
23. Diosady, L.L.; Tar, C.G.; Rubin, L.J.; Naczk, M. *Acta Alimentaria*, **1987**, *16*, 167-179.
24. Shahidi, F.; Naczk, M.; Myhara, R.M. *J. Food Sci.* **1990**, *55*, 1470.
25. Naczk, M.; Shahidi, F. Food Chem. **1988**, *31*, 159.
26. Shahidi, F.; Naczk, M. J. Food Sci. **1989**, *54*, 1082.
27. Shahidi, F.; Naczk, M. In *Canola and Rapeseed: Production, Chemistry, Nutrition, and Processing Technology*; Shahidi, F., Ed.; Van Nostrand Reinhold; New York, NY, **1990**, pp. 291-306.
28. Shahidi, F.; Gabon, J.E. *Lebensm.-Wisc., U.-tech.* **1990**, *23*, 154.
29. Shahidi, F.; Gabon, J.E. *J. Food Sci.* **1990**, *55*, 793.

Chapter 10

Cyanogenic Glycosides of Flaxseeds

Fereidoon Shahidi and P. K. J. P. D. Wanasundara

Department of Biochemistry, Memorial University of Newfoundland, St. John's, Newfoundland A1B 3X9, Canada

Cyanogenic glycosides are secondary metabolites that are found in various plant tissues and produce HCN upon hydrolysis. They are widely distributed in the plant kingdom and are synthesized during metabolism of aromatic amino acids such as phenylalanine and tyrosine and branched amino acids such as leucine, isoleucine and valine. Flaxseed contains linamarin, linustatin and neolinustatin. Cyanogenic glycosides can be quantified in the intact form by chromatographic methods or indirectly by determining the content of HCN released due to their decomposition. The ability of cyanogenic glycosides to release HCN is due to their enzymic hydrolysis which may cause cyanide poisoning. Therefore, removal of cyanogenic glycosides is necessary to improve the nutritional value and safety of cyanogen containing foods including flaxseeds.

Cyanogenesis is the ability of the living tissues to produce hydrocyanic acid (HCN) and widely observed in the plant kingdom. HCN does not occur in the free form in higher plants but is released from cyanogenic precursors due to enzymic hydrolysis. The cyanogenic compounds of plants are usually carbohydrate derivatives, specifically β-glycosides of α-hydroxynitriles (cyanohydrins). Cyanolipids are alternate sources of HCN in plants but are limited in nature. Cyanogenic glycosides of plants are nitrogen-containing secondary metabolites and are found in leaves, roots, seeds or other plant tissues. Several of the cyanophoric plants play important economic, dietary and nutritional roles and are extensively incorporated into human food and animal feed. Since the cyanogenic compounds are largely localized within the vacuoles of the cyanophoric plant cells, and their hydrolytic enzymes are cytoplasmic, damaging the cells leads to destruption of cellular integrity and results in enzymic hydrolysis of cyanogenic compounds and concomitant evolution of HCN. Therefore, cyanogenic glycosides are considered as part of the plant defence mechanism against pest/insect damage (*1*). There are also evidences that cyanogenic glycosides in seeds serves as a form of stored nitrogen which can be converted to amino acids when there is a great demand for nitrogen such as germination (*2*). Table I summarizes the sources of cyanogenic glycosides and their estimated content of HCN released upon acid hydrolysis.

Flaxseed is traditionally used for oil extraction for industrial purposes, but has recently been investigated for its potential use in value-added products. Presence of biologically-active phytochemicals such as α-linolenic acid, lignans and soluble fibre has generated new and increased interest about the nutritional and therapeutical value of flaxseed. Due to the presence of a wide spectrum of biologically active phytochemicals in flaxseed, this oilseed has also been identified as an item for the "Designer Foods" project in the United States (*3*). The Canadian grown flaxseed contains, on the average 41% oil (on a moisture-free basis), 26% protein (%N × 6.25), 4% ash, 5% acid detergent fibre and 24% nitrogen-free extract (*4*). However, presence of cyanogenic glycosides in flaxseed is a concern and limits its use of these seeds in large quantities in foods and livestock feed formulations. Therefore, it is desirable to remove cyanogenic glycosides from foods via processing or by employing biotechnological means.

Chemistry of Cyanogenic Glycosides

Both the vegetative parts and seeds of flax contain cyanogenic glycosides. Linamarin (2-[(6-O-β-D-glucopyranosyl)-oxy]-2-methylpropanenitrile) and lotaustralin ([(2R)-[(6-O-β-D-glucopyranosyl)-oxy]-2-methylbutanenitrile]) are the monosaccharide cyanogenic glycosides which may be present in flaxseed (Figure 1; *1* and *5*). Smith *et al.* (*6*) have reported that two disaccharide cyanogenic glycosides (Figure 1) may also be isolated from flaxseed meal, namely linustatin (2-[(6-O-β-D-glucopyranosyl-β-D-glucopyranosyl)-oxy]-2-methylpropanenitrile) and neolinustatin ([(2R)-[(6-O-β-D-glucopyranosyl-β-D-glucopyranosyl)-oxy]-2-methylbutanenitrile]). Recent studies by Oomah *et al.* (*7*) and Wanasundara *et al.* (*8*) have shown the presence of linamarin, linustatin and neolinustatin in the seeds. The content of these three glycosides depends on the cultivar, location and year of production, with cultivar having the most important effect. The predominant cyanogenic glycoside of the Canadian cultivars is linustatin (213 to 352 mg/100 g seed) which accounts for 54 to 76% of the total content of cyanogenic glycosides (Table II). The content of neolinustatin ranges from 91 to 203 mg/100 g seed and linamarin is present in less than 32 mg/100 g seed (*7*) and are not present in some cultivars. Table III presents cyanogenic glycoside content of commercial flaxseed products available in Canada.

Linamarin and lotaustralin are cyanohydrins of acetone and 2-butanone, respectively. The unstable cyanohydrin moiety is stabilized by a glycosidic linkage to D-glucose in linamarin and lotaustralin and to D-gentiobiose in linustatin and neolinustatin (Figure 1). Therefore, a close structural relationship exists among linamarin, linustatin, lotaustralin and neolinustatin. Linustatin and neolinustatin have been isolated as levorotatory crystalline solids. The IR spectra of these compounds show strong hydroxyl absorption (3400 cm^{-1}) and a weak absorption due to C≡N (2240 cm^{-1}) with no carbonyl absorption. The ^{13}C NMR spectra indicate that these

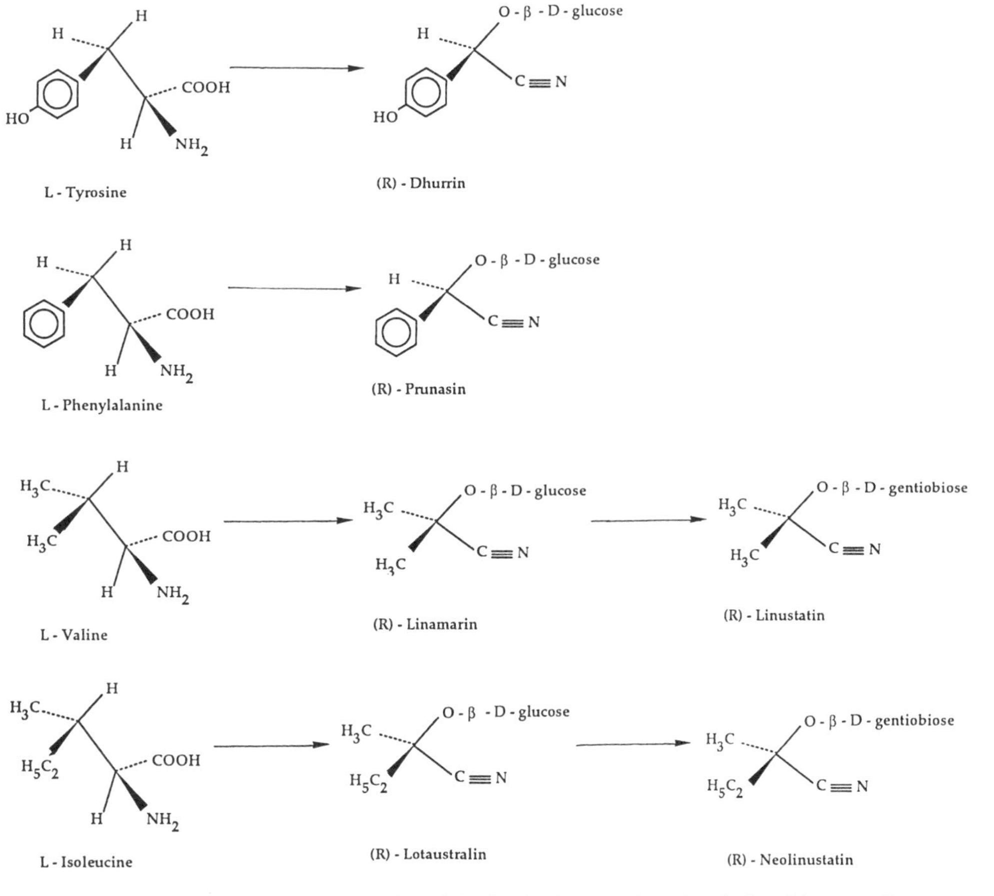

Figure 1. Cyanogenic glycosides found in food plants and their relationship to amino acids.

Table I. Cyanogenic food plants and their yield of HCN[1]

Plant	HCN yield (mg/100 g fresh weight)
Almond (amygdalin)	
bitter seed	290
young leaves	20
Apricot, seed (amygdalin)	60
Bamboo (taxiphyllin)	
stem, unripe	300
tops of unripe sprouts	800
Cassava (linamarin and lotaustralin)	
less toxic clones, bark of tuber	69
inner part of tuber	7
leaves	77
very toxic clones, bark of tuber	84
inner part of tuber	33
leaves	104
Flax, seedling tops (linamarin, linustatin and neolinustatin)	91
Lima bean, mature seed (linamarin)	
Puerto Rico, small black	400
Burma, white	210
American, white	10
Peach (prunasin)	
seed	160
leaves	125
Sorghum (dhurrin)	
mature seed	0
etiolated shoot tips	240
young green leaves	60
Wild cherry, leaves (amygdalin)	90-360

[1] Adapted from Ref. (*13*). Name of cyanogenic glycoside in each case is given in parenthesis.

Table II. Cyanogenic glycoside content of flaxseed cultivars[1]

Cultivar	Cyanogenic glycosides, mg/100 g seed			
	Linamarin	Linustatin	Neolinustatin	Total
Andro	16.7 ± 3.8	342 ± 38	203 ± 24	550 ± 53
Flanders	13.8 ± 3.7	282 ± 55	147 ± 22	432 ± 47
AC Linora	19.8 ± 5.4	269 ± 28	122 ± 20	402 ± 51
Linott	22.3 ± 8.2	213 ± 29	161 ± 25	396 ± 54
McGregor	25.5 ± 4.0	352 ± 56	91 ± 19	464 ± 76
Noralta	20.3 ± 3.4	271 ± 34	163 ± 18	455 ± 50
NorLin	ND	295 ± 46	201 ± 37	496 ± 81
Norman	ND	231 ± 63	135 ± 37	365 ± 97
Somme	27.5 ± 12.1	322 ± 46	149 ± 25	489 ± 78
Vimy	31.9 ± 8.3	262 ± 31	115 ± 21	409 ± 54

[1] Adapted from Ref. (*9*)

Table III. Cyanogenic glycoside content of commercial flaxseed products[1]

Product	Cyanogenic glycosides, mg/100 g product			
	Linamarin	Linustatin	Neolinustatin	Total
Bread mix	ND	9.4 ± 0.8	ND	9.4 ± 0.8
Bread (fresh)	0.3 ± 0.2	13.3 ± 0.5	ND	13.7 ± 0.6
Bread (7d at -20°C)	0.13 ± 0.04	18.0 ± 1.1	ND	18.0 ± 1.1
Bread (7d at 23°C)	2.1 ± 0.06	21.6 ± 0.1	ND	23.6 ± 0.5
Energy drink	20.6 ± 2.1	50.8 ± 3.5	35.0 ± 2.9	106.5 ± 7.9
Muffin mix	ND	4.5 ± 0.9	ND	4.5 ± 0.9
Stabilized flax	13.1 ± 2.1	126.8 ± 9.9	ND	263.1 ± 25.3
Waffle mix	ND	11.4 ± 2.4	123.2 ± 13.3	11.4 ± 2.4

[1] Adapted from Ref. (*9*)

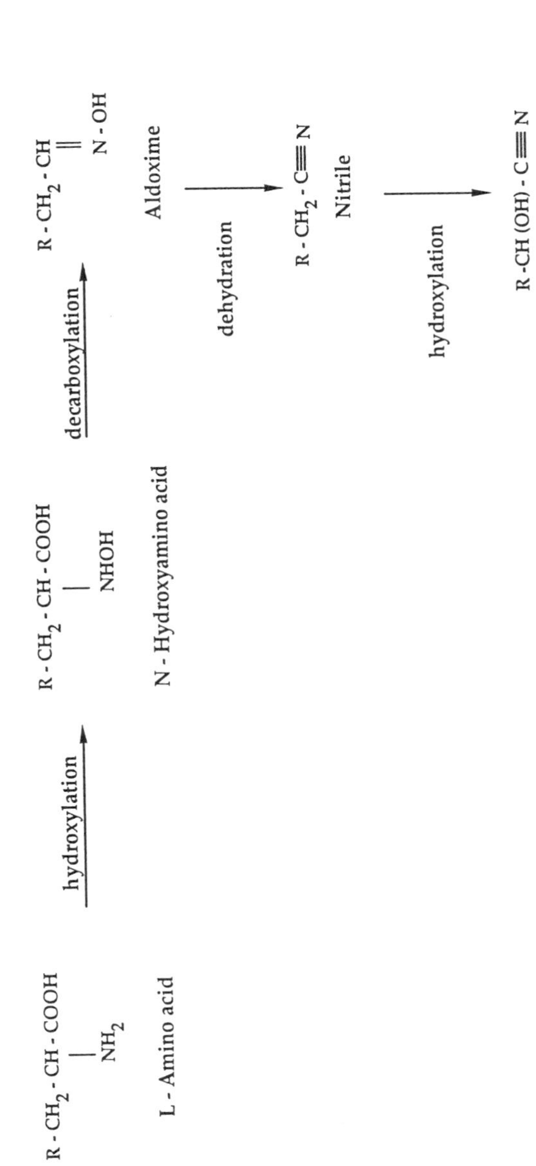

Figure 2. Biosynthesis of cyanogenic glycosides from amino acids.

compounds are cyclic polyols (*6*). Glucose is the only monosaccharide produced when linustatin and neolinustatin are completely hydrolysed with mineral acids. Linustatin cleaves regiospecifically into D-glucose and linamarin upon enzymatic hydrolysis with a β-glucosidase.

Biosynthesis of Cyanogenic Glycosides

Cyanogenic glycosides are β-glycosidic derivatives of α-hydroxynitriles. They are mainly derived by multi-step biosynthetic sequences from the hydrophobic amino acids such as L-phenylalanine, L-tyrosine, L-leucine, L-isoleucine and L-valine. It has been shown that synthesis of linamarin and lotaustralin (*5*) and linustatin and neolinustatin (*1*) in flaxseed is closely associated with the metabolism of valine and isoleucine, respectively (Figure 1). As the initial step of cyanogenic glycoside synthesis, the amine nitrogen of the precursor amino acid is oxygenated to form the corresponding aldoxime. Dehydration of aldoxime yields nitrile and stereospecific oxygenation of it produces cyanohydrin (Figure 2). These reactions are catalysed by a membrane-bound enzyme system (*10*) which is considered as a multienzyme complex. The last step of the biosynthetic sequence of cyanogenic glycosides is the glycosylation of the cyanohydrin catalysed by a uridine diphosphate-glucose-ketone cyanohydrin β-glucosyl transferase (*11*). Hahlbrock and Conn (*12*) have confirmed that the same enzyme is responsible for the final formation of both linamarin and lotaustralin.

The biochemically-related cyanogenic glycosides from valine, linamarin and linustatin, dominate the isoleucine series lotaustralin and neolinustatin (Figure 1) during the growth of whole flax fruits from anthesis to maturity (*2*). The ratio of monoglycosides to total cyanogenic glycosides shifts from 100% to 0% at maturity. The monoglycosides, linamarin and lotaustralin are the main glycosides (>90% of total cyanogenic glycosides) of young flax fruits (0 to 18 days after flowering). About 30% of the cyanogens in older fruits (20 to 40 days after flowering) are diglycosides linustatin and neolinustatin (*13*).

The synthesis of cyanogenic glycosides may occur in leaves and stems of the flax plant as monoglycosides (linamarin and lotaustralin) and then glycosylated and translocated into the growing seeds as the corresponding diglycosides linustatin and neolinustatin, respectively (*2*). According to the linustatin pathway postulated by Selmar *et al.* (*2*) the translocation of diglycoside occurs via vascular system to the endosperm tissues, where the terminal glucose is hydrolysed by linustatinase. The linamarin produced is stored temporarily in the endosperm and glycosylated again and translocated into the cotyledons to be stored as linustatin during seed maturation. The diglycosides would persist in the seed as such until the seed starts to germinate, then degradation and metabolism of the diglycosides would occur. Studies on flaxseed (variety, Somme) have shown that linustatin and neolinustatin levels are reduced during germination; 40 and 70% reduction in their contents was observed on day eight of the gemination, respectively (Wanasundara and Shahidi, unpublished data).

Catabolism and Toxicity of Cyanogenic Glycosides

The cyanogenic glycosides of flaxseed are referred to as bound cyanide. The molecular HCN is regarded as free or non-glycosidic cyanide. Under normal physiological conditions, tissues of cyanophoric plants contain little or no detectable HCN. However, when plant tissues are disrupted, HCN may be released rapidly from cyanogenic glycosides upon hydrolysis. The catabolism of cyanogenic glycosides is initiated by cleavage of their carbohydrate moiety by one or more β-glucosidases, thus yielding the corresponding α-hydroxynitrile. This intermediate may decompose either spontaneously or enzymatically in the presence of α-hydroxynitrile lyase to yield HCN and an aldehyde or a ketone (Figure 3; *14*).

Crude flaxseed extracts contain two distinct β-glucosidases which cooperate in stepwise removal of glucose residues. Linustatinase catalyses the hydrolysis of β-(*bis*-1,6)- and β-(*bis*-1,3)-glucosides but is inactive toward linamarin and cyanogenic disaccharides having terminal xylose or arabinose moieties. Therefore, linustatin and neolinustatin are hydrolysed to linamarin and lotaustralin, respectively, by linustatinase (Figure 4). These monosaccharides are further degraded to their corresponding α-hydroxynitriles by linamarase (*15*). Flaxseed linamarase shows only a moderate degree of substrate specificity with respect to the aglycone moiety and catalyses the hydrolysis of both aliphatic and aromatic cyanogenic monosaccharides but is virtually inactive towards cyanogenic disaccharides (*16*).

Potential toxic levels of cyanogenic glycosides for animals consuming plant materials containing cyanogenic glycosides depend on factors such as species and size of animal, the level of β-glucosidases in the plant, the length of time between tissue disruption and ingestion, the presence and nature of other components in the meal and the rate of detoxification of HCN by the animal (*14*). For acute toxicity to occur, enough plant materials must be ingested in a sufficiently short period by the animal. In humans, the minimum lethal dose of HCN taken orally is approximately 0.5 - 3.5 mg/kg body weight or 35 - 245 mg for a 75 kg person (*14*). Cyanide exerts an acute toxic effect by combining with metalloporphyrin-containing enzyme systems, most importantly cytochrome oxidase. Cyanide content of approximately 33 μmole can completely inhibit the mitochondrial electron transport system, thus swiftly preventing utilization of oxygen by the cells (*1*).

Analyses of Cyanogenic Glycosides

In order to isolate cyanogenic glycosides from the containing tissues it is necessary to inactivate their degradative enzymes. Extraction of dried, defatted and ground flaxseed with boiling 80% (v/v) ethanol (*8*) or 70% (v/v) methanol at room temperature (*7*) may inactivate the degradative enzymes and at the same time extracts cyanogenic glycosides into the alcohol. Following extraction, the volume is reduced by removal of solvent at low temperature and undesirable pigments (chlorophyll) and lipids could be removed by extraction with a nonpolar solvent such as chloroform.

Various chemical procedures have been described and used for quantification of total cyanogenic glycosides of plants. Most methods use acid hydrolysis to break down the glycoside to HCN which is then steam distilled into dilute alkali. The

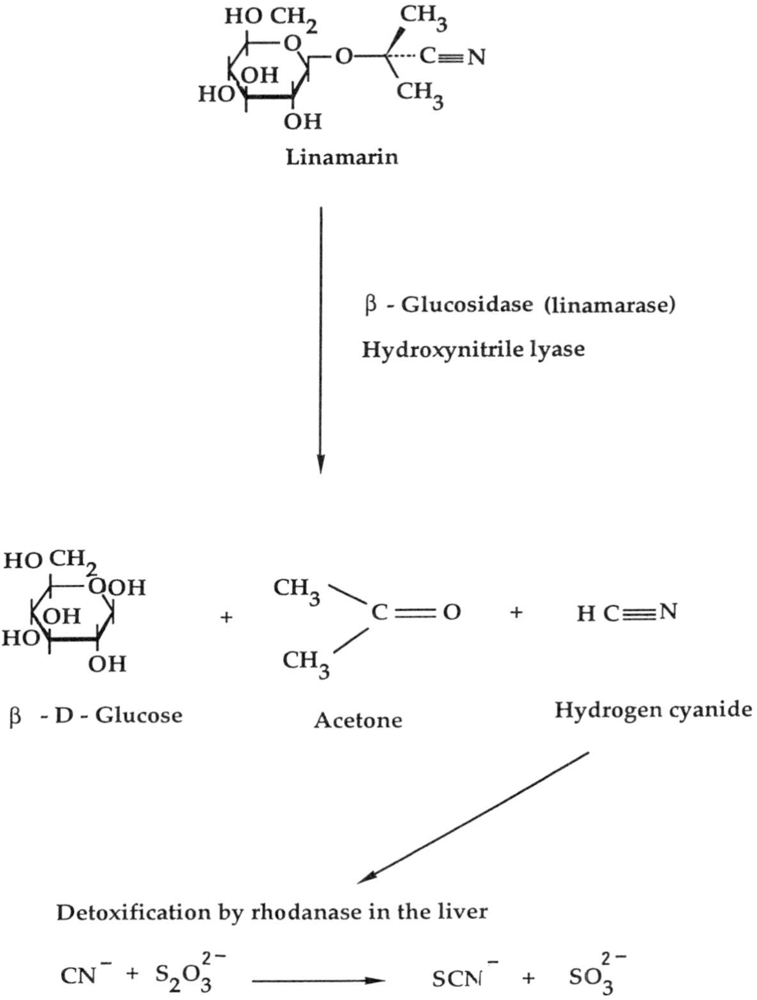

HO CH$_2$ CH$_3$ O C≡N OH HO OH CH$_3$

Linamarin

β - Glucosidase (linamarase)

Hydroxynitrile lyase

HO CH$_2$ OOH OH HO OH + CH$_3$ C=O CH$_3$ + H C≡N

β - D - Glucose   Acetone   Hydrogen cyanide

Detoxification by rhodanase in the liver

$$CN^{-} + S_2O_3^{2-} \longrightarrow SCN^{-} + SO_3^{2-}$$

Figure 3. Catabolism of cyanogenic glycosides

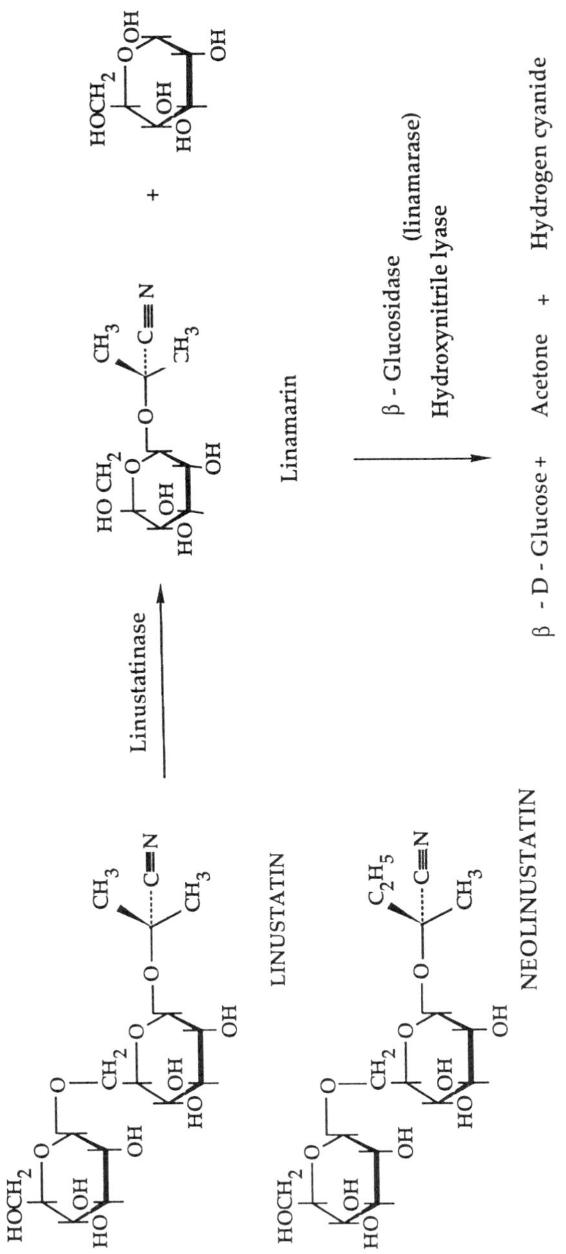

LINUSTATIN

NEOLINUSTATIN

Linustatinase

Linamarin

β - Glucosidase (linamarase)

Hydroxynitrile lyase

β - D - Glucose + Acetone + Hydrogen cyanide

Figure 4. Hydrolysis of linustatin and neolinustatin and their products.

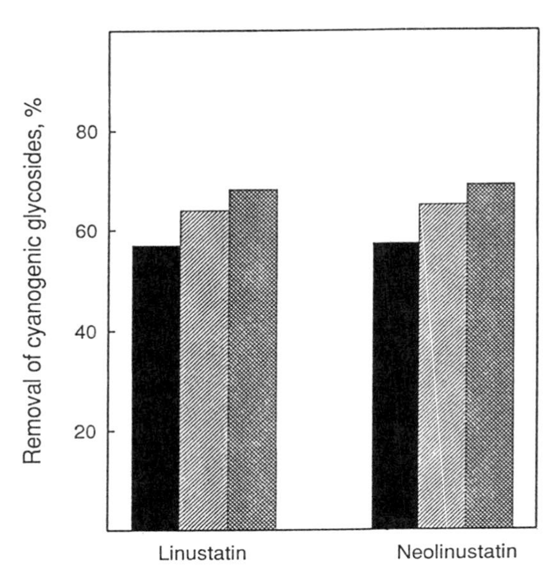

Figure 5. Removal of linustatin and neolinustatin of flaxseed meal as affected by the content of water in the polar phase of methanol-ammonia-water/hexane extraction: ■, 5%; ▨ ,10%; ▩, 15%.

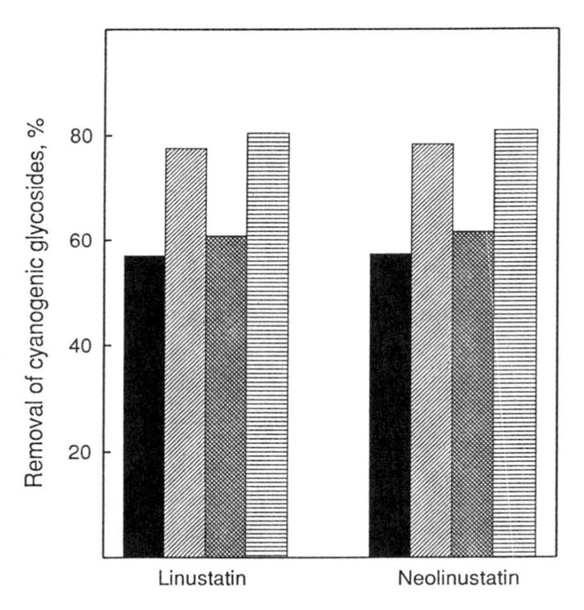

Figure 6. Removal of linustatin and neolinustatin of flaxseed meal as affected by the quiescent period and polar-phase volume to seed ratio (R, v/w): ■, 15 min, R=6.7; ▨,15 min, R=13.3; ▩, 30 min, R=6.7; ▤, 30 min, R=13.3.

colorimetric assay of the complex formed between HCN with barbituric acid-pyridine is used for quantification of cyanogenic glycosides (*17*). The most commonly used picrate method (*18* and *19*) has been improved to allow densitometric quantification of HCN formed upon enzymic hydrolysis on a thin layer chromatographic (TLC) plate (*20*). Quantification of released CN^- from homogenized flaxseed with water by HPLC equipped with an anion exchange column and an electrochemical (amperometric, oxidation) detector (*21*) was recently reported.

Chromatographic separation of cyanogenic glycosides of flaxseed by TLC (*22*) and high pressure liquid chromatography (HPLC, *6* - *8*) has been established. Unhydrolysed cyanogenic glycosides of flaxseed may be determined by reverse phase-HPLC equipped with a refractive index detector and a C_{18} (bonded silica) column and 15% (v/v) methanol (*8*) or water/methanol/phosphoric acid (94.95:5.00:0-.05, v/v/v; *7* and *23*). Authentic standards of cyanogenic glycosides are used as standards, however, linustatin and neolinustatin are not commercially available, they have to be isolated and purified from readily available sources (*6, 22* and *24*).

Removal of Flaxseed Cyanogenic Glycosides

Several attempts have been made to remove the cyanogenic compounds from flaxseed meal. These treatments include boiling in water, dry and wet autoclaving and acid treatment followed by autoclaving, among others (*25* and *26*). Soaking of flax cake in four times its weight of water and drying the residue after decantation, reduced cyanogenic glycoside content to one half of the original level however, autoclaving for 15 min at 1.055 kg/cm^2 pressure resulted in a maximum reduction (from 85 to 12 ppm HCN) of glycoside content (*26*). A dry milling process to remove flaxseed toxic compounds has been described by Dev and Quensel (*27*).

Furthermore, a simultaneous extraction procedure for the removal of cyanogenic glycosides from flaxseed and oil has been reported (*8*). A two-phase solvent system composed of methanol-ammonia-water (95:10:5, v/v/w)/hexane was employed in these studies. Presence of water or ammonia in the extraction medium largely enhanced the reduction of cyanogenic glycosides. The combination of water and ammonia in methanol resulted 57% reduction (Figure 5) of both linustatin and neolinustatin. Inclusion of water increased the polarity of the methanol phase and that may be one of the reasons for higher efficiency of the solvent system for the removal of polar components including cyanogenic glycosides. Increasing the water content in the polar phase of methanol-water-ammonia/hexane system removed more cyanogenic glycosides (Figure 5) but gave a sticky meal. Doubling the volume of the extraction solvent (R) and duration of the extraction period resulted in the removal of over 80% (Figure 6) of both linustatin and neolinustatin. A three stage extraction process afforded meal with greater than 10 fold reduction in the content of its cyanogenic glycosides (Figure 7). This process has the advantage of simultaneously extracting the oil and removing the cyanogenic glycosides of flaxseed meal. Removal of the cyanogenic glycosides of flaxseed provides a means for better utilisation of meals in feed and possibly for human food formulations. The choice of solvent system and extraction variables would allow efficient removal of cyanogenic glycosides.

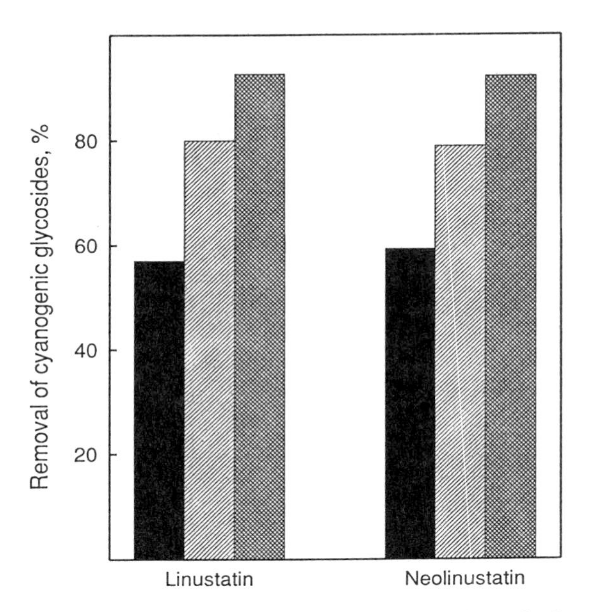

Figure 7. Removal of linustatin and neolinustatin of flaxseed meal as affected by the number of extraction stages: ▆ , one extraction; ▨ , two extractions; ▩ , three extractions.

Cyanogenic Glycosides of Flaxseed and Selenium Toxicity

The high selenium levels of soil in certain geographical locations of Canada and United States has been identified as a cause of selenium toxicity in farm animals grazing vegetation grown in these areas. Reduced weight gain and lowered reproductive rates are the major economic losses associated with selenium toxicity in farm animal although it is not fatal. Flaxseed meal at a dietary level of 25% (w/w) protects rats against adverse effects from 10 ppm Se (*29*). The complete isolation and identification of chemical compounds of flaxseed responsible for protection against Se toxicity in rats was achieved by Smith and co-workers (*6*). These authors were able to show that both linustatin and neolinustatin were involved. At 2 mg/g level both linustatin and neolinustatin were able to overcome the adverse effects of Se and showed a significant ($p<0.05$) growth increase in rats (*29*). The monoglycoside linamarin present in flaxseed was not as effective as the two diglycosides in preventing selenium toxicity.

Free cyanide has been shown to block Se stimulation of the thiol-induced swelling of liver mitochondria from Se deficient rats, possibly through the formation of $SeCN^-$ (*30* and *31*). It has also been reported that cyanide can inhibit the Se-containing enzyme glutathione peroxidase and release Se from the enzyme in the form of $SeCN^-$ (*32*). These observations suggest that CN^- derived from the cyanogenic glycosides must be responsible for the protective action of purified glycosides and/or flaxseed meal against Se toxicity (*29*).

Literature Cited

1. Conn, E.E. In *Secondary Plant Products*; Conn, E.E., Ed.; The Biochemistry of Plants; Academic Press: New York, 1981, Vol. 7; pp. 479-500.
2. Selmar, D.; Lieberei, R.; Bichl, B. *Plant Physiol.* **1988**, *86*,711-716.
3. Caragay, A.B. *Food Technol.* **1992**, *46(4)*, 65-68.
4. Oomah, B.D.; Mazza, G. *Food Chem.* **1993**, *48*, 109-114.
5. Butler, G.W. *Phytochem.* **1965**, *4*, 127-131.
6. Smith, C.R. Jr.; Weisidler, D.; Miller, R.W. *J. Org. Chem.* **1980**, *45*, 507-510.
7. Oomah, B.D., Mazza, G.; Kenaschuk, E.O. *J. Agric. Food Chem.* **1992**, *40*, 1346-1348.
8. Wanasundara, P.K.J.P.D.; Amarowicz, R.; Kara, M.T.; Shahidi, F. *Food Chem.* **1993**, *48*, 263-266.
9. Mazza, G. and Oomah, D.B. In *Flaxseed in Human Nutrition*, Cunnane, S.C. and Thompson, L.U. Eds.; American Oil Chemists' Society; Champaign, IL. 1995, pp. 56-81.
10. Cutler, A.J.; Sternberg, M.; Conn, E.E. *Arch. Biochem. Biophys.* **1985**, *238*, 272-279.
11. Hahlbrock, K.; Conn, E.E. *J. Biol. Chem.* **1970**, *245*, 917-922.
12. Hahlbrock, K.; Conn, E.E. *Phytochem.* **1971**, *10*, 1019-1023.
13. Frehner, M.; Scalet, M.; Conn, E.E. *Plant Physiol.* **1990**, *94*, 28-34.
14. Poulton, J.E. In *Food Proteins*, Kinsella, J.E. and Soucie, W.G. Eds.; American Oil Chemists' Society; Champaign. IL. 1989, pp. 381-401.
15. Fan, T.W.M.; Conn, E.E. *Arch. Biochem. Biophys.* **1985**, *243*, 361-373.
16. Butler, G.W.; Bailey, T.W.; Kennedy, L.D. *Phytochem.* **1965**, *4*, 369-381.
17. Nambisan, B.; Sundaresen, S. *J. Assoc. Off. Anal. Chem.* **1984**, *67*, 641-643.
18. Gilchrist, D.G.; Lueschen, W.E.; Hittle, C.N. *Crop. Sci.* **1967**, *7*, 267.
19. Cooke, R.D.; Blake, G.G.; Battershill, J.M. *Phytochem.* **1978**, *17*, 381-383.
20. Brimer, L.; Chritensen, S.B.; Molgarrd, P.; Nartey, F. *J. Agric. Food Chem.* **1983**, *31*, 789-793.
21. Chadha, R.K.; Lawrence, J.F.; Ratnayake, W.M.N. *Food Additives Contam.* **1995**, *12*, 527-533.
22. Amarowicz, R.; Wanasundara, P.K.J.P.D.; Shahidi, F. *Die Nahrung.* **1993**, *37*, 88-90.
23. Schilcher, V.H.; Wilkens-Sauter, M. *Fett. Seifen Anstrichnit*, **1986**, *88*, 287-290.
24. Amarowicz, R.; Chong, X.; Shahidi, F. *Food Chem.* **1994**, *48*, 99-101.
25. Madhusudhan, K.T.; Singh, N. *J. Agric. Food Chem.* **1985**, *33*, 1219-1222.
26. Madhusudhan, K.T.; Singh, N. *J. Agric. Food Chem.* **1985**, *33*, 1222-1226.
27. Deshmukh, S.V.; Netke, S.P.; Dabadghao, A.K. *J. Anim. Sci.* **1982**, *52*, 241.
28. Dev, D.K.; Quensel, E. *J. Food Sci.* **1988**, *53*, 1834-1837,1857.
29. Palmer, I.S. In *Flaxseed and Human Nutrition*, Cunnane, S.C. and Thompson, L.U. Eds.; American Oil Chemists' Society; Champaign, IL, 1995, pp. 165-173.
30. Levander, O.A.; Morris, V.C.; Higgs, D.J. *Biochem.* **1973**, *12*, 4586-4590.
31. Levander, O.A.; Morris, V.C.; Higgs, D.J. *Biochem.* **1973**, *12*, 4591-4595.
32. Prohaska, J.R.; Oh, S.H.; Hoekstra, W.G.; Ganther, H.E. *Biochem. Biophys. Res. Commun.* **1977**, *74*, 64-71.

Chapter 11

Nutritional Implications of Canola Condensed Tannins

Marian Naczk[1] and Fereidoon Shahidi[2]

[1]**Department of Human Nutrition, St. Francis Xavier University, P.O. Box 5000, Antigonish, Nova Scotia B2G 2W5, Canada**
[2]**Department of Biochemistry, Memorial University of Newfoundland, St. John's, Newfoundland A1B 3X9, Canada**

Tannins are complex phenolics with molecular weights of 500-3000 Daltons. Based on their structural features tannins are classified either as hydrolyzable or condensed. Tannins may form complexes with essential minerals, proteins, and carabohydrates, thus lowering the nutritional value of products. Canola hulls contained up to 2.3% condensed tannins and this is up to ten times more than those reported previously, using the vanillin assay. Canola tannins play an important role in iron-binding properties of canola meal and have a significant biological activity, *in vitro,* as determined by protein precipitation and dye-labelled asssays. The effect of protein type, pH, and tannin concentration on protein precipitating capacity of canola tannins is discussed. The interaction of canola tannins with wheat starch was the strongest while tapioca and rice starches were able to bind 2-3 times less tannins than wheat starch.

The global production of rapeseed, including canola varieties, ranks third amongst other oilseed crops. The meal obtained after oil extraction contains about 40% protein. The quality of rapeseed meal as represented by its amino acid composition is well-balanced for human use (*1*). Shahidi et al. (*2*) found that amino acids of canola were not sensitive to extraction with methanol-ammonia/ hexane as well as to commercial processing. Delisle et al. (*3*) have also reported that the protein efficiency ratios of rapeseed and soybean meals were 2.64 and 2.19, respectively. However, use of rapeseed as a source of food protein is still thwarted by the presence of undesirable components such as glucosinolates, phenols, phytates and hulls. The composition of rapeseed has been significantly altered by Canadian breeders who have developed double-low varieties of rapeseed, known as canola, which contain less than 2% erucic acid in the oil and no more than 30 μmol of any one combinations of two or more of four aliphatic glucosinolates per gram of moisture-free, defatted meal. Compared to the older varieties, the level of glucosinolates in canola has been significantly reduced (almost ten-fold), but the new varietes still contain too high levels of glucosinolates to be considered as a suitable protein source in the food products. This is due to the fact that hydrolytic decomposition of glucosinolates may lead to the formation of toxic isothiocyanates, nitriles, and thiocyanates depending on the raction conditions (*4*). Heating of crushed canola/rapeseed inactivates the enzyme myrosinase, which is responsible for the formation of glucosinolate breakdown products. However, glucosinolates in the meal may still undergo decomposition to toxic compounds by microorganisms in the lower gastrointestinal tract (*5-7*). Therefore, a number of new processes have been developed to reduce the content of glucosinolates in the meal. These methods involve chemical, microbial and physical treatments of meals or seeds (*5, 8, 9*).

Phenolic compounds are also important factors when considering canola/rapessed meal as a source of food-grade protein because they contribute to the dark color, bitter taste, and astrigency of rapeseed protein products (*10-13*). Phenolics and their oxidized products can also form complexes with essential amino acids and proteins, thus lowering the nutritional value of rapeseed products (*14-17*).

The content of phenolic compounds in rapeseed protein products is much higher than those in other oleaginous seeds (Table I). The predominant phenolics present in rapeseed/canola are phenolic acids and condensed tannins. Phenolic acids are present in the free-, esterified- and insoluble-bound forms. Rapeseed meals contain up to 1837 mg of phenolic acids per 100g defatted meal, on a dry weight basis. On the other hand, the content of phenolic acids in rapeseed flours ranges from 623.5 to 1280.9 mg per 100g sample, on a dry weight basis (*18-22*). Tannins are complex phenolics with molecular weights of 500-3000 Daltons. They are classified either as condensed or hydrolyzable based on their structural features and their reactivity towards hydrolytic agents, particularly acids. Hydrolyzable tannins are phenolics that upon acidic, alkaline or enzymatic hydrolysis produce a polyol moiety (usually D-glucose) and phenolic acids such as gallic acid and/or hexahydroxydiphenic acid. The latter acid, upon lactonization, produces ellagic acid. On other hand, condensed tannins are dimers, oligomers and polymers of flavan-3-ols which upon acidic hydrolysis (butanol-HCl) produce anthocyanidins. Therefore, they are also known as proanthocyanidins. Of these, only the condensed tannins were first identified in rapeseed hulls (*23*). This finding was later verified by Durkee (*24*).

Chemical Structure and Quantification of Tannins

Chemical Structure of Tannin. Durkee (*24*) identified cyanidin, pelargonidin and an artefact, n-butyl derivative of cyanidin, in the hydrolytic products of rapeseed hulls. Later, Leung et al. (*25*) reported that leucocyanidin was the basic unit of tannins isolated from rapeseed hulls (Figure 1).

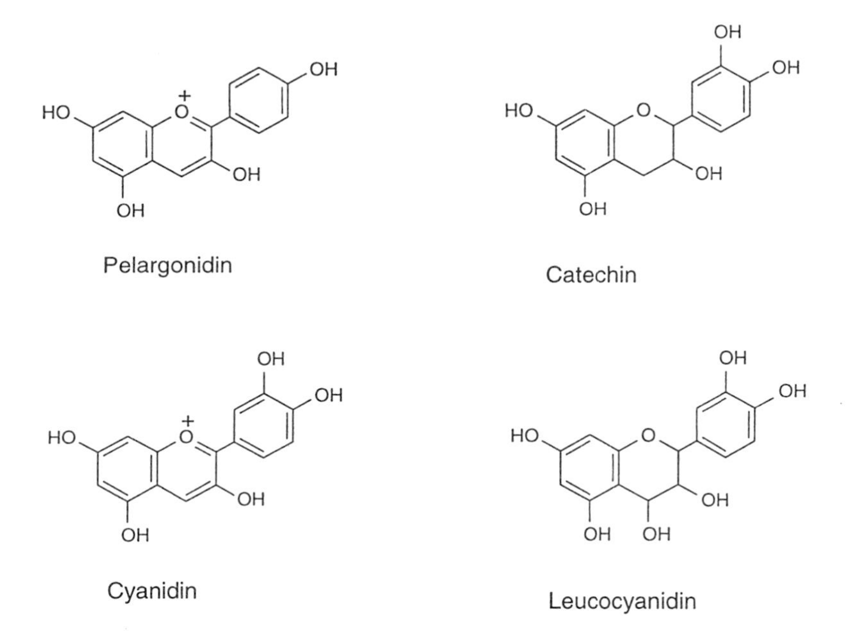

Figure 1. Structures of basic units of condensed tannins of rapeseed.

Table I. Total Content of Phenolic Acids in Oilseed Products

Product	*Total Phenolic Acids [mg/100g dry weight]*
soybean flour	23.4
cottonseed flour	56.7
rapeseed/canola flour	639-1280
canola meal	1542-1837
soybean meal	455

Adapted from ref. 14

Recovery of Tannins by Various Solvent Systems. Several solvent systems have been employed for the extraction of condensed tannins from plant tissues. The list includes absolute methanol, ethanol, acidified methanol, acetone, water or their combinations. Gupta and Haslam (*26*) have found that methanol is the most efficient solvent for the recovery of condensed tannins from sorghum grains. However, according to Maxson and Rooney (*27*) and Price et al. (*28*) addition of concentrated HCl to methanol enhances the extraction of tannins from sorghum. On the other hand, Leung et al. (*25*) used 70% acetone to recover tannins from rapeseed.

Table II. Effect of Water Content in Solvent on the Recovery of Tannins (mg catechin equivalents per 100g dry meal)

Water Content [%, v/v]	*Acetone*	*Methanol*	*N,N-dimethylformamide*
0	0	35.1	0
10	156.3	87.3	242.2
20	321.3	190.0	331.4
30	321.3	241.8	321.3
50	260.7	234.9	-

Adapted from ref. 29.

Recently Naczk et al. (*29*) have investigated the yield of tannin recovery as affected by various solvent systems and extraction conditions. Table II summarizes the effect of addition of water to methanol, acetone and N,N-dimethylformamide (DMF) on the recovery of canola tannins. Pure solvents provided poor extraction media for canola tannins, but addition of up to 30% (v/v) water markedly increased the yield of tannin recovery. Thus, 70% (v/v) acetone or DMF were most efficient solvent systems for the recovery of tannins. On the other hand, results presented in Table III indicate that addition of concentrated HCl to the

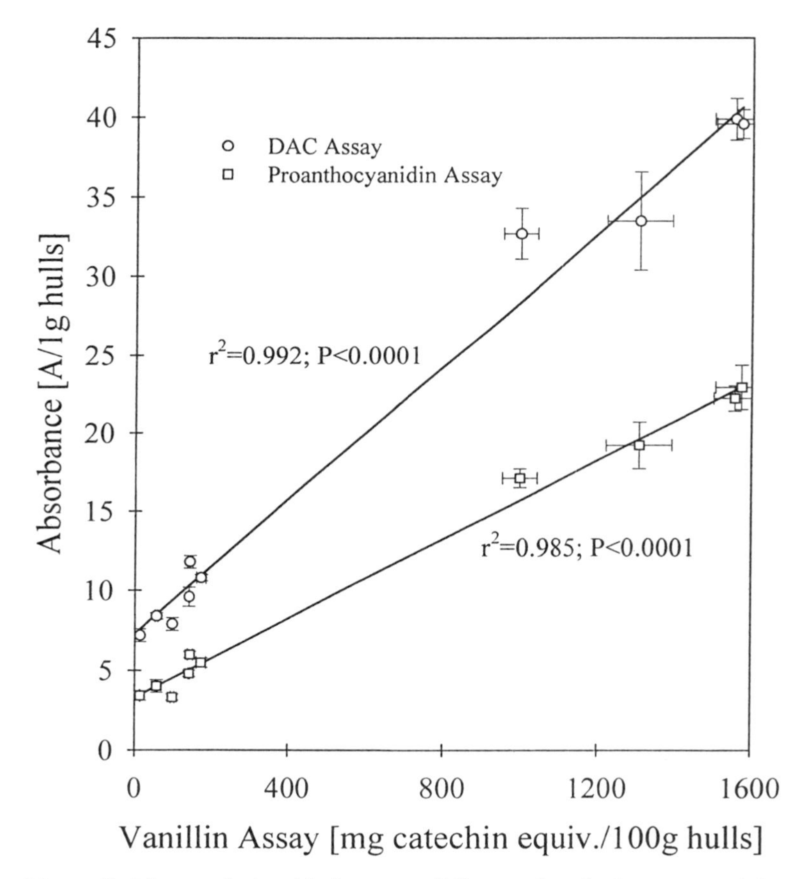

Figure 2. Linear relationship between different chemical assays used for quantification of canola tannins (adapted from ref.37).

extraction medium lowers the recovery of canola tannins. Thus, the data suggest that the chemical nature and solubility of canola tannins differ from those found in sorghum. Furthermore, Naczk et al. (*29*) found that two-stage extraction of canola meal with 70% (v/v) acetone at a meal to solvent ratio of 1 : 10 (w/v) was adequate for total recovery of their tannin constituents.

Table III. Effect of HCl Addition on the Recovery of Tannins

Solvent System	*Tannins [mg catechin equiv. per 100g meal]*
70% (v/v) methanol	241.8
1% conc. HCl in 70% (v/v) methanol	225.9
70% acetone (v/v)	321.3
1% conc. HCl in 70% (v/v) acetone	216.9

Adapted from ref. 29.

Quantification of Tannins by Chemical Assays. Several colorimetric methods have been developed for the quantification of tannins, but only a few are specific towards condensed tannins. These include the modified vanillin assay of Price et al. (*28*), the proanthocyanidin assay (*30, 31*) and the 4-(dimethylamino) cinnamaldehyde (DAC) assay (*32,33*). Of these, the vanillin method is widely used for quantification of condensed tannins due to its specificity for flavanols and dihydrochalcones, both of which possess a single bond at the 2,3-position of the pyran ring and free hydroxyl groups at positions 5 and 7 of the benzene ring (*34*). Methanol is usually used for carrying out the vanillin assay. In this solvent, the reaction is less sensitive to the monomeric than polymeric tannins (*35*). On the other hand, DAC reacts only with the terminal flavan-3-ol residues, but is sensitive to both monomeric and polymeric tannins (*32*). This leads to overstimation of tannin content as determined by the DAC assay. The proanthocyanidin assay depends on acid hydrolysis of interflavan bonds of condensed tannins to produce anthocyanidins. The yield of the reaction depends on the concentration of HCl, reaction time and temperature, proportion of water in the reaction mixture, and presence of transition metal ions in the reaction medium (*36*) as well as the chain length of the proanthocyanidin molecule (*31*).

Naczk et al. (*37*) have evaluated the usefulness of the aforementioned assays for quantification of crude tannins isolated from canola hulls of Westar, Cyclone, Excel, and Delta varieties. Canola hulls used in this study were separated from cotyledons by air aspiration as described by Sosulski and Zadernowski (*38*). Figure 2 shows the data for the DAC and proanthocyanidin assays plotted against the vanillin assay results. Statistically significant ($P<0.001$) linear relationships (correlation coefficients, $r^2>0.96$) existed between the tannin content determined by either the DAC or the proanthocyanidin method versus those obtained using the vanillin assay. Therefore, any of these chemical assays can be used for quantification of crude condensed tannins in extracts isolated from canola/rapeseed hulls.

Tannin Contents. Mitaru et al. (*39*) reported that rapeseed hulls contain 0.02 - 0.22% tannin. According to Leung et al. (*28*) rapeseed hulls contain no more than 0.1% condensed tannin extractable by solvent systems commonly used for isolation of polyphenols. Recently, Naczk et al. (*37, 40*) have determined the content of condensed tannins in several canola hulls of Westar, Cyclone, Excel, Vanguard, and Delta varieties using the modified vanillin assay of Price et al. (*27*). The content of tannins in hulls ranged from 14 to 2300 mg catechin equivalents per 100g hulls. The differences in tannin contents within canola varieties were between 9- and 15-fold (Table IV). These data indicate that both cultivar differences and environmental growing conditions may influence the content of condensed tannins in canola hulls.

Table IV. Content of Condensed Tannins in Canola Hulls

Canola Cultivar	*Tannin Content [mg catechin equiv. per 100g hulls]*
Westar	
sample 1	1556 ± 54
sample 2	173 ± 12
sample 3	142 ± 2
sample 4	98 ± 5
Cyclone	
sample 1	1307 ± 85
sample 2	994 ± 45
sample 3	1574 ± 58
sample 4	695 ± 12
sample 5	2318 ± 82
sample 6	372 ± 10
Delta	14 ± 1
Excel	144 ± 7
Vanguard	1059 ± 15

Adapted from ref. 37 and 40.

Rapeseed meals contain approximately 3% tannins as assayed by the method of tannin determination in cloves and allspice (*41, 42*); however, this value includes sinapine (*43*). Fenwick et al. (*44*) have later reported that whole and dehulled

Tower meals contain 2.71 and 3.91% tannins, respectively. On the other hand, the content of tannins assayable by the modified vanillin method of Price et al. (*27*) ranged from 0.09 to 0.39% in the defatted rapeseed cotyledons and from 0.23 to 0.54% in the defatted canola cotyledons (*45*). Shahidi and Naczk (*46, 47*) have reported that canola meals contain from 0.68 to 0.77% of extractable condensed tannins. Existing discrepancies in tannin contents may originate from differences in solvent systems employed for tannin recovery as well as methods used for their quantification.

Sensory Effects

Some phenolic substances present in plants are able to bring about a puckering and drying sensation over the whole surface of the tongue and the buccal mucosa, called astringency (48, 49). This sensantion is related to the ability of the substance to precipitate salivary proteins (50). According to Haslam (51), only tannins with a molecular weights of 500 to 3,000 Daltons may bring about the astringency sensation. Delcour et al. (52) have determined the taste thresholds of astringency for tannic acid, (+)-catechin, procyanidin B-3, and a mixture of trimeric and tetrameric proanthocyanidins dissolved in deionized water. These threshold values range from 4.1 to 46.1 mg/ml. These authors also found that substances with higher molecular weights had lower threshold values. Rapeseed hulls contained up to 2,300 mg condensed tannins/100g sample (37, 40). Thus, condensed tannins present in rapeseed hulls may contribute to the astringent taste of the meals as reported by Malcomson et al.(12).

Nutritional Effects

Phenolics compounds and their oxidized products may form complexes with minereals essential amino acids, proteins and carbohydrates, thus lowering the nutritional value of rapeseed products. The nutritional effects of tannins on rapeseed protein products are summarized below.

Interaction with Minerals. Phenolic compounds have been identified as possible inhibitors of iron absorption (53, 54). This inhibition may be due to the formation of insoluble iron-phenol complexes in the gastrointestinal tract which make the iron unavailable for absorption. Brune et al. (55) have suggested that phenolic compounds with galloyl and/or catechol groups are responsible for inhibition of iron absorption. Futhermore, a relationship existed between the content of galloyl groups in foods and the degree of inhibition of iron absorption.

Naczk and Shahidi (40) have determined the content of condensed tannins and iron-binding phenolics in selected canola varieties developed within the Canadian breeding program. The contents of tannins and iron-binding phenolics in defatted canola meals were determined by the modified vanillin assay of Price et al. (27) and the ferric ammonium sulfate (FAS) assay described by Brune et al. (56), respectively. The content of condensed tannins ranged from 358.3 to 692.3

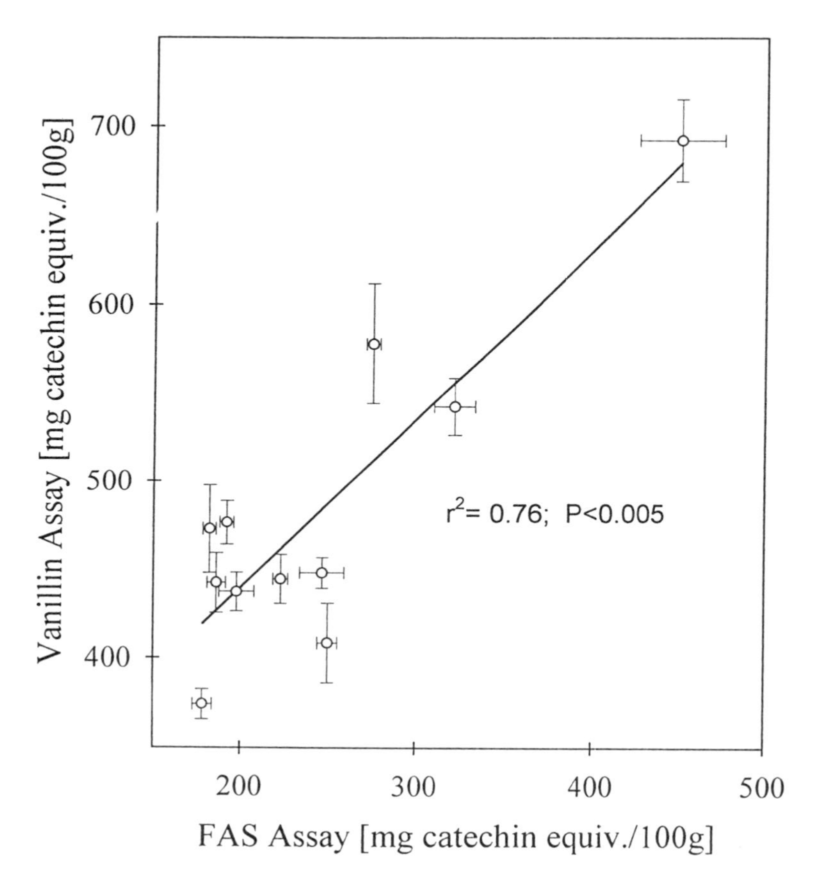

Figure 3. Linear relationship between the contents of tannin and iron-binding phenolics (adapted from ref. 40).

mg catechin equivalents per 100g meal, while the content of iron-binding phenolics ranged from 157 to 451 mg catechin equivalents per 100g meal. A statistically significant ($P<0.005$) linear relationship ($r^2= 0.76$) existed between the content of condensed tannins and iron-binding phenolics (Figure 3). This relationship indicates that condensed tannins of rapeseed/canola play an important role in iron-binding properties of the meal and therefore their removal improves the nutritional value of protein products from canola/rapeseed.

Interaction with Proteins. The available information on rapeseed tannin-protein interactions is still diverse and fragmentary. Leung et al. (28) have reported that tannins isolated from rapeseed hulls form a white precipitate after addition to a 1% gelatin solution. Later, Mitaru et al. (39) reported that condensed tannins isolated from rapeseed hulls were unable to inhibit the activity of α-amylase *in vitro*. Futhermore, Fenwick et al. (57, 58) and Butler et al. (59) have reported that tannins in rapeseed meal may be responsible for tainting of eggs. They postulated that this tainting effect may originate from the formation of tannin - trimethylamine (TMA) oxidase complex. This enzyme converts TMA to the odorless and water-soluble TMA oxide. On the other hand, addition of tannins (extracted from rapeseed meal using water) to chicks diets containing soybean resulted in a reduction in their metabolizable energy, but did not have any apparent effect on absorption of proteins by chicks (60).

Molecular Intractions between Proteins and Polyphenols. Loomis and Battaile (61) suggested that phenols may reversibly complex with proteins via hydrogen bonding of hydroxyl groups of phenols and carbonyl functionalites of peptide bonds in proteins, or irreversibly by oxidation to quinones which may in turn combine with reactive groups of protein molecules. Wade et al. (62) have found that the extent of binding of serum albumin to simple phenols correlated well with PK_a values of phenols. Thus, hydrogen bonding between the phenol molecule and proteins is stronger for more acidic phenols. The phenol-protein complex may also be stabilized by other types of molecular interactions such as ionic bonds between the phenolate anion and the cationic site of the protein molecules (63) or/and hydrophobic interactions between aromatic ring structure of tannins and hydrophobic region of proteins (63-66). However, it is believed that phenol-protein complexation is usually due to the formation of hydrogen bonds and hydrophobic interactions (65), particularly in the acidic pH range (67).

Methods of Quantification of Tannin-Protein Interaction. The biological and ecological role of phenolics is attributed to their abilities to bind/precipitate proteins or to inhibit enzymatic activities. No satisfactory correlations have been found between the tannin content and degree of inhibition of various enzymes (68). Therefore, estimation of biological activity of tannins to bind and precipitate proteins *in vitro* seems to be a matter of choice. Several methods are available for determination of the protein-precipitating capacity of tannins (69). Of these methods, the dye-labelled BSA (Bovine Serum Albumin) assay developed by

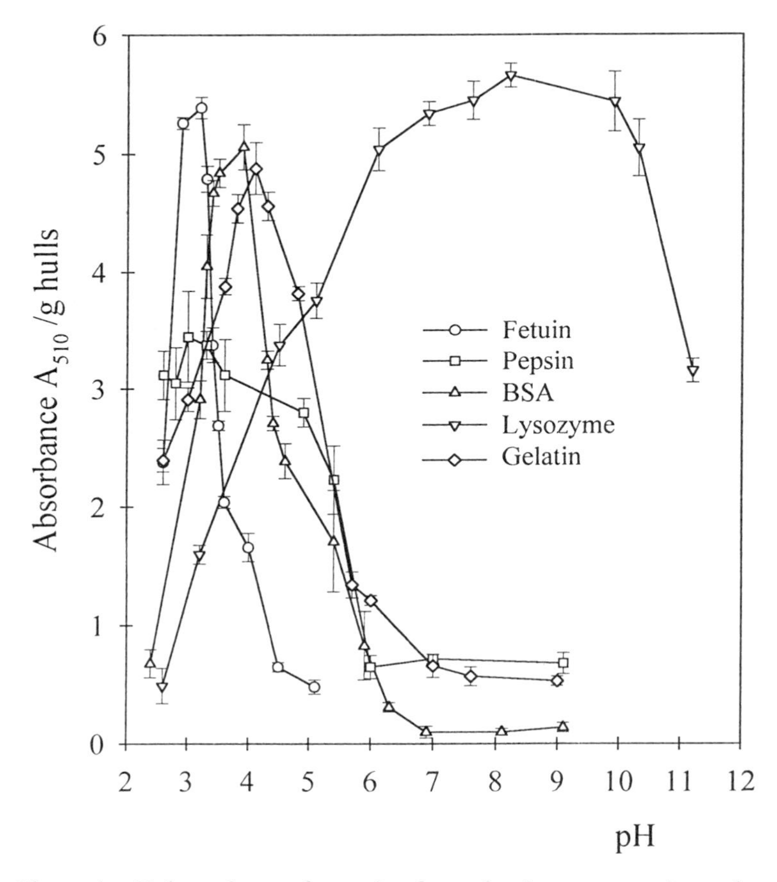

Figure 4. pH dependence of complex formation between canola tannins and several proteins (adapted from ref. 73)

Asquith and Butler (68) and protein precipitation assay developed by Hagerman and Butler (70) were selected for quantification of protein-precipitating activity of tannins isolated from canola hulls. The dye-labelled BSA assay provides a direct quantitation method for proteins precipitated by tannins, while protein-precipitation assay allows for estimation of the amount of phenolics bound to proteins.

Effect of pH. The effect of pH on precipitation of condensed tannins with selected proteins, as determined by the protein precipitation assay of Hagerman and Butler (70) is shown in Figure 4. Crude tannin extracts, isolated from hulls of Westar canola, (sample # 1; Table IV) were used in this study. BSA, fetuin, gelatin and pepsin were effectively precipitated by crude tannin extracts at pH 3.0 - 5.0. However, maximum precipitation of lysozyme occured at pH > 8.0. Thus, precipitation of proteins by canola tannins not only depends on the availability of unionized phenolic groups for hydrogen bonding, but also on other variables such as protein type (71). The optimum pH for performing the protein precipitation assay of Hagerman and Butler (70) was 4.0. A statistically significant ($P < 0.0004$) linear relationship ($r^2 = 0.99$) existed between the isolectric point of the protein and pH optimum for its precipitation by condensed tannins (Figure 5).

The effect of pH on the precipitation of dye-labelled BSA by crude canola tannins is shown in Figure 6. Tannins were isolated from hulls of Excel (low-tannin) and Cyclone (sample #1; high-tannin) canola (71; Table IV). Dye-labelled BSA was largely precipitated by canola tannins at pH = 3.5. Table V shows the precipitation of dye-labelled BSA at pH 3.5 and 4.8. The data indicate that pH=3.5 was optimum for carrying out the dye-labelled BSA assay of Asquith and Butler (68) and not 4.8 as indicated by these authors.

Effect of Tannin Concentration. Figure 7 exhibits the titration curves of a known amount of protein versus increasing concentrations of tannins (73). The crude extracts of canola tannins were partially purified by extraction of their aqueous solutions with ethyl acetate (65). Crude extracts contained approximately 20% proanthocyanidins which were soluble in ethyl acetate, as determined by the modified vanillin assay of Price et al. (27). According to Porter (72), only monomeric and dimeric proanthocyanidins are highly soluble in ethyl acetate. The protein precipitating capacities of crude and partially purified tannins were determined by the dye-labelled BSA and the protein precipitation assays. These assays were carried out at optimum pH values reported by Naczk et al. (73).

Figure 7 indicates that canola tannins did not delay the binding of dye-labelled BSA but did show a definitive threshold prior to binding with unlabelled BSA. The titration curves exhibit a linear relationship between the amount of precipitated tannin-protein complex and the amount (up to 0.8 mg catechin equiv./mL for low-tannin hulls and up to 1.5 mg tannins/mL for tannin extracts from high-tannin hulls) of canola tannins added to the system, both in the crude and partially purified forms. Thus ethyl acetate-soluble proanthocyanidins also contribute to protein-precipitating capacity of crude canola tannins. Furthermore,

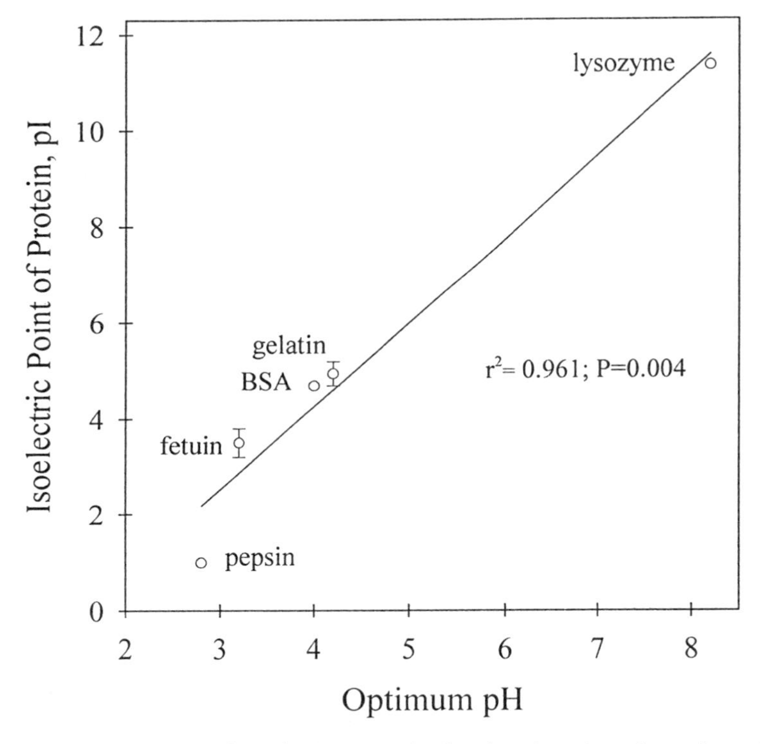

Figure 5. Linear relationship between the isoelectric point of protein and pH optimum for its precipitation (M.Naczk, unpublished results).

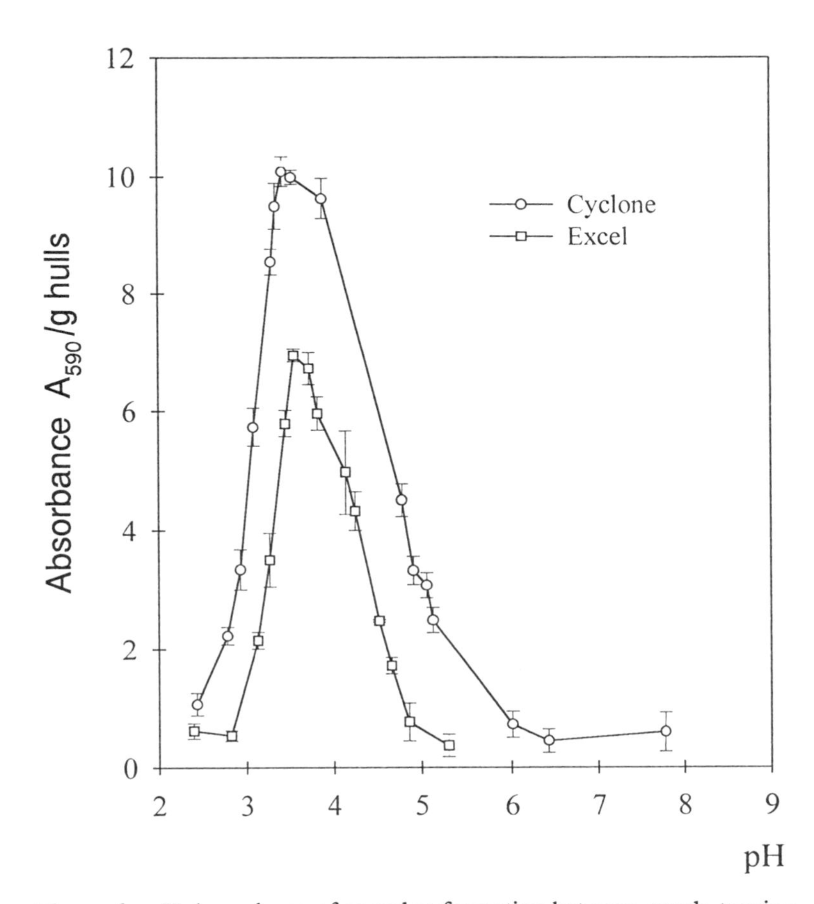

Figure 6. pH dependence of complex formation between canola tannins and dye labelled BSA (adapted from ref. 73).

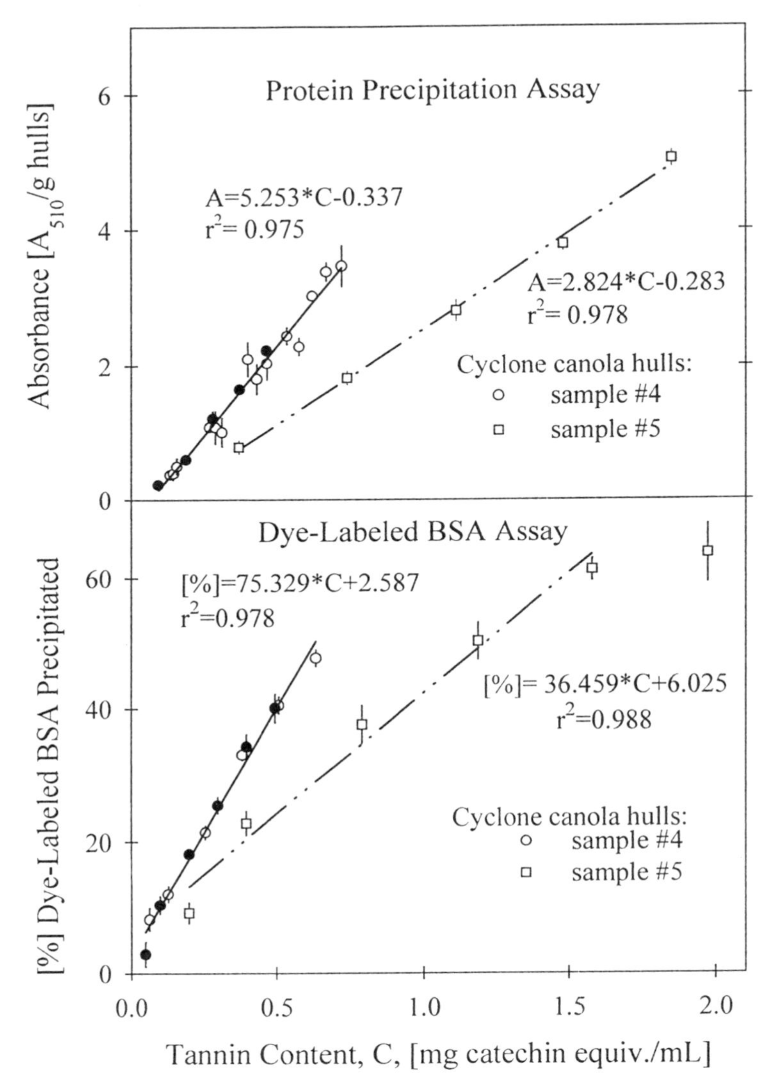

Figure 7. Titration curves of a known amount of protein with increasing tannin concentration (adapted from ref. 73).

these titration curves showed a saturation effect at higher concentrations of tannins obtained from high-tannin Cyclone hulls, using labelled BSA assay. The observed differences between the titration curves obtained for crude tannins isolated from low- and high-tannin hulls may be due to differences in their affinities for proteins.

Protein-Precipitating Capacity of Canola Tannins. Protein-precipitating capacity of tannins isolated from canola hulls, as determined by the dye-labelled BSA assay of Asquith and Butler (68) and the protein precipitation assay of Hagerman and Butler (70) is shown in Table VI. The data, as determined at optimum pH values shown in Table VI, are similar to those reported by Asquith and Butler (68) and Hagerman and Butler (70) for condensed tannins isolated from sorghum. Naczk et al. (37) proposed to characterize the affinity of canola tannins for proteins as protein precipitation index (P_I), expressed as mg dye-labelled BSA

Table V. Effect on pH on Precipitation of Dye-Labelled BSA by Canola Tannins

Cultivar	*%, Dye-Labelled BSA Precipitated at*	
	pH = 4.8	*pH = 3.5*
Westar	26.6 ± 1.1	73.5 ± 2.3
sample 1	4.0 ± 0.5	38.9 ± 1.6
sample 2	4.0 ± 0.5	29.8 ± 1.7
sample 3	5.1 ± 0.8	33.0 ± 2.2
sample 4		
Cyclone		
sample 1	27.2 ± 1.2	65.0 ± 1.2
sample 2	28.3 ± 1.5	55.7 ± 2.1
sample 3	33.4 ± 0.8	65.8 ± 1.5

precipitated by 1 mg tannins (as mg catechin equivalents) determined by the modified vanillin assay (27). It was found that P_I values for tannins isolated from high-tannin canola hulls did not exceed 5.0 mg BSA/mg tannins, but for low-tannin canola hulls ranged from 17.7 to 40.7 mg BSA/mg tannins. According to Porter and Woodruffe (74) protein-precipitating capacity of tannins depends on their molecular weight. Therefore, differences in P_I values may be due to the existing differences in molecular weights of tannins extracted from low- and high-tannin canola hulls.

Figure 8 shows the relationship between tannin content, detemined by the DAC (32, 33) and the modified vanillin assays (27) and the corresponding protein

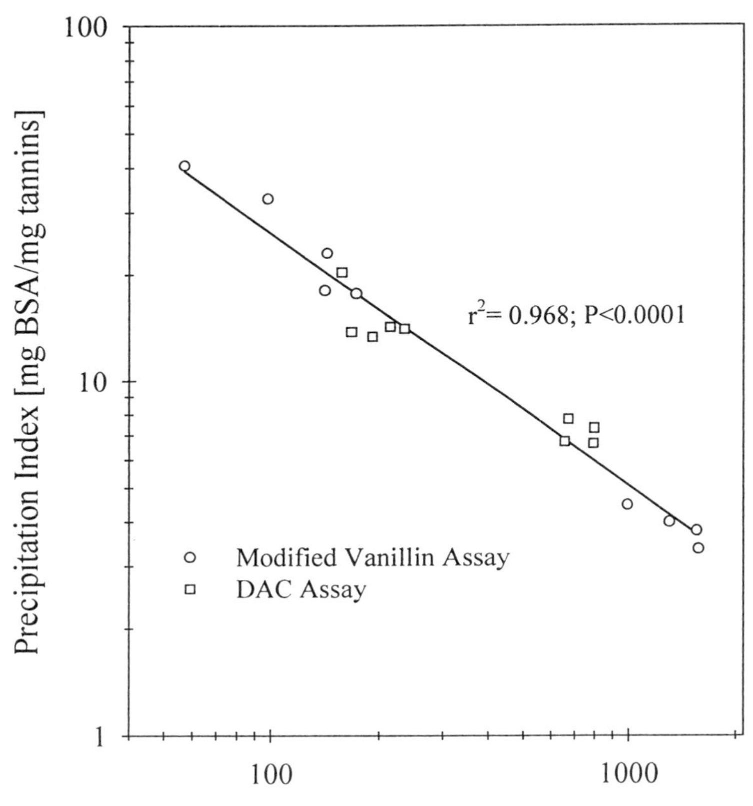

Figure 8. Logarithmic relationship between precipitation index, P_I, and tannin content determined by the vanillin and the DAC assays (adapted from ref. 37).

precipitation indices, P_I. A statistically significant ($P < 0.001$; $r^2 = 0.968$) linear logarithmic relationship existed between the P_I values and tannin contents determined by either modified vanillin or DAC assay (37). Thus, both the DAC and modified vanillin assays can be used for estimation of protein-precipitating capacity of canola tannins.

Interaction with Carbohydrates. The affinity of tannins for carbohydrates depends on their molecular weight, conformational mobility, and shape, as well as solubility of tannins in water (75). Desphande and Salunkhe (76) have studied the interaction of tannic acid and catechin with different starches in model systems. These authors reported that starches were associated with up to 652 μg of tannic acid and up to 586 μg of catechins per 100 mg of sample. Recently, Naczk (unpublished results) has studied binding of canola tannins to selected starches using the method of Desphande and Salunkhe (76). In this, 1 ml of methanolic solution of crude canola tannins (containing 890 μg/ml of tannin expressed as catechin equivalents) was added to a suspension of 100 mg of starch in 9 ml of distilled water. This slurry was stirred for two hours at room temperature, centrifuged for 15 min at 5000xg and the residual tannins were

Table VI. Protein Precipitating Capacity of Canola Condensed Tannins

Cultivar/Assay	*Protein Precipitation [A_{510}/g hulls]*	*Dye-Labelled BSA [mg BSA/g hull]*
Westar		
sample 1	5.0 ± 0.5	58.6 ± 1.9
sample 2	1.1 ± 0.05	30.7 ± 1.4
Cyclone		
sample 1	4.9 ± 0.3	52.2 ± 1.0
sample 2	4.5 ± 0.2	44.2 ± 1.4
Excel	2.0 ± 0.1	33.2 ± 0.5

Adapted from ref. 37 and 70.

determined using the DAC assay (32, 33). Figure 9 shows the affinity of selected starches for canola tannins. Wheat starch had the strongest affinity for canola tannins, while tapioca, corn starch, rice starch, amylose and amylopectin were able to bind 2-3 times less tannins than wheat starch.

Effect of Processing

Naczk and Shahidi (21) have reported that methanol alone extracted only 16% of tannins present in rapeseed meals. Addition of 5% (v/v) water to methanol

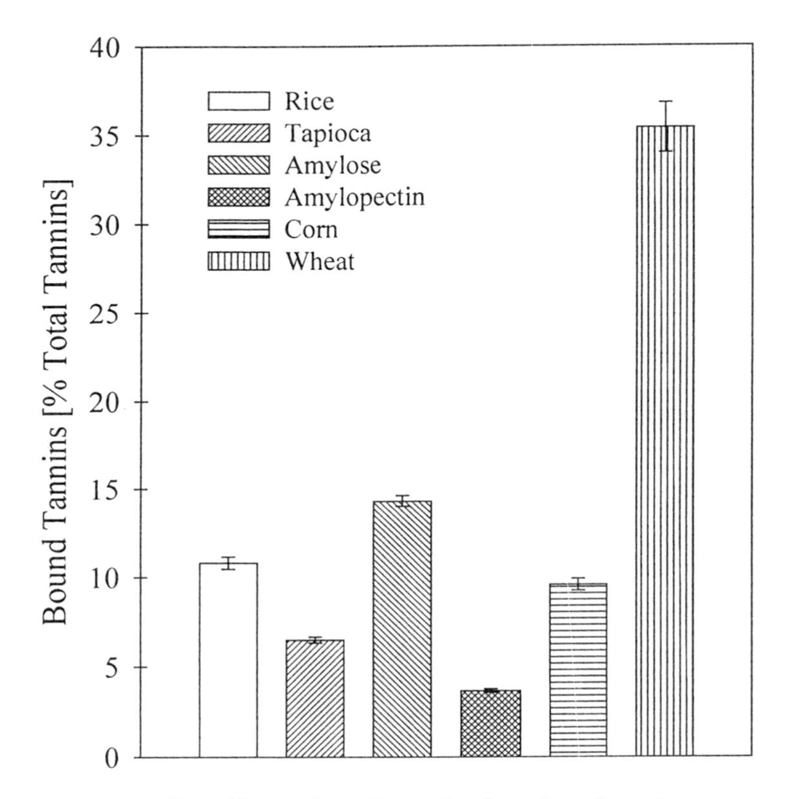

Figure 9. Affinity of canola tannins for selected starches.

increased the efficiency of tannin extraction to 36%. Presence of ammonia in absolute or 95% methanol greatly enhanced the extraction of condensed tannins from rapeseed. Methanol-ammonia-water/ hexane, however, was the most effective solvent system for the removal of tannins as the resultant meals contained 4 - 33% of condensed tannins originally present in the meals. This decrease in tannin content may be due to the extraction of tannins out of the seeds and into the polar phase and/or decomposition of tannins to products that are insensitive to the vanillin reagent. Ghandi et al. (77) have found that ammonia depolimerizes the tannins present in salseed meal. Processed meals so prepared were nontoxic and palatable. Moreover, tannins upon alkali treatment may form phlobaphenes, which are both chemically and nutritionally unreactive compounds (78). On the other hand, Fenwick et al. (43) have reported that treatment of *Brassica napus* meals with ammonia or lime does not affect the tannin contents appreciably.

Acknowledgments

This work was supported, in part, by a grant from the Natural Sciences and Engineering Research Council of Canada to Marion Naczk.

Literature Cited

1. Ohlson, R. *Proc. 5th Intern. Rapeseed Congress,* Malmo, Sweden 1978, Vol.2; pp. 152-167.
2. Shahidi, F.; Naczk, M.; Hall, D.; Synowiecki, J. *Food Chem.* **1992**, *44*, 283-285.
3. Delisle, J.; Amiot, J.; Goulet, G.; Simard, C.,Brisson, G.J.; Jones, J.D. *Qual. Plant.-Plant Foods Human Nutr.* **1984**, *34*, 243-251.
4. Nishie, K.; Daxenbichler, M.E.. Food Cosmet.Toxicol. **1980**, *18*, 159-172.
5. Fenwick, G.R., Spinks,E.A., Wilkinson,A.P., Heaney,R.K.; Legoy, M.A. *J. Sci. Food Agric.* **1986**, 37, 735-741.
6. Oginsky, E.L.; Stein, A.E.; Greer, M.A. *Proc. Soc. Experiment. Biol. and Medicine* **1965**, *119*, 360-364.
7. Rabot, S.; Guerin, C.; Nugon-Baudon, L.; Szylit, O. *Proc. 9th Intern. Rapeseed Congress,* Cambridge, UK, 1995, vol 1; pp. 212-214.
8. Rubin, L.J., Diosady,L.L., Naczk,M.; Halfani, M. *Can. Inst. Food Sci. Technol. J.* **1986**, *19*, 57-61.
9. Shahidi, F.; Naczk, M. In *Canola and Rapeseed: Production, Chemistry, Nutrition, and Processing Technology;* Shahidi, F., Ed.; An Avi Book: New York, 1990; pp. 291-306.
10. Clandinin, D.R. *Poultry Sci.* **1961**, *40*, 484-487.
11. Ismail, F. Vaisey-Genser, M.; Fyfe,B. 1981. *J. Food Sci.* **1981**, *46*, 1241-1244.
12. Malcolmson, L; Vaisey-Genser, M.; Walker, B. 1978. *Proc. 5th Intern. Rapeseed Congress,* Malmo, Sweden, 1978, Vol 2; pp. 147-152.
13. Sosulski, F.W. *J. Am. Oil Chem. Soc.* **1979**, *56*, 711-715.

14. Kozlowska, H.; Naczk, M.; Shahidi, F; Zadernowski, R.; In *Canola and Rapeseed: Production, Chemistry, Nutrition, and Processing*; Shahidi, F., Ed; An Avi Book: New York, 1990; pp 193-210.
15. Smyk, B.; Drabent, R. *Analyst* **1989**, *114*, 723-726.
16. Smyk, B., Amarowicz, R., and Zadernowski, R. Materialy 22 Sesji Naukowej: Procesy Technologiczne a wartosc odzywcza zywnosci. Olsztyn Poland 1991; p. 143; (in polish).
17. Zadernowski, R. *Acta Acad. Agric. Techn. Olst.* **1987**, 21(F), 1-55; (in polish).
18. Kozlowska, H., Sabir, M.A., Sosulski, F.W.;Coxworth, E. **1975**. *Can. Inst. Food Sci.Technol. J.* **1975**, *8*, 160-63.
19. Kozlowska, H.; Rotkiewicz, D.A.; Zadernowski, R.; Sosulski, F.W. *J. Am. Oil Chem. Soc.* **1983**, *60*, 1119-1123.
20. Krygier, K.; Sosulski, F.W.; Hogge, L. *J. Agric. Food Chem.* **1982**, *30*, 334-336.
21. Naczk, M.; Shahidi, F. *Food Chem.* **1989**, *31*, 159-164.
22. Shahidi, F.; Naczk, M. *Bull. de Liason Groupe Polyphenols* **1990**, *15*, 236-239.
23. Bate-Smith, E.C.; Ribereau-Gayon, P. *Qual. Plant. Mater. Vegetable.* **1959**, *5*, 189-198.
24. Durkee, A.B. *Phytochemistry* **1971**, *10*, 1583-1585.
25. Leung, J.; Fenton, T.W.; Mueller, M.M.; Clandinin, D.R. *J. Food Sci.* **1979**, *44*, 1313-1316.
26. Gupta, R.K.; Haslam, E. In *Polyphenols in Cereals and Legumes,* Hulse, J.H., Ed.; International Development Research Center: Ottawa, Canada, 1980; pp. 15- 25.
27. Maxson, S.D.; Rooney, L.W. *Cereal Chem.* **1972**, *49*, 719-729.
28. Price, M.L.; Van Scoyoc, S.; Butler, L.G. *J. Agric. Food Chem.* **1978**, *26*, 1214-1218.
29. Naczk, M.; Shahidi, F.; Sullivan, A. *Food Chem.* **1992,** *45*, 51-55.
30. Mole,S.; Waterman, P.G. *Oecologia*, **1987**, *72*, 137-140.
31. Porter, L.J.; Hrstich, L.N.; Chan, B.G. *Phytochemistry* **1986**, *23*, 223-230.
32. McMurrough, I.; McDowell, J. *Anal. Biochem.* **1978**, *91*, 92-100.
33. Thies, M.; Fischer, R. *Mikrochim. Acta* **1979**, 9-13.
34. Sarkar, S.K.; Howarth, R.E. 1976. *J. Agric. Food Chem.* **1976,** *24*, 317-320.
35. Butler, L.G. *J. Agric. Food Chem.* **1982**, *30*, 1090-1994.
36. Scalbert, A. In *Plant Polyphenols: Synthesis, properties, Significance;* Hemigway, R.W.; Laks, P., Eds.; Plenum Press: New York, 1992; pp. 259-281.
37. Naczk, M.; Nichols, T.; Pink, D.; Sosulski, F. *J. Agric. Food Chem.* **1994**, *42*, 2196-2200.
38. Sosulski, F.W.; Zadernowski, R. *J. Am. Oil Chem. Soc.* **1981**, *58*: 96-98.
39. Mitaru, B.N.; Blair, R.; Bell, J.M.; Reichert, R.D. *Can. J. Animal Sci.* **1982**, *62*, 661-663.

40. Naczk, M.; Shahidi, F. *Abstracts of Papers*; 86th AOCS Annual Meeting, San Antonio, Texas, 1995.
41. AOAC *Official Methods of Analysis*. 10th ed. Association of Official Agricultural Chemists: Washington, DC, 1965.
42. Clandinin, D.R.; Heard, J. *Poultry Sci.* **1968**, *47*, 688-689.
43. Fenwick, R.G.; Hoggan, S.A. *Br. Poultry Sci.* **1976**, *17*, 59-62.
44. Fenwick, G.R.; Curl, C.L.; Pearson, A.W.; Butler, E.J. *J. Sci. Food Agric.* **1984**, *35*, 757-761.
45. Blair, R.; Reichert, R.D. 1984. *J. Sci. Food Agric.* **1984**, *35*, 29-35.
46. Shahidi, F.; Naczk, M. *J. Food Sci.* **1989**, *54*, 1082-1083.
47. Shahidi, F.; Naczk, M. *Bull.de Liason Groupe Polyphenols*. **1988,** *14*, 89-92.
48. Haslam, E.; Lilley, T.H. *CRC Crit. Rev. Food Sci. Nutr.* **1988**, *27*, 1-40.
49. Lea, A.G.H.; Arnold, G.M. *J. Sci. Food Agric.* **1978**, *29*, 478-483.
50. Bate-Smith, E.C. *Phytochemistry* **1973**, *12*, 907-912.
51. Haslam, E. In The *Biochemistry of Plants*; Stumpf P.K.; Conn, E.E., Eds., Academic Press: London, 1981, Vol. 7; p. 527-556.
52. Delcour, J.A., Vandenberghe, M.M., Corten, P.F.; Dodeyne, P.C. *Am. J. Enol. Vitic.* **1984**, *35*, 134-136.
53. Gillooly, M., Bothwell, T.H., Torrance, J.D., MacPhail, A.P, DErman, D.P., Bezwoda, W.R., Mills, W.,Charlton, R.W. ;Mayet, I. *Br. J. Nutr.* **1983**, *49*, 331-342.
54. Hallberg, L. and Rossander, L. *Human Nutr.: Appl. Nutr.* **1982**, *36A*, 116-123.
55. Brune, M.; Rossander, L.; Hallberg, L. *Eur. J. Clin. Nutr.* **1989**, *43*, 547-557.
56. Brune, M.; Hallberg, L.; Skanberg, A.B. *J. Food Sci.* **1991**, *56*, 128-131.
57. Fenwick, G.R.; Pearson, A.W.; Greenwood, N.M.; Butler, E.G. *Anim. Feed Sci. Technol.* **1981**, *6*, 421-423.
58. Fenwick, G.R.; Curl, C.L.; Butler, E.J.; Greenwood, N.M.; Pearson, A.W. *J. Sci. Food Agric.* **1984**, *35*, 749-757.
59. Butler, E.J.; Pearson, A.W.; Fenwick, G.R. *J. Sci. Food Agric.* **1982**, *33*, 866-875.
60. Yapar, Z.; Clandinin, D.R. *Poultry Sci.* **1972,** *51*, 222-228.
61. Loomis, W.D.; Battaile, J. 1966. *Phytochemistry* **1966**, *5*, 423-438.
62. Wade, S.; Tamioka, S.; Moriguchi, I. *Chem. Pharm. Bull.* **1969**, *17*, 320-323.
63. Loomis, W.D. *Methods Enzymol.* **1974**, *31*, 528-544.
64. Goldstein, J.; Swain, T. *Phytochemistry* **1965**, *4*, 185-192.
65. Hagerman, A.E.; Butler, L.G. *J. Agric. Food Chem.* **1980**, *28*, 947-952.
66. Oh, H.I.; Hoff, J.E.; Armstrong, G.S.; Haff, L.A. *J. Agric. Food Chem.* **1980**, *28*, 394-398.
67. McManus, J.P.; Davis, K.G.; Beart, J.E.; Gaffney, S.H.; Lilley, T.H.; Haslam, E. *J. Chem. Soc. Perkin Trans.* **1985**, *2*, 1429-1438.
68. Asquith, T.N.; Butler, L.G. *J. Chem. Ecol.* **1985,** *11*, 1535-1543.

69. Makkar, H.P.S. *J. Agric. Food Chem.* **1989**, *37*, 1197-1202.
70. Hagerman, A.E.; Butler, L.G. *J. Agric. Food Chem.* **1978**, *26*, 809-812.
71. Naczk; M.; Oickle, D.; Shahidi, F. *Proc. 6th Intern. Rapeseed Congress,* Cambridge, UK, Vol. 3, 1995; pp. 882-884.
72. Porter, L.J. In *Methods in Plant Biochemistry,* Harborne, J.B., Ed.; Academic press: San Diego, 1989, Vol. 1; pp. 389-420.
73. Naczk, M.; Oickle, D.; Pink, D.; Shahidi, F. *J. Agric. Food Chem.* **1996**, *44*, 1444-1448.
74. Porter, L.J.; Woodruffe, J. *Phytochemistry,* **1984**, *23*, 1255-1256.
75. Cai, Y.; Lilley, T.H.; Haslam, E. In *Chemistry and Significance of Condensed Tannins;* Hemingway, R.W.; Karchesy, J.J., Eds.; Plenum Press: New York, 1988; pp 307-323.
76. Desphande, S.S.; Salunkhe, D.K. *J. Food Sci.* **1982**, *47*, 2080-2082.
77. Gandhi, V.M.; Cheriyan, K.K.; Mulky, M.J.; Menon, K.K.G. 1975. *J. Oil Technol. Assoc. India (Bombay).* **1975,** *7*, 44-50.
78. Swain, T. In *Herbivores: their interaction with secondary plant metabolites;* Rosenthal, G.A.J.; Janzen, D., Eds; Academic Press: New York, 1979; pp.557-682.

Chapter 12

Methods for Determination of Condensed and Hydrolyzable Tannins

Ann E. Hagerman, Yan Zhao, and Sarah Johnson

Department of Chemistry, Miami University, Oxford OH 45056

Tannins are natural products which are characterized by their phenolic nature and their ability to precipitate protein. Their diverse biological effects, and their common occurrence in plants used to make foods, beverages, herbal medicines, and animal feeds, have created widespread interest in methods for the analysis of tannins. However, their chemical complexity and diversity impose constraints upon the application of many of the analytical methods that have been developed. After briefly reviewing the chemistry of tannins, we describe some of the methods available for analysis of tannins, and outline the limitations of those methods. The discussion of each method emphasizes the utility of the method for analysis of mixtures of tannin, like those found in many foods. A procedure for preparative scale purification of a standard gallotannin, pentagalloyl glucose, is included. A new method for simultaneous determination of condensed and hydrolyzable tannins is summarized.

Distribution and Roles of Tannins

The natural products known as tannins are found in many plants, including some used as foods, herbal medicines or in the production of beverages (*1*,*2*,*3*,*4*,*5*). Many plants consumed by wild or domestic animals contain tannin (*6*,*7*,*8*). The taxonomic distribution of tannins in certain plant families has been surveyed (*9*,*10*,*11*,*12*). Tannins are common in the roots, flowers, leaves and wood of various Gymnosperms and Dicots. Although at one time tannins were not thought to be found in the vegetative tissues of Monocots (*13*), recent reports have demonstrated their presence in leaves of grasses (*14*). Up to 2% by weight of the cereal grains sorghum and barley may be tannin (*15*). Marine brown algae contain compounds known as phlorotannins, which have reactivities similar to those of the tannins of higher plants (*16*). The widespread occurrence of tannins, and their pronounced biological effects, have led some investigators to postulate that they serve to protect plants against herbivores and pathogens (*17,18,19*).

Tannins contribute to the taste of foods and beverages by providing astringency, a sensation of dryness that is especially pronounced in some red wine, tea, and unripe fruit. The sensation apparently results from the interaction between tannins and proteins of the saliva and the mucous tissues of the mouth (*20,21*). Beyond taste, consequences of tannin consumption often include apparent depression in protein utilization (*17,22,23*). A variety of other effects have been reported in some experimental systems (*24,25,26*). Variability in the reported effects of tannins can be attributed in part to the variety of sources and types of tannin used in feeding experiments and the variable responses of different experimental animals (*27,28,29,30,31,32*). Amount of protein in a diet, or steps in diet preparation such as heating (*33*) influence the effects of tannin.

Tannins may contribute to the therapeutic effects of certain herbal medicines (*2,3*). *In vitro* effects of some tannins on some microbes have been noted (*34,35,36*). Consumption of certain tannin-rich beverages, including red wine and tea, has been associated with reduced incidence of heart disease or cancer. Like some smaller phenolics, tannins may serve as dietary antioxidants, and may thus have beneficial effects (*37*).

The common occurrence of tannins in foods, beverages and feeds, and the diverse biological effects of ingested tannin, make qualitative and quantitative analysis of tannins important. The structural chemistry of the tannins, and widely used methods for their analysis, are reviewed here. Before attempting analysis of tannins, it is necessary to develop appropriate techniques for tissue collection, preservation and extraction (*38,39,40,41,42,43*).

Chemistry of Tannins

The term "tannin" was historically used to describe a chemically heterogeneous group of compounds which precipitate proteins. The name derives from the tradition of "tanning" animal hides with infusions of oak or chestnut bark to make leather (*44*). The term tannin was somewhat more rigorously defined by Bate-Smith in 1962 when he stated that tannins are "water-soluble phenolic compounds having molecular weights between 500 and 3,000 and, besides giving the usual phenolic reactions, they have special properties such as the ability to precipitate alkaloids, gelatin and other proteins" (*45*). Modern methods of separation chemistry and spectroscopic structure determination have advanced our knowledge of tannins substantially. It is now realized that several groups of compounds have in common the general properties of tannins as defined by Bate-Smith, but that these groups are quite distinct from one another in their detailed structural chemistry (*44,46*).

Proanthocyanidins. The condensed tannins, or proanthocyanidins (*12,44,47,48*), are flavanol-based compounds. Condensed tannins react in alcoholic solutions of strong mineral acid to release the corresponding anthocyanidin, which has a characteristic color. Structural diversity in the proanthocyanidins is a consequence of the substitution patterns and stereochemistry of the flavanol subunits. Structural complexity also results from the diversity of positions for interflavan bond formation, and from the stereochemical variation in the interflavan bond. The procyanidin (**1**) found in sorghum grain (*49*) is representative of the simplest condensed tannins. It is a linear (4→8) polymer of epicatechin with catechin terminal units. The only commercially available condensed tannin is the profisetinidin (**2**) found in preparations of Quebracho tannin. Quebracho tannin, a crude extract of the bark of *Schinopsis spp.*, contains a complex mixture of phenolics including a branched chain condensed tannin comprised of (4→6)- and (4→8)-linked monomers of the 5-deoxyflavan-3-ol fisetinidol (*50*).

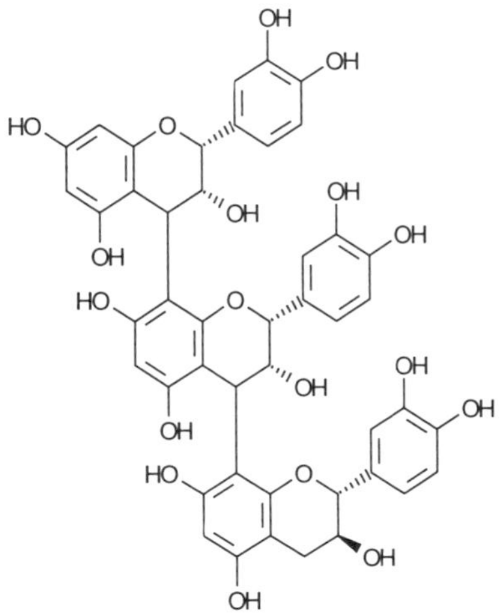

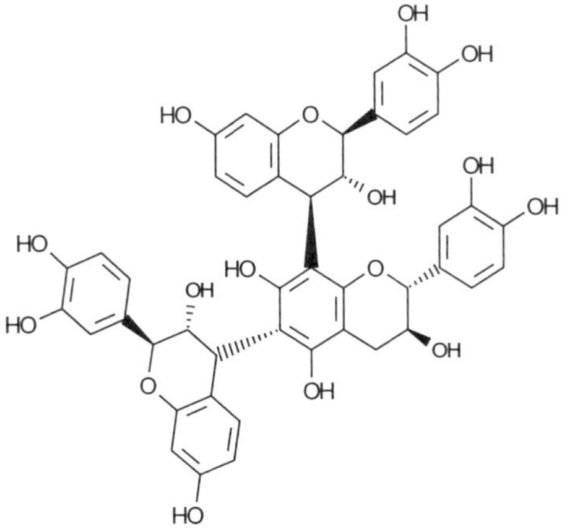

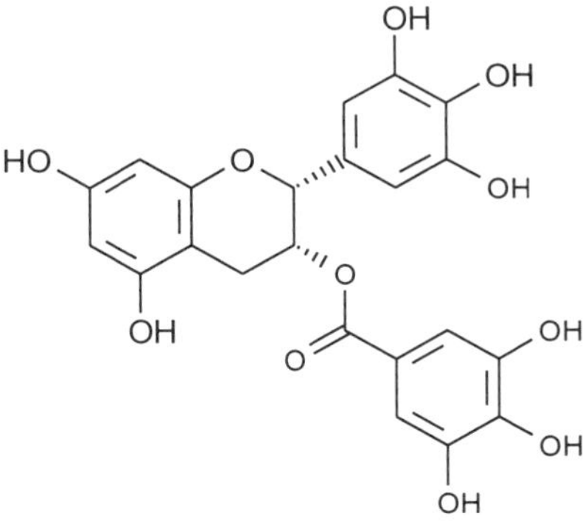

The galloylated catechins are related to the condensed tannins. In these compounds, the 3-hydroxyl group of the flavanol subunit is esterified by gallic acid. Addition of this galloyl group significantly alters reactivity; for example, epigallocatechin does not precipitate protein, but epigallocatechin gallate (**3**) is the principle astringent component of green tea (*51*). Higher molecular weight galloylated catechins have not been characterized.

Hydrolyzable Tannins. The hydrolyzable tannins are comprised of phenolic acids, such as gallic acid or hexahydroxydiphenic acid (HHDP), esterified to a polyol such as glucose. The simple gallotannins, comprised only of galloyl esters of glucose or quinic acid, are relatively rare in nature but are the principal constituents of commercial tannic acid (*52,53*). In the simple gallotannins, the core (e.g. pentagalloyl glucose, **4**) has been further elaborated by addition of galloyl groups

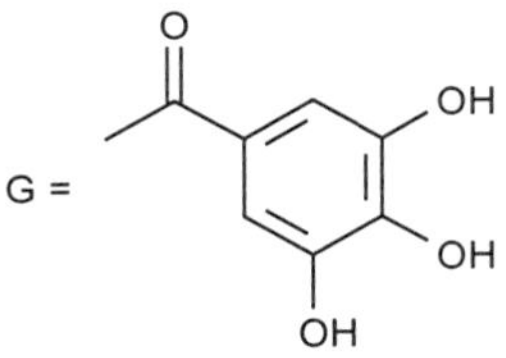

gallic acid 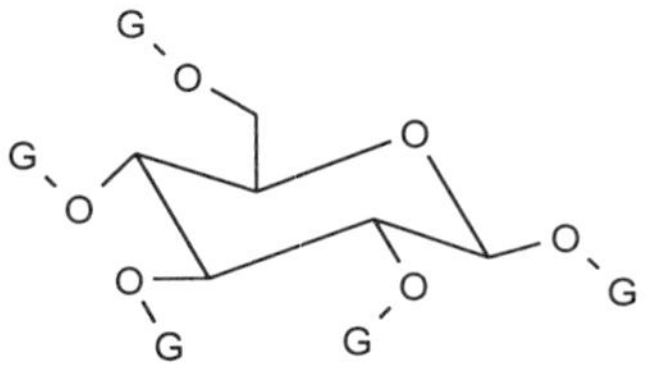 **4**

as "depsides", or esters of phenolic hydroxyls, yielding polyesters containing up to 12 galloyl groups. The ellagitannins such as tellimagrandin II (**5**) are biosynthetically derived from the gallotannins by oxidative coupling of adjacent galloyl residues to yield HHDP esters (*54*). *In vitro* hydrolysis of the ellagitannins

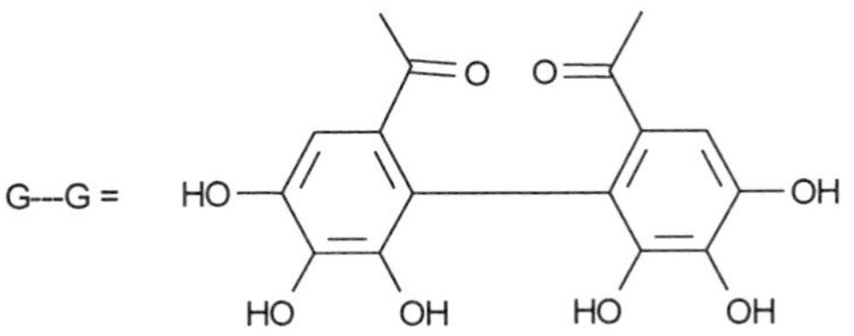

hexahydroxydiphenic acid 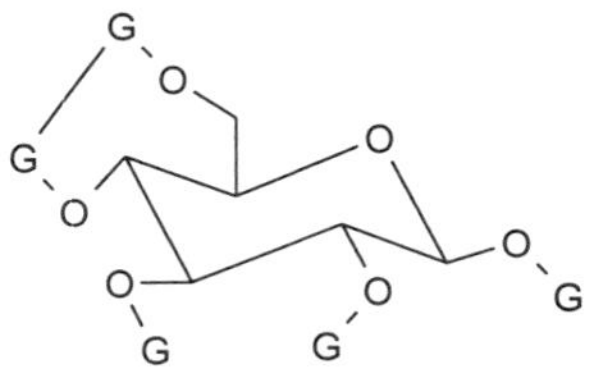 **5**

releases HHDP, which spontaneously lactonizes to yield ellagic acid. Structural diversity in the gallotannins and ellagitannins is a consequence of various degrees and patterns of esterification, and of various patterns of oxidative coupling (*44,54*).

Oligomeric hydrolyzable tannins such as oenothein B (**6**) were unknown before 1982 but may be the most common type of hydrolyzable tannin (*55*). The oligomeric hydrolyzable tannins have at least two ester subunits comprised of polyols esterified with HHDP and/or gallic acid. The ester subunits are linked by oxidative coupling between a phenolic hydroxyl group from one subunit and an aromatic ring carbon from another subunit. The order of linkage of the units, and the stereochemistry of the crosslinking bond contribute to significant structural diversity within this class of tannins.

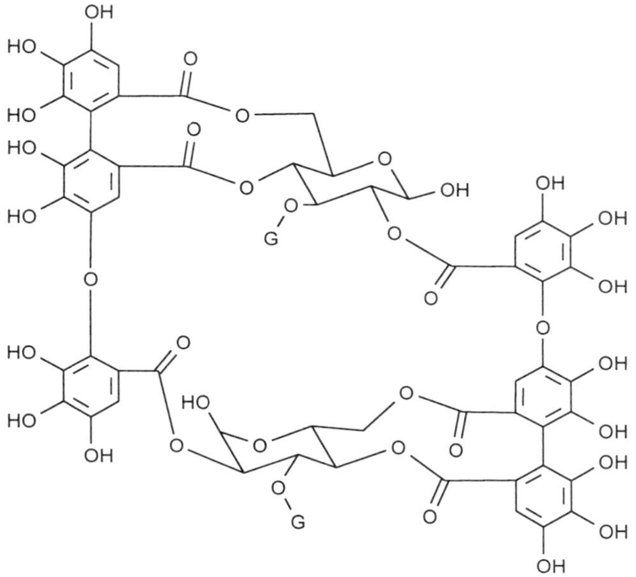

6

Phlorotannins. A final class of compounds which have typical characteristics of tannins are the phlorotannins. These compounds, comprised of aryl-aryl or ether-linked phloroglucinol subunits (e.g. **7**) , are found only in marine brown algae. Like the more familiar terrestrial tannins, they precipitate proteins, and may contribute to both the taste and nutritional quality of brown algae for marine herbivores. Details of their chemistry, analysis and reactions with protein (*16,56*) will not be discussed further here.

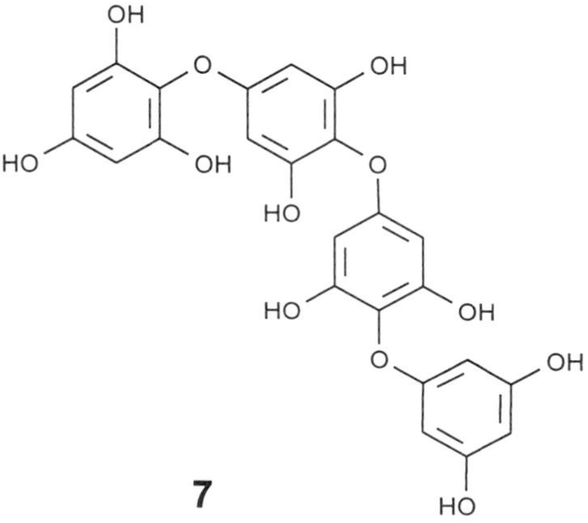

7

Nontannin phenolics. In addition to tannins, plants contain a wide variety of nontannin phenolics. These phenolics, which are usually relatively low molecular weight (<500), are distinguished from the tannins by their inability to precipitate proteins. Dietary nontannin phenolics have different metabolic fates from tannins because of their different reactivity.

Analytical Approaches

Numerous methods for quantitative analysis of tannins have been described (*43,46,57,58,59*). In planning an analysis, it is important to realize that each type of tannin responds differently in each of these assays, consistent with the diversity of structural chemistry of the tannins. This variability in response makes it impossible to use any single method to accurately assesses the "tannin" content of a complex mixture. Instead, several methods must be employed to obtain a qualitative and quantitative picture of the tannins present in the mixture. Complex mixtures are of particular importance since typical higher plants, including those used as foods, beverages and herbal medicines, contain one or more types of tannins and a variety of nontannin phenolics. For example, strawberries (*Fragaria ananassa*) contain both condensed and hydrolyzable tannins (*5*).

In general, the methods used for tannin analysis are based on either general phenolic reactions; protein precipitation; functional group reactions; or HPLC. Several of the most widely used methods in analysis of tannins are discussed below, with emphasis on the utility of the methods for analyzing tannins in complex mixtures. For each method of analysis, proper standards must be selected to allow meaningful interpretation of the data (*59*).

Total Phenolics Methods. The Prussian blue method (*60*) or the Folin-Denis method (*61*) are used to measure both tannin and nontannin phenolics in plant extracts. In both methods, the phenolic analyte is oxidized and the reagent is reduced to form a chromophore. Easily oxidized nonphenolics such as ascorbic acid interfere with the methods. The Folin method is plagued by formation of precipitates which interfere with spectrophotometric determination, and many variations on the original method have been developed in attempts to minimize this problem (*43*). Interferences are rare with the Prussian blue method, and a recent modification to diminish time dependence of color formation simplifies the method (*62*).

Since the pattern of substitution affects the redox chemistry of phenolics (*60*), significant differences in response in these total phenolic methods must be expected for different tannins. For example, with the Prussian blue assay, the color yield for epicatechin is only 70% of the color yield for gallic acid. Furthermore, the redox methods provide neither a means to distinguish tannins (protein precipitating phenolics) from nontannin phenolics, nor a means to identify specific types of tannins in a mixture.

Methods based on formation of colored phenolic-metal ion complexes are also useful for measuring total phenolics. Several methods dependent on formation of the ferric-phenolate complex at high pH have been described (*63,64*). These methods are useful for any phenolic containing *ortho*-substituted phenolics, and interference from nonphenolics is unlikely. Nontannin phenolics cannot be distinguished from tannin with this method, and specific types of tannin cannot be identified. It has been suggested that condensed tannin can be distinguished from hydrolyzable tannin by formation of different colored ferric ion complexes in neutral solution (*64*) but that idea has not been thoroughly evaluated, and must be used with caution (*65*).

Precipitation Methods. The defining reaction of tannins is their ability to precipitate protein (*66,67*), and many methods for determining tannin have been developed based on reactions with protein (*59,68*). Among the simplest of these methods is one in which the phenolic is determined after precipitation by a standard protein. The precipitated phenolic is redissolved in basic solution, and reacted with ferric ion to form a colored complex (*63*). This method discriminates tannin from nontannin phenolics, but does not allow selective determination of various types of tannin.

Several methods for determining the amount of protein precipitated by tannins have been described. Since tannin and protein are very similar in their reactivity to many reagents, it is essential to either use protein that is labeled in some fashion or to isolate the protein after precipitation. Proteins can be labeled with a radioisotope (*69*) or with a chromophore (*70,71*). Separation achieved chromatographically (*72*) or with a membrane (*73*) can be followed by conventional analysis of the protein. None of these methods allow selective determination of various types of tannin.

All precipitation methods are sensitive to reaction conditions. For example, the amount of protein precipitated by a given amount of tannin may depend on the pH (*63*), the presence of organic solvent (*74*), the ionic strength (*73*), and the time allowed for precipitation (*74*). The ratio of tannin to protein influences the precipitation reaction, with soluble complexes forming preferentially when protein is in excess and precipitable complexes forming when tannin is in excess (*74*). The interaction between tannin and protein is quite specific (*75,76*), so that structural variations in either tannin or in protein may alter the stoichiometry and solubility of the complex.

Although the ability of tannins to precipitate proteins is widely recognized, under some conditions soluble complexes between tannin and protein may form. Attempts to analyze soluble complexes electrophoretically (*28*) are limited by their qualitative nature. More success has been obtained using competitive binding assays (*75*).

Functional Group Methods. These methods depend on the specific chemical reactivity of the functional group characteristic of a given type of tannin. As a result, these methods are often very specific and provide both qualitative and quantitative information on the tannins found in a plant sample.

Condensed Tannins. The vanillin assay (*77*) is a method for determining proanthocyanidins. Aromatic aldehydes such as vanillin react with *meta*-substituted phenolics to yield highly conjugated colored products. The method has been used to determine condensed tannin because many proanthocyanidins have an appropriate *meta*-substitution on the "A" ring (**1**) and thus react with vanillin. However, interpretation of the results is complicated by the fact that vanillin reacts with nontannin flavanols such as catechin, but does not react with proanthocyanidins based on 5-deoxyflavanols, such as the profisetinidin in Quebracho tannin (**2**). Additional problems stemming from the complex kinetics of the reaction with vanillin, and from the sensitivity of the reaction to trace amounts of water, limit the utility of the method.

The acid butanol method is a simple, specific method for determining proanthocyanidins (*78*). The method depends on the oxidative cleavage of the interflavan bond in alcoholic solutions of hot mineral acid to yield the colored anthocyanidin. Reaction conditions must be carefully controlled; for example, traces of iron catalyze the reaction while water inhibits the reaction (*78*). The response is dependent on the structure of the tannin, with profisetinidins yielding less color because of the increased stability of the (4→6) interflavan bond typical of 5-deoxyflavanols (*58*). A method based on similar chemistry but utilizing sulfuric acid instead of HCl/butanol (*79*) is apparently less sensitive to the presence of water.

The acid butanol method is satisfactory for selectively determining condensed tannins in the presence of hydrolyzable tannins. Most nontannin phenolics do not interfere with the method, although anthocyanidins and other pigments can interfere. A modification of the acid butanol method for determining proanthocyanidins in the presence of chlorophyll has been described (*80*). That method can also be used to determine the unusual 3-deoxy proanthocyanidins (*12*) and the nontannin leucoanthocyanidins (flavan-3,4-diols and flavan-4-ols).

None of these methods provide unambiguous information on the composition or size of the condensed tannins. Although NMR and mass spectroscopy are powerful tools for analysis of proanthocyanidins (*46,81*), the spectra of polymeric condensed tannins are difficult to interpret. Chemical methods for evaluating composition and estimating molecular weight are more accessible. The degree of polymerization of procyanidins can be estimated by comparing the results from the vanillin/glacial acetic acid assay (in which only terminal units react to form chromophore) with the results from the acid butanol assay (in which only extender units react to form chromophore) (*82,83*). The method can only be used to compare chemically similar tannins, since reactivity in the acid butanol assay is a function of the reactivity of the interflavan bond. Degree of polymerization can also be determined by gel permeation chromatography of acetylated condensed tannins (*84*). Proanthocyanidins can be oxidatively cleaved in the presence of a nucleophile such as phloroglucinol to yield the underivitized terminal flavanol and the phloroglucinol adducts of the extender flavanols. These products can be separated and quantitated by HPLC to yield information on the composition and degree of polymerization of the parent tannin (*85*).

Hydrolyzable Tannins. The iodate assay (*86*) relies on the reaction of potassium iodate with gallate esters to yield a chromophore. The chemistry of the color-producing reaction is poorly understood. The method cannot be employed with samples containing complex mixtures of tannins, because samples turn brown instead of the characteristic pink (*87*). A method for determining intact ellagitannins relies on color changes in the presence of nitrous acid (*88*).

The rhodanine assay and nitrous acid assays were devised to allow estimation of gallotannins or ellagitannins in mixtures (*89,90*). The basis for each assay is selective determination of the phenolic acid released by hydrolysis of the parent hydrolyzable tannin. Successful hydrolysis is critical to the methods. The acid hydrolysis must be carried out in the absence of oxygen, since gallic acid and HHDP are sensitive to oxygen. Following hydrolysis, gallic acid can be reacted with rhodanine to form a chromophore. Ellagic acid can be isolated using pyridine and then reacted with nitrous acid to form a chromophore. Each of the methods is very selective, allowing determination of the esterified forms of the specific acid of interest. Unfortunately, sample hydrolysis is difficult and inconvenient for large numbers of samples. In addition, the results do not provide information on the degree of esterification or structure of the parent tannin. Degree of esterification can be a critical determinant of biological activity of hydrolyzable tannins (*29*).

The degree of esterification of an unknown mixture of gallotannins can be determined using either normal phase (*29,52*) or reversed phase HPLC (*91*). In each of these chromatographic systems, the log of the retention time is a linear function of the degree of esterification. The methods can be standardized using commercially available methyl gallate and pentagalloyl glucose.

Pentagalloyl glucose is prepared from commerical tannic acid using a method modified from Yoshizawa *et al.* (*92*). Tannic acid from *Rhus semialata* galls ("Chinese" tannin) is a suitable starting material since it is comprised of higher esters of pentagalloyl glucose (*54*). However, "Turkish" or "Aleppo" tannins, from *Quercus infectoria* galls, are based on a tetragalloyl glucose core and do not yield pentagalloyl glucose (*52,53*). The tannic acid (0.5 g) is dissolved in 10 mL of 70% methanol/30% acetate buffer (0.1 M acetate, pH 5.0) and is incubated at 65°C for 15 h. The solution is then stirred while adjusting the pH to 6.0 with 0.25 M NaOH, and is evaporated under reduced pressure while maintaining the temperature at <30°C. As the methanol is removed, water is added to maintain the volume. The aqueous solution is extracted three times with diethyl ether, and is then extracted three times with ethyl acetate. The ethyl acetate fractions are combined and evaporated under reduced pressure. Water is added to maintain the volume as the ethyl acetate is removed. When all the organic solvent has been removed, the aqueous suspension is freeze dried to yield pure pentagalloyl glucose. Proton NMR is used to confirm the identity of the purified material [from TMS: Glucose C-1,

6.3 ppm (d, 1H); glucose C-2, C-4, 5.6 ppm (q, 2H); glucose C-3, 6.0 ppm (t, 1H); glucose C-5, 4.5 ppm (d, 1H); glucose C-6, 4.4 ppm (dd, 1H); galloyl group, between 6.9-7.2 ppm, 5 singlets (2H)].

Many of the smaller ellagitannins are easily crystallized to yield pure compounds which can be identified by NMR and MS (*57*). Methods for HPLC of ellagitannins have been described (*93,94,95*) but are not useful for identification of unknowns.

None of the functional group methods are satisfactory for the oligomeric hydrolyzable tannins. Although gallic acid and ellagic acid may be released from oligomeric hydrolyzable tannins, other complex phenolic acids are major products of these compounds. For example, valoneic acid (**8**), a hydrolysis product of oenothein B (**6**), is unreactive in all these functional group assays. The oligomeric hydrolyzable tannins can be separated using the HPLC methods described for gallotannins and ellagitannins, but the results are not useful for identification of unknowns. NMR and MS can be used to establish structures of purified complex tannins (*57*).

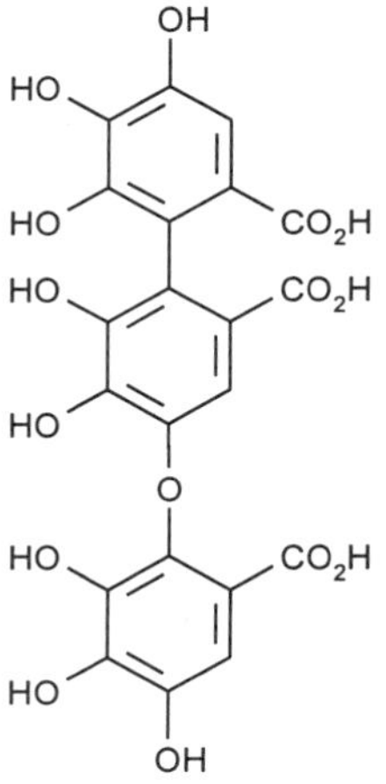

8

Evaluation of Mixtures of Tannins. It is possible to use the array of chemical and spectroscopic methods described here to qualitatively and quantitatively describe the tannins in an extract. In many instances, assessment of the gross composition of the tannin is of more importance than detailed chemical evaluation. The acid butanol method provides a quick, reliable method for determining proanthocyanidins in extracts. Because it is more difficult to perform the functional group assays necessary for hydrolyzable tannins, this group of tannins is often neglected (*96*).

In an effort to simplify screening of samples for hydrolyzable tannins, we have developed a new method for rapid assessment of crude tannin extracts. The extracts are analyzed with the radial diffusion assay (*97*), a simple protein precipitation method in which the plant extract is allowed to diffuse through a protein-containing gel. Tannins in the extract precipitate the protein in the gel to form a visible ring. The diameter of the ring is proportional to the amount of tannin in the extract. This method as originally developed is not useful for discriminating condensed tannins from hydrolyzable tannins, since both types of tannin form rings of precipitate.

Hydrolyzable tannins, like other esters, are susceptible to cleavage under mild conditions with hydroxylamine hydrochloride (*98*). Upon reaction, the hydrolyzable tannins yield the core polyol and the hydroxamic acid of the phenolic residues. Simple gallotannins are completely decomposed by reaction with hydroxylamine hydrochloride under mild conditions (pH 5.5, 70°C, 48h) while proanthocyanidins are unchanged (*99*).

To screen samples for the presence of hydrolyzable tannins, the radial diffusion method is performed twice for each extract. One sample of the extract is assayed directly, and a second sample of the extract is assayed after treatment with hydroxylamine hydrochloride. The precipitation ring obtained with the first sample represents all of the tannin in the extract. If a smaller precipitation ring is obtained with the hydroxylamine-treated extract, it suggests that the extract contained some hydrolyzable tannins which were destroyed by the hydroxylamine hydrochloride. If the size of the ring is not changed by the hydroxylamine hydrochloride treatment, the sample did not contain hydrolyzable tannins.

Although this method is not a substitute for complete chemical characterization of tannins, it does provide a simple method for screening for hydrolyzable tannins. Availability of such a simple method should ensure that this important group of tannins receives more attention in future studies.

Literature Cited

1. Singleton, V. L. In *Plant Polyphenols*; Hemingway, R. L.; Laks, P. E., Eds.; Plenum Press: NY, NY, 1992; pp 859-880.
2. Haslam, E.; Lilley, T.H.; Cai, Y.; Martin, R.; Magnolato, D. *Planta Medica* **1989**, *55*, 1-8.
3. Okuda, T.; Yoshida, T.; Hatano, T. *Planta Medica* **1989**, *55*, 117-122.
4. Pierpoint, W.S. In *Flavonoids in Biology and Medicine III. Current Issues in Flavonoid Research*; Das, N.P., Ed.; National University of Singapore: Singapore, 1990; pp 497-514.
5. Foo, L.Y.; Porter, L.J. *J. Sci. Food Agric.* **1981**, *32*, 711-716.
6. Degen, A.A.; Becker, K.; Makkar, H.P.S.; Borowy, N. *J. Sci. Food Agric.* **1995**, *68*, 65-71.
7. Foo, L.; Jones, W.T.; Porter, L.J.; Williams, V.M. *Phytochemistry* **1982**, *21*, 933-935.
8. Kumar, R. *J. Agric. Food Chem.* **1983**, *31*, 1364-1366.
9. Bate-Smith, E.C. *Phytochemistry* **1978**, *17*, 1945-1948.
10. Mole, S. *Biochem. Syst. Ecol.* **1993**, *21*, 833-846.
11. Okuda, T.; Yoshida, T.; Hatano, T.; Iwasaki, M.; Kubo, M.; Orime, T.; Yoshizaki, M.; Naruhashi, N. *Phytochemistry* **1992**, *31*, 3091-3096.
12. Stafford, H.A. *Flavonoid Metabolism;* CRC Press: Boca Raton, FL, 1990; 298 p.
13. Jung, H.G.; Batzli, G.O.; Seigler, D.S. *Biochem. Syst. Ecol.* **1979**, *7*, 203-209.
14. Iason, G.R.; Hodgson, J.; Barry, T.N. *J. Chem. Ecol.* **1995**, *21*, 1103-1112.
15. Butler, L.G. In *Toxicants of Plant Origin*; Cheeke, P.R. Ed.; CRC Press: Boca Raton, FL, 1989; pp 95-121.
16. Ragan, M.A.; Glombitza, K. *Prog. Phycologic. Res.* **1986**, *4*, 177-241.
17. Bernays, E.A.; Cooper-Driver, G.; Bilgener, M. *Adv. Ecol. Res.* **1989**, 19, 263-302.

18. Harborne, J.B. In *Plant Defenses against Mammalian Herbivory*; Palo, R.T.; Robbins, C.T. Eds.; CRC Press: Boca Raton, FL, 1991; pp 45-60.
19. Schultz, J.C. *Ecology* **1988**, *69*, 896-897.
20. Luck, G.; Liao, H.; Murray, N.J.; Grimmer, H.R.; Warminski, E.E.; Williamson, M.P.; Lilley, T.H.; Haslam, E. *Phytochemistry* **1994**, *37*, 357-371.
21. Mehansho, H.; Butler, L.G.; Carlson, D.M. *Ann. Rev. Nut.* **1987**, *7*, 423-440.
22. Mole, S.; Waterman, P.G. In *Tannins as Allelochemicals in Agriculture, Forestry and Ecology*; Waller, G.R. Ed.; American Chemical Society: Washington, DC, 1987; pp 572-587.
23. Robbins, C.T.; Hanley, T.A.; Hagerman, A.E.; Hjeljord, O.; Baker, D.L.; Schwartz, C.C.; Mautz, W.W. *Ecology* **1987**, *68*, 98-107.
24. Makkar, H.P.S.; Singh, B.; Dawra, R.K. *Brit. J. Nut.* **1988**, *60*, 287-296.
25. Mehansho, H.; Hagerman, A.; Clements, S.; Butler, L.; Rogler, J.; Carlson, D.M. *Proc. Nat. Acad. Sci. U. S. A.* **1983**, *80*, 3948-3952.
26. Mole, S.; Rogler, J.; Butler, L. In *Flavonoids in Biology & Medicine III. Current Issues in Flavonoid Research*; Das, N.P., Ed.; National University of Singapore: Singapore, 1990; pp 581-589.
27. Mehansho, H.; Ann, D.K.; Butler, L.G.; Rogler, J.C.; Carlson, D.M. *J. Biol. Chem.* **1987**, *262*, 12344-12350.
28. Hagerman, A.E.; Robbins, C.T. *Can. J. Zool.* **1993**, *71*, 628-633.
29. Hagerman, A.E.; Robbins, C.T.; Weerasuriya, Y.; Wilson, T.C.; McArthur, C. *J. Range Manag.* **1992**, *45*, 57-62.
30. Mole, S.; Rogler, J.C.; Butler, L.G. *Biochem. Syst. Ecol.* **1993**, *21*, 667-677.
31. Robbins, C.T.; Hagerman, A.E.; Austin, P.J.; McArthur, C.; Hanley, T.A. *J. Mammal.* **1991**, *72*, 480-486.
32. VanAltena, I.A.; Steinberg, P.D. *Biochem. Syst. Ecol.* **1992**, *20*, 493-499.
33. Dietz, B.A.; Hagerman, A.E.; Barrett, G.W. *J. Mammal.* **1994**, *75*, 880-889.
34. Sakagami, H.; Nakashima, H.; Murakami, T.; Yamamoto, N.; Hatano, T.; Yoshida, T.; Okuda, T. *J. Pharmacobiodyn.* **1992**, *15*, S5.
35. Scalbert, A. *Phytochemistry* **1991**, *30*, 3875-3883.
36. Tsai, Y.J.; Aoki, T.; Maruta, H.; Abe, H.; Sakagami, H.; Hatano, T.; Okuda, T.; Tanuma, S. *J. Biol. Chem.* **1992**, *267*, 14436-14442.
37. Okuda, T.; Yoshida, T.; Hatano, T. In *Active Oxygens, Lipid Peroxides, and Antioxidants*; Yagi, K. Ed.; CRC Press: Boca Raton, FL, 1993; pp 333-346.
38. Cork, S.J.; Krockenberger, A.K. *J. Chem. Ecol.* **1991**, *17*, 123-134.
39. Hagerman, A.E. *J. Chem. Ecol.* **1988**, *14*, 453-462.
40. Makkar, H.P.S.; Singh, B. *J. Sci. Food Agric.* **1991**, *54*, 323-328.
41. Orians, C.M. *J. Chem. Ecol.* **1995**, *21*, 1235-1243.
42. Torti, S.D.; Dearing, M.D.; Kursar, T.A. *J. Chem. Ecol.* **1995**, *21*, 117-125.
43. Waterman, P.G.; Mole, S. *Analysis of Phenolic Plant Metabolites;* Blackwell: Oxford, U. K., 1994; 238 p.
44. Haslam, E. *Plant Polyphenols: Vegetable Tannins Revisited;* Cambridge University Press: Cambridge, U. K., 1989; 230 p.

45. Bate-Smith, E.C.; Swain, T. In *Comparative Biochemistry*; Mason, H.S.; Florkin, A.M. Eds.; Academic Press: NY, NY, 1962; pp 755-809.
46. Porter, L.J. In *Methods in Plant Biochemistry. Plant Phenolics*; Harborne, J.B. Ed.; Academic Press: NY, NY, 1989; pp 389-419.
47. Hagerman, A.E.; Butler, L.G. *Meth. Enz.* **1994**, *234*, 429-437.
48. Porter, L.J. In *The Flavonoids*; Harborne, J.B. Ed.; Chapman and Hall Ltd.: London, UK, 1988; pp 21-62.
49. Gupta, R.K.; Haslam, E. *J. Chem. Soc. Perk. Trans. I* **1978**, 892-896.
50. Hemingway, R. W. In *Chemistry and Significance of Condensed Tannins*; R. W. Hemingway; J. J. Karchesy, Eds.; Plenum Press: NY, NY, 1989; pp 265-284.
51. Sakata, I.; Ikeuchi, M.; Maruyama, I.; Okuda, T. *Yakugaku Zasshi-J Pharm Soc J* **1991**, *111*, 790-793.
52. Beasely, T. H.; Ziegler, H. W.; Bell, A. D. *Anal. Chem.* **1977**, 49, 238-243.
53. Armitage, R.; Bayliss, G.S.; Gramshaw, J.S.; Haslam, E.; Haworth, R.D.; Jones, K.; Rogers, H.J.; Searle, T. *J. Chem. Soc.* **1961**, 1842-1853.
54. Haddock, E.A.; Gupta, R.K.; Al-Shafi, S.M.K.; Layden, K.; Haslam, E.; Magnolato, D. *Phytochemistry* **1982**, *21*, 1049-1062.
55. Okuda, T.; Yoshida, T.; Hatano, T. *Phytochemistry* **1993**, *32*, 507-521.
56. Stern, J.L.; Hagerman, A.E.; Steinberg, P.D.; Mason, P.K. Phlorotannin-protein interactions. *J. Chem. Ecol.* **1996**, in press.
57. Okuda, T.; Yoshida, T.; Hatano, T. *J. Nat. Prod.* **1989**, *52*, 1-31.
58. Scalbert, A. In Plant Polyphenols; Hemingway, R. W.; Laks, P. E. Eds.; Plenum Press, NY, NY, 1992; pp. 259-280.
59. Hagerman, A. E.; Butler, L. G. *J. Chem. Ecol.* **1989**, *15*, 1795-1810.
60. Price, M.P.; Butler, L.G. *J. Agric. Food Chem.* **1977**, *25*, 1268-1273.
61. Ribereau-Gayon, P. *Plant Phenolics*. Oliver & Boyd: Edinburgh, U.K., 1972; 254 p.
62. Graham, H.D. *J. Agric. Food Chem.* **1992**, *40*, 801-805.
63. Hagerman, A.E.; Butler, L.G. *J. Agric. Food Chem.* **1978**, *26*, 809-812.
64. Mole, S.; Waterman, P.G. *Oecologia* **1987**, *72*, 137-147.
65. Grove, J.F.; Pople, M. *Phytochemistry* **1979**, *18*, 1071-1072.
66. Hagerman, A.E. In *Phenolic Compounds in Food and Their Effects on Health I. Analysis, Occurrence, and Chemistry*. Ho, C.T.; Lee, C.Y.; Huang, M.T. Eds.; American Chemical Society: Washington, DC, 1992; pp 236-247.
67. Spencer, C.M.; Cai, Y.; Martin, R.; Gaffney, S.H.; Goulding, P.N.; Magnolato, D.; Lilley, T.H.; Haslam, E. *Phytochemistry* **1988**, *27*, 2397-2409.
68. Makkar, H.P.S. *J. Agric. Food Chem.* **1989**, *37*, 1197-1202.
69. Hagerman, A.E.; Butler, L.G. *J. Agric. Food Chem.* **1980**, *28*, 944-947.
70. Asquith, T.N.; Butler, L.G. *J. Chem. Ecol.* **1985**, *11*, 1535-1544.
71. Schultz, F.C.; Baldwin, I.T.; Nothnagle, P.J. *J. Agric. Food Chem.* **1981**, *29*, 823-826.
72. Martin, J.S.; Martin, M.M. *J. Chem. Ecol.* **1983**, *9*, 285-294.
73. Martin, M.M.; Rockholm, D.C.; Martin, J.S. *J. Chem. Ecol.* **1985**, *11*, 485-494.

74. Hagerman, A.E.; Robbins, C.T. *J. Chem. Ecol.* **1987**, *13*, 1243-1259.
75. Hagerman, A.E.; Butler, L.G. *J. Biol. Chem.* **1981**, *256*, 4494-4497.
76. Asquith, T.N.; Butler, L.G. *Phytochemistry* **1986**, *25*, 1591-1593.
77. Price, M.L.; Van Scoyoc, S.; Butler, L.G. *J. Agric. Food Chem.* **1978**, *26*, 1214-1218.
78. Porter, L.J.; Hrstich, L.N.; Chan, B.C. *Phytochemistry* **1986**, *25*, 223-230.
79. Bae, H..D.; McAllister, T.A.; Muir, A.D.; Yanke, L.J.; Bassendowski, K.A.; Cheng, K..J. *J. Agric. Food Chem.* **1993**, *41*, 1256-1260.
80. Watterson, J.J.; Butler, L.G. *J. Agric. Food Chem.* **1983**, *31*, 41-45.
81. Self, R.; Eagles, J.; Galletti, G.C.; Mueller-Harvey, I.; Hartley, R.D.; Lea, A.G.H.; Magnolato, D.; Richli, U.; Gujer, R.; Haslam, E. *Biomed. Environ. Mass Spec.* **1986**, *13*, 449-468.
82. Butler, L.G.; Price, M.P.; Brotherton, J.E. *J. Agric. Food Chem.* **1982**, *30*, 1087-1089.
83. Butler, L.G. *J. Agric. Food Chem.* **1982**, *30*, 1090-1094.
84. Williams, V.M.; Porter, L.J.; Hemingway, R.W. *Phytochemistry* **1983**, *22*, 569-572.
85. Koupai-Abyazani, M.R.; McCallum, J.; Bohm, B.A. *J. Chrom.* **1992**, *594*, 117-123.
86. Bate-Smith, E.C. *Phytochemistry* **1977**, *16*, 1421-1426.
87. Haslam, E. *Phytochemistry* **1965**, *4*, 495-498.
88. Bate-Smith, E.C. *Phytochemistry* **1972**, *11*, 1153-1156.
89. Inoue, K.H.; Hagerman, A.E. *Anal. Biochem.* **1988**, *169*, 363-369.
90. Wilson, T.C.; Hagerman, A.E. *J. Agric. Food Chem.* **1990**, *38*, 1678-1683.
91. Barbehenn, R.V.; Martin, M.M. *J. Insect Physiol.* **1992**, *38,* 973-908.
92. Yoshizawa, S.; Horiuchi, T.; Suganuma, M.; Nishiwaki, S.; Yatsunami, J.; Okabe, S.; Okuda, T.; Muto, Y.; Frenkel, K.; Troll, W.; Fujiki, H. In *Phenolic Compounds in Food and their Effects on Health II. Antioxidants and Cancer Prevention*; Huang, M.T.; Ho, C.T.; Lee, C.Y. Eds.; American Chemical Society: Washington, DC, 1992; pp 316-325.
93. Hatano, T.; Yoshida, T.; Okuda, T. *J. Chrom.* **1988**, *435*, 285-295.
94. Okuda, T.; Mori, K.; Seno, K.; Hatano, T. *J. Chrom.* **1979**, *171*, 313-320.
95. Scalbert, A.; Duval, L.; Peng, S.; Monties, B.; DuPenhoat, C. *J. Chrom.* **1990**, *502*, 107-119.
96. Martin, J.S.; Martin, M.M. *Oecologia (Berlin)* **1982**, *54*, 205-211.
97. Hagerman, A.E. *J. Chem. Ecol.* **1987**, *13*, 437-449.
98. March, J. *Advanced Organic Chemistry: Reactions, Mechanisms, and Structure*, 3rd ed. Wiley: NY, NY, 1985; pp 370-375.
99. Zhao, Y. *Simultaneous Determination of Hydrolyzable Tannin and Condensed Tannin*; M.S. Thesis; Miami University, Oxford, OH, 1995; 90 p.

Chapter 13

Lawsone: Phenolic of Henna and Its Potential Use in Protein-Rich Foods and Staining

Rashda Ali and Syed Asad Sayeed

Department of Food Science and Technology, University of Karachi, Karachi–75270, Pakistan

The leaves of *Lawsonia inermis* have been used since ancient times in the Indian Subcontinent for decorating and dyeing hands, soles, beard and hair and to impart beautiful shades of dark red color. An aqueous mixture is often used for medicinal purposes focusing various skin ailments. The lawsone, a brown color powder, was isolated from the leaves of *L. inermis* and was used as a staining agent for the resolved proteins in polyacrylamide gel electrophoresis (PAGE). It exhibited an excellent affinity to bind with a majority of the known proteins such as albumin, blood serum protein, keratin and casein.

The leaves of henna *(Lawsonia inermis)* are widely used in Pakistan and India for decorating the hands and feet of brides and ladies on special occasions such as wedding and Eid festivals. The application of henna gave us the novel idea that the extract must have some dye which binds to proteins (*i.e.*, keratin and collagen) of hair, wool, skin and nail. As expected, the modified leaf extract when applied to resolved proteins in polyacrylamide gel electrophoresis (PAGE) produced very successful results.

Ali and Sayeed (*1*) first reported the interaction between lawsone and protein molecules, and used lawsone for staining of known proteins such as bovine serum albumin and casein in electrophoresis. The results showed the superiority of the dye as compared with those of Coomassie brilliant blue R-250 or amido black which are traditionally used for detection of resolved protein bands in electrophoresis (*2*). The protein bands are sharper in appearance, last longer and have clear boundaries; moreover, the procedure is simple, rapid and dependable as results can be easily reproduced.

Origin of Lawsone

The slurry of henna leaves, commonly known as "Mehndi", has been used since ancient times. It seems that the extract of leaves used for dyeing wool, silk and animal skin was first reported in the literature by Hoffstein (*3*) and Tommasi (*4*). The wool is generally immersed in lawsone (*i.e.*, a coloring constituent of an aqueous extract) for less than half an hour and is stained quite effectively. Moreover, it has been found that the rate of dye absorption depends upon pH and the time allocated for diffusion of the pigment (*5*).

Although the specific binding site in the structure of the wool and the active site of the dye itself has not yet been explored, it is evident that the interaction of the two molecules takes place at variable pH and that both the intensity and color retention are pH dependent. The extract imparts a fast color in an acid bath while the alkaline medium, although it intensifies the color, limits the binding capacity of the fiber and dye (*6*).

The aqueous mixture is presently employed for dyeing hair as well as for decorative purposes (*7*). Furthermore, the mixture offers good medicinal benefits as it is believed to cure some skin disorders and dandruff. The application of the extract provides cooling and soothing effects in the summer (*8*). Keratin, the protein of skin, hair and nail, is responsible for interaction with lawsone, the coloring material of henna.

Occurrence and Growth

Lawsonia inermis, known as henna, is a bush-like tree with numerous branches, thickly surrounded by small (2-4 cm) green leaves; it appears more like a shrub. In view of the dense accumulation of leaves, it is grown as a hedge in many tropical countries like Egypt, Sudan, India, Pakistan and Australia. Once rooted, the plant continues to survive for over a 100 years or more (*9*).

Mikailov and Allakhverdiev (*10*) suggested that the growth conditions affected the lawsone content of the leaves which varied from region to region. The highest content of lawsone is estimated to be 0.8-0.95%, on a dry weight basis, in leaves from plants grown in Nakhichevan (part of Russia). The amount decreased to 0.62-0.64% for henna grown in Shirvan (also part of Russia) and the yield of the lawsone dropped even further to only 0.5-0.56% (*10*) when grown in Sheron.

Karawya *et al.* (*11*) reported that the lawsone content of henna leaves varies from 0.75 to 0.95% in Edfo and Assuit (a region of Egypt where the temperature is higher), while in the other regions such as Cairo, Mett and Kanana, the leaves contain only 0.7, 0.63 and 0.55% of lawsone, respectively. The areas mentioned above are located in the southern part of Egypt where the temperature is comparatively low. Markaryn (*12*) reported the effect of dimethyl sulfoxide on the yield of lawsone. The bark contains 0.21-0.24%, on a dry weight basis, of lawsone which is the next highest amount after the leaves; other parts of the plant contain a negligible amount of the dye. We have found that the pericarp of the plant and the seeds also contain lawsone at 0.05 and 0.13%, respectively. The dye has also been isolated from old roots of the plants.

The maximum yield of lawsone from the leaves is reached during June and July; however, the amount gradually decreases to about 0.78-0.88% in the winter season. During our investigation, it was observed that the protein active compound (PAC) of leaves, harvested in Multan, contained a slightly higher amount of dye, 1.32-1.37%, because atmospheric temperature of Multan is comparatively higher than that of Karachi. Thus it seems that atmospheric temperature may be one of the factors affecting variation in the quantity of lawsone as the surrounding temperature in Karachi ranges between 37 and 42°C in the months of June and July, and from 8 to 32°C in August to February. The quantitative variation of lawsone in various parts of the plant is presented in Table I.

The three major countries which export henna are India, Egypt and Sudan. However, India has taken the lead by producing approximately 65,000 - 70,000 metric tons per year which constitutes 85% of the total export production. The leaves are pressed and packed as bales, while powder in gunny bags weighing over 250 lbs is exported to various countries and only 15% of the total production is used in India. More than 17 varieties or brands of henna powder are commercially processed. The powder is graded on the basis of the color intensity of the extract. Different varieties of henna powder is available, according to variation in the color, particle size and purity.

Use in edible products

Although all parts of the shrub have been explored, the leaves of henna have been more extensively studied. The essential oil from henna contains 90% α- and β-ionone with a major quantity of β-ionone which is present at a concentration four times higher than that of α-ionone. The oil from henna has a pleasent aroma that has already been commercially exploited by the perfume industry and carries tremendous scope in the food industry as a flavoring agent. Some characteristics of the oil from flower and seeds of henna are given in Table II.

Recently, Ahmad *et al.* (*13*) investigated the fatty acid composition of the seed oil. It consisted of a large proportion of omega-3 and omega-6 fatty acids. The present work, based on Lawsone Protein Adduct (LPA), further demonstrated the use of lawsone in trace amounts in ice cream, yogurt, cheese and bakery products. Apart from imposing itself as a colorant, LPA has emerged as an effective preservative as it extends the shelf life of treated product.

Uses in non-food products

Textile Industry. The textile industry has successfully applied the henna extract for dyeing of fibers and has demonstrated that inorganic salts, if properly manipulated, may be effective in producing a variety of colors which may overshadow its own reddish brown color. Thus, addition of traces of potassium dichromate ($K_2Cr_2O_7$), ferrous sulfate ($FeSO_4$), tin chloride ($SnCl_3$), sulfuric acid (H_2SO_4) and alum ($K_2SO_4.Al_2(SO_4)_3.24H_2O$) to henna results in the production of a wide range of colors (*14*) from brown to cement grey. The pre- and post-soaking of the fiber in acidic

Table I. Quantitative Distribution of Lawsone

Plant part	Percentage on dry weight basis	Season	Place	Reference
Leaves	0.80-0.95	Summer	Nakhichevan[a]	*10*
Leaves	0.62-0.64	Winter	Shirvan[a]	*10*
Leaves	0.50-0.56	Winter	Sheron[a]	*10*
Leaves	0.97-0.95	Summer	Edfo & Assuit[b]	*11*
Leaves	0.55-0.63	Winter	Cairo & Fayed[b]	*11*
Leaves	1.00-1.12	Summer	Karachi[c]	*This work*
Leaves	0.78-0.88	Winter	Karachi	*This work*
Bark	0.21-0.24	Summer	Karachi	*This work*
Pericarp	0.13-0.25	Summer	Karachi	*This work*
Leaves	0.92-1.37	Summer	Multan[c]	*This work*

[a] Nakhichevan, Shirvan and Sheron: parts of Russia
[b] Edfo, Assuit, Cairo and Fayed: parts of Egypt
[c] Karachi and Multan: parts of Pakistan

solutions makes the staining permanent, resistant to soap washing, chlorine bleaching and sunlight.

Cosmetic Industry. The extract has also been exploited for use as a hair dye, perfume, and medicine. The leaf powder has recently been patented in Canada as a hair color conditioner and a hair dye cream from henna has also been commercially prepared in seven different shades as well as for use in liquid shampoo, cream rinse, premix color base, natural conditioner, and as a dye concentrate (*15*); nearly nine different shampoos have been prepared on a large scale using henna. In the USA, a clay base hair conditioner from henna has been patented (*16*).

The use of certain additives not only intensifies the usual color of henna, but even changes the resultant shades. Henna when mixed with indigo, logwood or other phytomass provides various shades of hair color. The light brown shade for hairs may be obtained by applying the commercial brand of "Henna-Reng" which is a mixture of henna and indigo in the ratio of 1:2, while the hair may be shaded dark brown by increasing the amount of indigo. A suitable ratio of 1:3 between dye and the additive is generally recommended. The mixture of the metallic salt and henna is known as "Henna Rasticks" and by varying the quantities of inorganic compounds and phenolics such as pyrogallol, aminophenol and tannins, a variety of shades from brown to black may be easily produced.

Organic compounds such as acetic acid (CH_3COOH) are also effective in slightly altering the shade of the color. The method of using oil for isolation of pigments suggested by Rehsi and Daruvala in 1957 (*14*) is still used even though it affords a poor yield. The methodology involves extraction of lawsone with chloroform from an aqueous sodium carbonate solution at pH 4-6 with a yield of less than 1%. The powder from leaves of henna when mixed with chestnut, milk, citric acid and water produced a slurry which is commonly applied for dyeing of hair and wool (*17*). The chemical structure of hair plays a key role in determining the absorption and retention of the dye.

Pharmaceutical Industry. The prophylactic action of henna leaves in curing skin disorders such as boils, burns, bruises and other inflammations or rashes has long been known to the rural population of the Indian Subcontinent. For curing dermatitis, the Japanese have prepared a lotion containing lawsone from henna leaves. The astringent phenolics are partially responsible for the action of henna and its extracts against microbes; mild antibiotic effects have been observed for alcoholic extracts of leaves against *Micrococcus pyrogenus, Staphylococcus aureus* and *Escherichia coli*. The antimicrobial nature of lawsone has created further interest for its utilization by the pharmaceutical industries (*18-20*). The plant has also been exploited as a traditional medicine in Saudia Arabia (*8*). The aqueous extract or liquor containing most of the essence provides relief to sore throats and is recommended in medicines for gargle and seed oil has been used for production of a muscle soothing and strain releasing agent in Uganda.

Table II. Some Characteristics of the Oil from Flower and Seeds of Henna

Characteristic	Flower	Seed
Essential oil	0.01-0.02 (%)	-
Color	brown	-
Specific gravity	0.9423	0.9545
Odour	tea rose	-
Major component	α- and β-ionone (1:4 ratio)	-
Acid value	6.8	18.2
Saponification value	-	149.0
Iodine value	-	60.0
Unsaponifiable matter	-	10.5 (%)

Table III. Solvent Systems Used for Chromatography

Solvent number	Solvent system	Abbreviation and relative concentration
1	Butanol-acetic acid-water	BAW (4:1:5)
2	tert-Butanol-acetic acid-water	TBAW (30:1:1)
3	Ethyl acetate-formic acid-water	EFW (14:3:3)
4	Acetic acid-hydrochloric acid-water	FORESTAL (30:3;10)
5	Acetic acid-water	AW (15:85)
6	Chloroform-acetic acid-water	CAW (30:15:2)

Isolation and Purification

The leaves of *L. inermis* (1 kg) were mixed with 2.5 L of a 1.5 M calcium hydroxide solution (pH 10) to a fine slurry using a blender, and then stirred for about 5 h using a magnetic stirrer. The mixture was transferred to a round bottom jar (henna jar) having a tap at the bottom. The jar was placed into a 60°C water bath for 15 min; this helped the solid particles to float to the top of the liquid. The clear solution containing natural dyes was collected through the tap at the bottom of the container, leaving the suspended solids behind. Residual solids present in the liquid were removed by centrifugation and the supernatant was sucction filtered through Whatman No. 1 filter paper. The extract was concentrated by freeze drying. Alternatively, the mixture was concentrated using a rotary evaporator at 50°C. The concentrated mixtures were extracted separately with chloroform (3 times) in a separatory funnel, filtered and dried using a rotary evaporator.

The residue was then taken into a mixture of chloroform, petroleum ether and acetic acid (4:6:0.5, v/v/v; solvent A). The precipitated lawsone was collected on a filter paper and further purified using a flash chromatography apparatus. The dye was identified by two dimensional TLC in solvents A and B (BuOH/AcOH/H_2O, 4:1:5, v/v/v). The major red fraction which had the affinity for the protein was further purified by thin layer chromatography. Both silica gel and cellulose supports were used as well as mobile phases listed in Table III. The purity of the compound was further confirmed by HPTLC in solvent systems A and B. The isolated compound in methanol was analyzed at 330 nm using a UV spectrophotometer.

Identification by Nuclear Magnetic Resonance Spectroscopy (NMR) and Mass Spectrometry (MS)

Nuclear magnetic resonance spectra at 300 or 400 MHz (^{1}H-NMR) were recorded in deuterated methanol using tetramethyl silane as internal standard.

The mass spectra were determined using a double focusing Finnigan mass spectrometer connected to a data handling assembly and a computer system. Peak matching, linked scan field desorption (FD), field ionization (FI) and fast atom bombardment (FAB) measurement were also performed on the MAT 312 mass spectrometer. FABMS was measured in glycerol-water at a ratio of 1:1 (v/v) in the presence of potassium iodide. Accurate mass measurements were made with the FAB source using an internal standard. High resolution mass spectra (HRMS) were also recorded. Ultraviolet, NMR and MS data were used to confirm that the red dye was indeed lawsone.

Lawsone: A Phenolic of Henna

The green leaves of *L. inermis* were found to contain 1-1.2% lawsone, a phenolic and PAC, on a dry weight basis. The quantitative estimation revealed that stem, bark and seeds also contain PAC, but only in trace quantities. Karawya *et al.* (*11*) and Verma & Rai (*21*) had used a TLC method for the isolation of lawsone from the leaves.

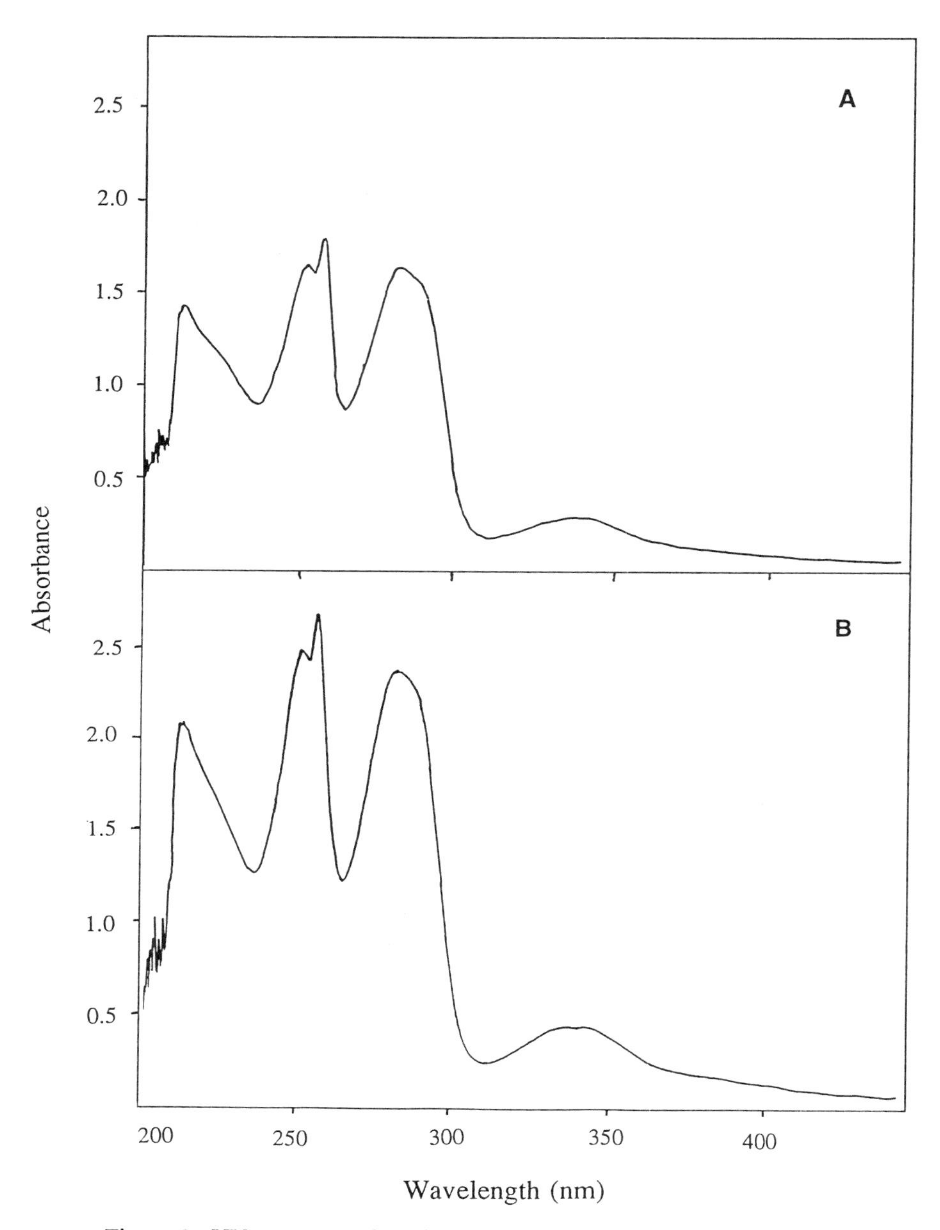

Figure 1. UV spectrum of (A) isolated and (B) commercial standard of lawsone.

Lawsone with R_f value of 0.72 was also separated (*22*) on a 10% acetylated cellulose powder mixed with 95% ethanol and without any binder. Kharbade (*23*) isolated five natural dyes from henna and demonstrated that the TLC system using benzene-ethyl formate-formic acid (74:24:1,v/v/v) has the best resolving power amongst 14 other mobile systems examined.

The present investigation used a novel mobile phase "A" (chloroform-petroleum ether-acetic acid, 4:6:0.5, v/v/v) after examining numerous solvent systems. The system "A" was the most effective as eleven bands were resolved on a silica gel column. The system also has several advantages, namely: It takes no more than 6-7 h for complete resolution on a column; Resolution efficacy is quite high; It is economical; It gives very clear visualization; Sharp boundaries of the coloring compounds are obtained; Least overlapping of bands occured; Immediate binding of fraction six with protein molecules is noted.

Fraction six, when further separated using the same solvent system, produced two subfractions. The pure fraction from round disc centrifuge chromatography (RDCC) produced a distinct red colored component that demonstrated its ability to bind to proteins; therefore, it is referred to as a PAC and was identified as lawsone. The PAC in methanol exhibited an absorption maximum at 331 nm, corresponding to that of lawsone, and it also corresponds to that of dihydrokaempferol (*24*) (Figure 1).

Chemical Nature of Lawsone

Lawsone belongs to the widely distributed family of natural quinones or more specifically 1,4-napthaquinones. Quinones are best known for their contribution to the color of plants and animal kingdom where their intensity varies with the position of the ketonic groups as demonstrated for *ortho* and *para*-quinones (Figure 2). Quinone are produced by microbes, plants and animals. The metabolic reactivity of quinone is attributed to various functions such as enhancing and inhibiting respiration, growth and excretion. Because lawsone is a napthaquinone, the discussion below is limited only to some of the napthaquinones found in nature.

1,4-Napthaqinones

Juglone. The 5-hydroxynapthaquinone is found in the skin of walnut (*Juglone nigra*) (*25*). The name juglone has been derived from the name of the genera in view of the fact that it is the major quinone present in the fruit.

Plumbagin. This 2-methyl-5-hydroxynapthaquinone of medicinal importance is isolated from the plumbago plant. It has been found to be toxic to larvae of many insects and may be used as an insecticide.

Phythiocol. An isomer of plumbagin has an interesting history as *Mycobaterium tuberculosis* bacteria treated with alcoholic alkali produced phythiocol from its precursor vitamin K_1 after hydrolysis. Lapadiol, menadione, droserone, lomatiol, alkanin, dunnione and many others have been isolated and identified. A pentahydroxy-1,4-napthaquinone with a red color was reported long ago in the eggs of sea urchins.

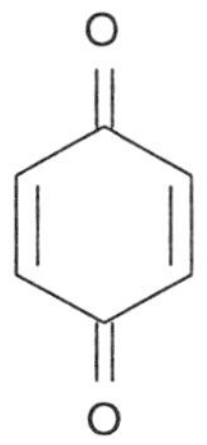

para-Benzoquinone (yellow) 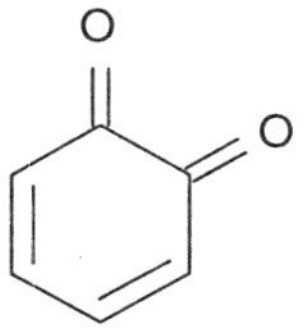   *ortho*-Benzoquinone (red)

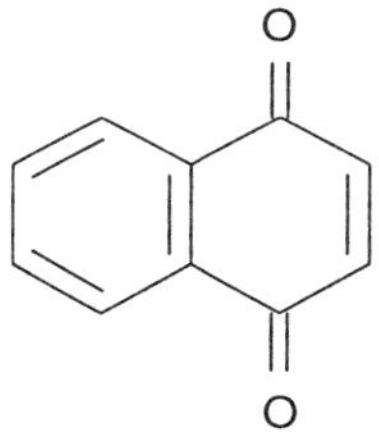

1,4-Napthaquinone

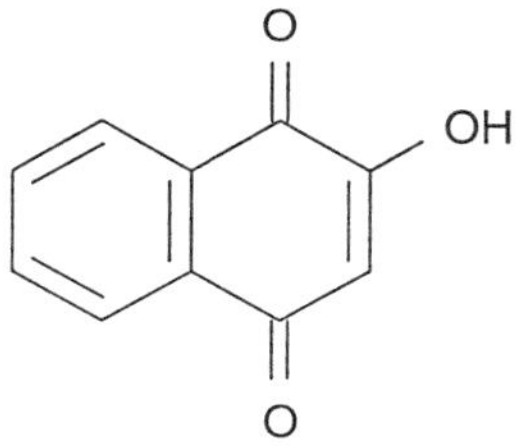

2-Hydroxy-1,4-Napthaquinone (Lawsone)

Figure 2. Chemical structures of benzoquinone, 1,4-napthaquinone and lawsone.

During the present investigations, one of the PACs, the dye from henna leaves was identified as lawsone with an empirical formula of $C_{10}H_6O_3$ and a chemical structure as given below.

2-Hydroxy-1,4-Napthaquinone (Lawsone)
The chemical structure (Figure 2) shows its ability to associate itself with protein at certain pH conditions of the compound. Lawsone being a 2-hydroxy-1,4-napthaquinone is expected to: (i) from an equilibrium in electrochemical reduction of quinone to hydroxyquinone in aqueous solution; (ii) undergo 1,4- addition reactions followed by enolization which may easily take place to form new isomers. The glutathione in reduced form adds easily to the quinoid ring. Similarly a number of other addition reactions are possible (Figure 3); and (iii) undergo substitution reactions such as alkylation or acetylation. which are less expected as formation of a free radical is also involved.

Interaction of lawsone with Protein Molecules

The 2-hydroxy-1,4-napthaquinone has been known to stain proteins by linking itself either electrostatically or covalently. The complex formation may occur via electrostatic interaction or covalent bonding.

Electrostatic interaction. As there are two ketonic groups present in lawsone, hydrogen bonds are formed with protein and lawsone is held in the folding of the protein molecule due to its secondary, tertiary and quaternary structures.

Covalent bonding forming conjugated proteins. (i) Reactions with SH group; As glutathione forms a covalent bond through the sulfhydryl group (SH), proteins having SH groups may react with lawsone to form conjugated chromoproteins. It also explains the staining reaction of lawsone to hair, skin and nail which contain keratin, a protein rich in cysteine. (ii) Reaction with NH_2 groups; Lawsone may also react with free NH_2 groups of basic amino acids such as lysine to form a Schiff base (Figure 3).

Possible Mode of Amino Acid Linkage with Lawsone

It is important to know whether the SH group of cysteine, or the NH_2 group of lysine, or the H-bonding through a ketonic group is involved in the formation of a complex between lawsone and proteins. An enzymic degradation of LPA was carried out. All peptide bonds present in "Lawsone Conjugated Protein" (LCP) should have been hydrolysed by the proteolytic enzymes therefore leaving the lawsone linked to only one amino acid, or a few amino acids, depending on whether partial or complete hydrolysis of peptide linkages has taken place. However, as none of the amino acids were detected, it was obvious that H-bond and van der Waals forces were involved in holding lawsone molecules between gaps of the protein structure.

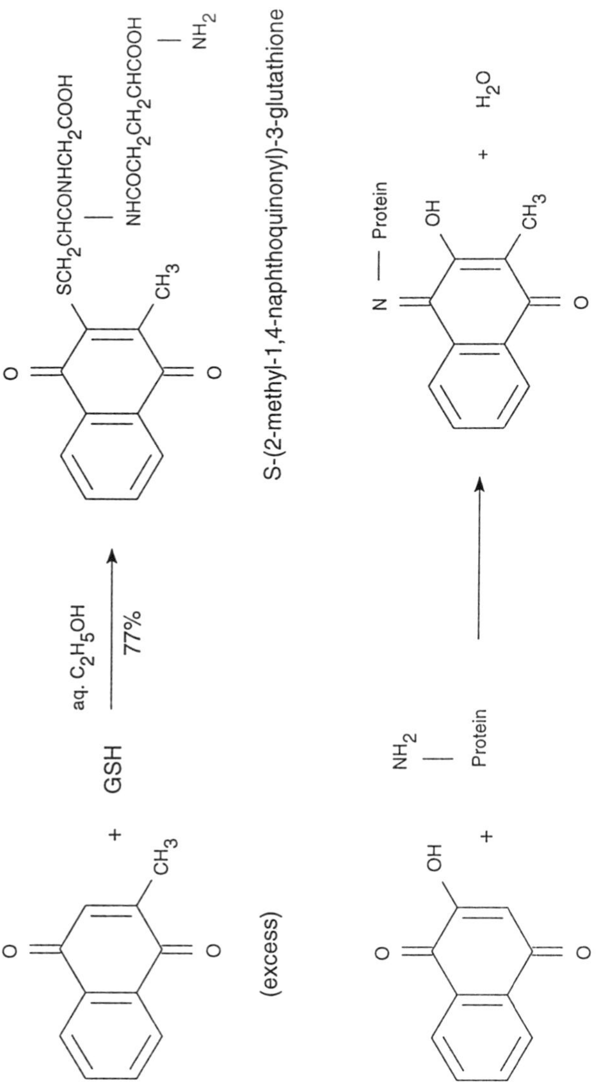

(excess) + GSH —aq. C_2H_5OH, 77%→ S-(2-methyl-1,4-naphthoquinonyl)-3-glutathione

Figure 3. Reaction of lawsone with reduced glutathione and free amino group.

Table IV. Methods of Fixing, Staining and Destaining by Various Dyes

Stain	Fixing	Staining	Destaining
Coomassie	1% TCA in 40% EtOH; 1-4 hr	0.2% dye in MeOH-H_2O-AcOH (2:5:2); ≥ 10 hr	EtOH-AcOH-H_2O(93:1:6) or 5% MeOH and 7% AcOH; ≥ 24 hr
Amido Black-10	7% AcOH at 200°C; ≥ 1 hr	2% dye in 7% AcOH at 100°C; ≥ 1 hr	7% AcOH at 100°C; ≥ 24 hr
Pakarli	Not necessary	2% dye in 7% AcOH 25°C; ½ hr	only H_2O at 25°C; ½-2 hr[a]

[a] bands visible even after 30 min and complete washing takes about 2 hours.

Lawsone as a Staining Agent for Proteins (Pakarli)[a]

Preparation of Pakarli Stain. Henna leaves (1 kg) in 1M KOH were mixed in a blender and then stirred for 24 hr using a magnetic stirrer. The mixture was acidified to pH 4.5, filtered and extracted with $CHCl_3$. The dye was reextracted with a 1% (w/v) KOH solution from the $CHCl_3$ layer. The staining solution was preparedby dissolving 0.1 g of it in 100 ml of (1:1, v/v) alcohol and 7% (v/v) aqueous acetic acid as shown in Table IV.

Suitable (pH) for staining. When sprayed on caseins, albumin or blood serum protein, the Pakarli stain at pH 2, 4, 7, 8 and 10 showed that it was effective over a broad pH range; however, the protein spots appeared to be darker in the alkaline pH range. The albumin staining capacity of the dye over a broad pH range reveals that it easily binds acidic proteins which are poorly stained by Indian ink and colloidal gold (*26,27*). The results also suggest that the carbohydrate moiety of glycoproteins has no adverse effect on the dye-binding capacity of the dye molecule. Another advantage of using Pakarli is that it could be directly used without any prior
fixing to protein bands with acid or alcohol. However, better results are obtained if the dye is dissolved in or mixed with a fixing solution. Pakarli shows immediate and rather permanent interaction with proteins without any prefixing treatment.

Experiments with casein, albumin and egg proteins demonstrated that although color intensity is best at pH 10, lawsone binds with proteins over a wide range of pH and with a great variety of proteins. Cellulose acetate membrane electrophoresis also produced better results when Pakarli was used.

Albumin, resolved on cellulose acetate membranes, stained with amido black and Pakarli produced equally good results. However, Pakarli is highly sensitive and long lasting even after a storage period of 18 months. The stainability of the dye with proteins on cellulose acetate membranes also suggests its possible application in immunological identification of proteins by blotting (*28*).

Staining of Electrophoretically Resolved Proteins with Pakarli

Six tubes (10 x 0.8 cm) containing 7.5% acrylamide gel with cross linking of 2.5% were used to resolve blood serum proteins, casein and albumin (*29*). Each set of gel was fixed, stained and destained with Pakarli, Coomassie brilliant blue R-250 and amido black-10 as given in Table IV.

Comparison of Pakarli with Other Staining Agents

The crude dye was an excellent staining agent for resolved proteins in PAGE because

[a]Pakarli represents Pakistan by its first three letters, A and R stands for Asad and Rashda (first names of the authors) and L and I are for the name of the plant *Lawsonia inermis*.

it contained lawsone, the PAC in reasonable quantity. Casein, albumin and a few blood serum proteins, were stained in Coomassie brilliant blue R-250 and Pakarli. The procedure of using Pakarli is rapid and does not require heating. The stained protein bands become dark brown, if heated, otherwise they appear as a bright reddish brown color at room temperature.

The results of the crude extract of Pakarli dye and purified stain of Coomassie brilliant blue R-250 after interaction with various blood serum proteins were compared. The crude dye produced better results, once purified.

Retention of Pakarli and its Superiority as a Staining Agent

The above studies have revealed that intensity of color of bands stained with Pakarli (*2*), did not fade after 2 years, even if the gel was stored in water. This illustrates the long lasting interaction of protein and the dye at neutral pH. In preliminary experiments, human blood serum proteins were separated by rod gel electrophoresis and stained with Pakarli and Coomassie brilliant blue R-250. The gels, stained with Pakarli and Coomassie brilliant blue R-250, were stored in water and 7% acetic acid, respectively, for a period of two and a half years. It was observed that some of the bands in the middle and in the lower regions faded and eventually disappeared after six months if stained with Coomassie brilliant blue R-250, but the color intensity of Pakarli treated samples remained unchanged. Moreover, the bands were completely diffused and disappeared in the case of Coomassie after a year while the protein bands with distinct and sharp boundaries were visible even after almost 3 years. Results demonstrate the superiority of the PAC present in the leaves.

As time progressed, the bands at the top of the gel began to lose their color intensity in the case of Coomassie and disappeared at the end of 18 months. In fact, most of the Coomassie stained bands lost their blue color before 30 months whereas the bands stained with Pakarli were visible at this point; thus, demonstrating the strong binding affinity of the dye to protein, even when stored in water. The stability of bands can be further increased by adjusting the pH to maximize the interaction between the dye and the protein.

The interaction of Pakarli with various proteins appear to be fast as the staining time was less than 30 min. Pakarli shows strong affinity towards protein molecules and binds to them even at room temperature. The strong interaction between the molecules of the dye and the protein is further supported by the prolonged stability of the color of bands in the gels when stored in water. In view of the above facts, it appears that the methodology may be adopted for routine electrophoretic analysis of proteins.

Lawsone Protein Adduct - LPA

Lawsone interacts with proteins such as keratin, gluten, gelatin, milk caseins, yogurt, caseins, and bovine albumins to form conjugated chromoprotein complexes. The LPA is resistant to microbial invasion as it shows prolonged stability. Furthermore, LPA appears to be temperature independent since lawsone is uniformly distributed throughout frozen milk products.

Table V. Absorbance of the Hydrolysates

Solvent number	Enzyme system	Absorbance at 280 nm
1	M(10ml) + L(1ml, 1% solution) + Pepsin	0.581
2	M + L + Trypsin	0.673
3	M + L + Papain	0.020
4	M + L + Bromelain	0.091
5	M + L + Proteinase-K	0.363
6	M + L	0.090
7	M	0.226

M = Cows's milk (tetra pack) purchased from market.
L = Lawsone 1% aqueous solution at pH 10.6.

Specificity of Protein Interaction

The wheat flour was allowed to react with lawsone before electrophoresis. Electrophoretic experiments further revealed that LPA of gluten was stable in an electrical field while a non-protein lawson adduct (NPLA) was also formed; this latter one was not stained with Coomassie.

The specificity of the interaction of τ-casein with lawsone was also demonstrated where a single yellow band or LPA was visible before staining while staining with Coomassie produced three distant bands of α, β and τ-casein. Thus, lawsone may be used in the identification of certain proteins.

The strong affinity of LPA was further demonstrated as four sharp yellow bands were observed during the electrophoric mobility of Yogurt Lawsone Adduct (YLA). Interestingly, only two bands were stained with Coomassie, thus three distinct factors may contribute to lawsone adduct complex formation. Firstly, lawsone forms adducts with molecules other than proteins as only two of the four yellow bands were identified as proteins by Coomassie, the others are non-protein lawsone adducts (NPLA) which are formed during fermentation. Secondly, there was a protein other than τ-casein of milk which is produced during fermentation and forms an adduct with lawsone, and this was visible in the electrophoretic pattern of caseins. The electrophoretic pattern of milk exhibited a single yellow band representing τ-casein-lawsone adduct after staining. Thirdly, lawsone may be used to identify specific proteins and non-protein organic substances even before staining of the gels.

The Albumin (bovine) from Serva which produced two yellow bands; only one band was stained with Coomassie, thus a non-protein-lawsone-adduct (NPLA) was also formed.

The LPA with gelatin also offered further verification of the results. The two yellow bands of the gelatin lawsone adducts (GLA) were stained with Coomassie while other gelatins were also present in the medium. Thus, lawsone reacted only with certain proteins and it was also capable of forming adducts with non-protein molecules.

Effect of pH on LPA

The LPA exhibited sharp color changes at different pH values, being pale yellow or colourless at pH 3.5-5.5, yellowish/orange at pH 5.5-9.0, dark red at pH 9.0-11.0, reddish brown at pH 11.0-13.0 and pale red at pH>13.0. The solubility of pure lawsone was maximum between pH 7.6 and 10, while it was insoluble in acidic media. The absorption of samples containing lawsone at various pH measured at 331 nm is shown in Figure 1.

In-vitro Digestibility of LPA

Several enzymes were selected for evaluation of digestibility of enzymes on the LPA. The optical density of the hydrolysates (Table V) demonstrate that LPA was digestible in *vitro*. The order of digestibility of LPA by various enzymes was found as trypsin

> pepsin > proteinase K > papain > bromelain. In fact bromelain and papain were nearly completely ineffective. It also appears that lawsone forms stable adducts with some of the enzymes as well. Moreover, it is likely that lawsone binds with the enzyme at its active site, so lawsone may also act as an enzyme inhibitor.

Toxicity of Lawsone

Although skin permeability of lawsone in humans is not well understood, the interaction of lawsone with keratin has long been established. The present studies revealed that lawsone may complex with many other proteins. In view of the long shelf life of LPA and the possibility of using lawsone as a food colorant and preservative, the subject of lawsone toxicology needs to be addressed.

The toxic role of 1,4-napthaquinone in cellular redox cycling and its involvement with glutathione, which reduces the oxidised quinone, has been investigated by several authors (*30,31*). Mild toxicity of lawsone seems to be due to three distinct reasons.

Firstly, being a hyroxynapthaquinone, lawsone participates in the intestinal redox cycling where favorable alkaline conditions exist, but constant interconversion of quinone to semiquinone will affect the extent of its toxicity. The presence of glutathione or other reducing agents will also be responsible for a reduced toxicity level (*32*) that is also directly related to electrophilicity depending on the position of OH group and presence of other substituents.

Secondly, the electron transfer theory for simultaneous existence of quinone and hydroquinone, though proposed much earlier by Emster *et al.* (*33*), elaborates the detoxifying of quinone. It suggests that quinone, or partially reduced quinone, forms complexes with a number of metabolites such as macromolecules, sulfate, glucouronides, and of course glutathione as studied in vertebrates.

Thirdly, feeding of moth with a 1,4-napthaquinone containing diet has proven that lawsone acts as a mild toxicant as compared to methylated hydroxyquinones (plumbagin). The well known dye juglone, 5-hydroxy-1,4-napthaquinone, is less toxic than lawsone. However, lawsone has often been reported as being the least toxic of several 1,4-napthaquinone (*34,35*).

Antimicrobial Activity of Lawsone and Henna Extract (HE)

Lawsone has been found active against many microorganisms. The lawsaritol and isoplumbagin from the stem, bark and roots of the plant have shown anti-inflammatory activity in rats (*36*).

Antimicrobial activity of henna extract (HE) has recently been studied against a number of gram positive and gram negative bacterial and fungi such as *Bacillus anthracis, B. cereus, B. subtilis, Enterobacer cloacae. P. mmirabilis and Shigella dysentry*. It was revealed that the henna extract effectively controlled the microbial growth. A virus inhibitory activity was also observed in *L. alba* which may be highly useful as a medicine (*37*).

Abd-el-Malek (*18*) has shown the prominent antimicrobial activity of lawsone against a wide range of microbes and has also observed that it was inactive against *Pseudomonas aeruginosa* and *Candida albicans*. However, later reports indicated that *Pseudomonas aeruginosa* growth was retarded by lawsone treatment (*38*). Bacterial growth was resisted by the presence of lawsone and leaf extract of *L. inermis*. The inhibition of various pathogenic and non-pathogenic bacteria have also been examined. *Staphylococcus aureus* was resisted equally by lawsone and the crude extract; the least degree of inhibition was recorded against *Streptococcus faecalis*. The *Citrobacter fruhdt, Salmonella typhi and Enterobacter cloacae* were controlled by lawsone as well as by the leaf extracts.

Malekzadeh & Shabestari (*38*) have also reported that the extract of henna was effective against both gram positive and gram negative bacteria *in-vitro*, such as *Bacillus anthracis, B. cereus, B. subtilis, Enterobacter aerugenes, Escherichia coli 0128-87, E. coli 055-B5, Proteus mirabilis, Pseudomonas aeruginosa, S. paratyphi, Shigella dysentry* type 1 and 7. Another compound 1,2-dihydroxy-4-glucosyloxynapthalene was also isolated from *L. inermis* and was active against *Bacillus subtilis* and *Saccharomyces pastorianus* (*20*). The present work demonstrated that the activity of the extract is due to the lawsone as it has the highest affinity for protein molecules. Lawsone was also used as a fungicide against *Achlaya orion, Isoachlaya anisospora, Aphanomyces helicoides* and *Saprolegnia parasitica* (*21*). We also examined antibacterial activity of aqueous, alcohol and chloroform extracts against different microbes as summarized in Table VI.

Future Research Prospects

The easy formation of lawsone protein complex at room temperature in absence of specific reactants, the stability of the complex LPA and its mild or nontoxic nature has unravelled many unknown facts that may allow its future utilization in food and allied industries. As lawsone reacts with the SH group of the cysteine of glutathione and probably of proteins, it may also inactivate sulfhydryl enzymes through SH group present at their active sites and thus may be useful in the food processing and preservation.

Lawsone was found to be toxic towards certain microorganisms. The mechanism of its antimicrobial action may be due to the complex formation between lawsone and membrane proteins, thus restricting the permeability of oxygen and nutrients for the living cell. The reaction of lawsone with sulfhydryl group of cysteine may have significance in bakery products and requires further investigation. The reaction suggests that addition of lawsone will decrease loaf volume as it will limit the formation of S-S bridges during kneading, thus may have an impact on the trapping of CO_2 during heating.

The experiments on electrophoresis have demonstrated that lawsone reacts with certain proteins and the colored proteins are mobile in the electric field on the gel. Thus, some of the proteins may be identified during electrophoresis before staining. It is important to learn the characteristics of these proteins and factors responsible for such interactions and electromobility. The specificity of LPA formation needs more

Table VI. Inhibition of Leaf Extract and Isolated Lawsone Against Various Microorganisms

Bacteria	Leaf extract			Isolated lawsone			Time (h)
	1	2	3	1	2	3	
Citrobacter	+	++	+++	-	++	+++	20
fruhdt	-	++	+++	-	+	++	43
	-	+	++	-		++	72
Staphylococcus	+	++	+++	+	++	+++	20
aureus	-	++	+++	-	++	+++	43
	-	++	+++	-	++	+++	72
Streptococcus	-	+	++	-	++	+++	20
faecalis	-	+	++	-	+	++	43
	-	-	+	-	+	+	72
Salmonella	-	++	+++	-	++	+++	20
typhi	-	+	++	-	++	++	43
	-	-	+	-	+	++	72
Enterobacter	-	+	++	-	++	+++	20
cloacae	-	+	++	-	++	+++	43
	-	-	-	-	+	++	72
Bergey	+	+++	+++	+	++	+++	20
maanula	+	+++	+++	+	++	+++	43
	-	++	+++	-	+	++	72

1-aqueous extract, 2-alcohol extract, 3-chloroform extract
+ Diameter of inhibition 1 to 10 mm
++ Diameter of inhibition 11 to 19 mm
+++ Diameter of inhibition ≥20 mm

research in order to identify such proteins. However, it is clear that lawsone may be used for identification of certain proteins, such as τ-casein of bovine milk. It will also be interesting to explore the binding of lawsone with various proteins with different numbers or no S-S bridges.

Conclusions

The present study has shown that the extract from leaves of *L. inermis* consists of 11 coloring compounds and Lawsone represents the major constituent showing affinity towards different proteins and macromolecules. The extract has been used specifically as a staining agent for electrophoretically resolved proteins. In view of the rapid methodology involved, clear sharp boundaries of the colored protein bands and long color retention, the dye shows superiority in comparison to the other common staining agents. Although lawsone forms LPA with a majority of proteins, however, the specificity of lawsone interaction with certain proteins shows that it may have potential for protein identification.

The strong binding between protein and the dye shows that it may be useful in food processing; lawsone is thought to be wrapped within the folding of proteins and this may be one of the reasons for the color getting darker with time as more and more lawsone molecules get attached within the secondary structure of proteins.

The dye has also been demonstrated to inhibit a wide range of bacteria and fungi. As some of microbes are food spoilage organisms lawsone or even leaf color extract may be used, not only as colorant, but, also as a food preservative.

Literature Cited

1. Ali, R.; Sayeed, S.A. *Protein Struct.-Funct. Relat. Proc. Int. Symp.* **1986**, pp.15-26.
2. Ali, R.; Sayeed, S.A. *Electrophoresis*, **1990**, *11*, 343-344.
3. Hoffstein, B.H. *Am. J. Pharm.* **1920**, *92*, 543-547.
4. Tommasi, G. *Gazz. Chim. Ital.* **1920**, *50*, 263-272.
5. James, K.C.; Spanouchi, S.P.; Turner, T.D. *J. Soc. Costmet. Chem.* **1986**, *37*, 359-367.
6. Singh, S.; Brin, P.J.; Hart, G.J. *Tetrahedron.* **1968**, *20*, 495-497.
7. Goldwell, A.G. *Eur. Patent Appl.* **1992**, EP 572-768.
8. Al-Yaya, M.A. *Col. Pharm. Fitoterapia*, **1986**, *57*, 179-182.
9. Roberts, M; Singh, K. *Marketing Henna Brochure,* **1955**, pp. 495-496.
10. Mikailvo, M.A.; Allakhvevdiev, S.R. *Az. SSR.* **1983**, *39*, 54-60.
11. Karawya, M.S.; Kharbade, B.V.; Agrawal, O.P. *J. Chromatogr.* **1969**, *347*, 447- 454.
12. Markaryan, S.A.; Vartanyam, M.K.; Tatevosyan, A.O. *Fiziol Biokim Kult Rast*, **1992**, *24*, 169-171.
13. Ahmed, M.B.; Rauf, A.; Osman, S.M. *J. Oil Technol. Asooc. India*, **1989**, *21*, 46-47.
14. Rehsi, S.S.; Daruvala, E.D. *J. Sci. Ind. Res. India,* **1957**, *16A*, 428.

15. William, A.P. *Perfumes, Cosmetics and Soaps,*Chapman and Hall: New York, **1969**, Vol. 2; p.17.
16. Gus, S.K. In *The Chemistry and Manufacture of Cosmetics*; Maison de Navorre, Ed.; Continental Press: Paris, 1975, Vol IV; pp. 908-909.
17. Moore, E.R.; Riddle, E.O. *Eur. Patent Appl.*, **1980**, *11*, 511-520.
18. Abd-el-Malek, Y.; El-Leithy, M.A.; Reda, F.A.; Khalil, M. *Zentralbl. Bakteriol. Parasitenk. Infektionskr. Hyg. Abt.*, **1973**, *128*, 61-67.
19. Tripathi, R.D.; Mekrani, S.; Dixit, S.N.; Srivastava, G.C. *Current trends Life Sci.*, **1982**, *9*, 55-58.
20. Afzal, M.; Al-Oriquat, G.; Al-Hassan, J.M.; Nazar, M. *Heterocycles*, **1984**, *22,* 813-816.
21. Verma, M.R.; Rai, J. *Indian Standard Institute Bulletin*, **1968**, *20*, 495-497.
22. Masschelein, K. *Mikrochim Acta*, **1967**, *2*, 1080-1085.
23. Kharbade, B.V.; Agrawal, O.P. *J. Chromatogr.* **1985,** *347*, 447-454.
24. Markham, K.R. *Techniques of Flavonoid Identification*, Academic Press: New York, 1982, p.32.
25. Robert, L.T.; Lindroth, R.L.; Tracy, J.W. *J. Chem. Enol.* **1994**, *20*, 1631-1641.
26. Muillerman, H.G.; Ter Hart, H.G.J.; Vandigk, W. *Anal. Biochem.* **1982**, *120*, 46-51.
27. Righetti, P.G.; Drysdale, J.W. *J. Chromatogr.* **1974**, *98,* 271-321.
28. Greshani, J.M.; Polade, G.E. *Anal. Biochem.* **1983**, *131*, 1-15.
29. Mauerer, H. R. In *Disc Electrophoresis and Related Techniques of Polyacrylamide Gel Electrophoresis.* **1971,** Walter de Gruyter: New York, 1971, pp. 224-225.
30. Ollinger, K.; Brinmark, K. *J. Biol. Chem.* **1991**, *266*, 21496-21503.
31. Ahmed, S. *Biochem. Syst. Ecol.* **1992**, *20*, 269-296.
32. Felton, G.W.; Duffey, S.S. *J. Chem. Ecol.* **1991**, *17*, 1821-1836.
33. Emster, L.; Danelson, L.; Lungren, L. *Biochem. Biophys. Acta.* **1962**, *58*, 171-188.
34. Doherty, M.; Roogers, A.; Cohem, G.M. *J. Appl. Toxicol.* **1987**, *7*, 123-129.
35. Buffington, G.D.; Ollinger, K.; Brudmax, A.; Cadenas, F. *Biochem. J.* **1989**, *257*, 561-571.
36. Gupta, S.; Ali, M.; Pillai, K.K.; Alam. M.S. *Fitotetrapia*, **1993**, *64*, 365-366.
37. Khan, M.M.; Abid, A.; Jain, D.C.; Bhakuni, R.S.; Zain, M.; Thakur, R.S. *Plant Sci.* **1991**, *75*, 161-165.
38. Malekzadeh, F.; Shabestari, P.P. *J. Sci. Islamic Repub. Iran.* **1989**, *1*, 7-12.

Chapter 14

Anticarcinogenic Activities of Polyphenols in Foods and Herbs

Ken-ichi Miyamoto[1], Tsugiya Murayama[2], Takashi Yoshida[3], Tsutomu Hatano[3], and Takuo Okuda[3]

[1]**Faculty of Pharmaceutical Sciences, Hokuriku University, Ho–3 Kanagawa-machi, Kanazawa 920–11, Japan**
[2]**Department of Microbiology, Kanazawa Medical University, Ishikawa 920–02, Japan**
[3]**Faculty of Pharmaceutical Sciences, Okayama University, Okayama 700, Japan**

The host-mediated anticancer activity, which is an anticarcinogenic activity found for polyphenols besides the inhibition of mutagenicity and tumor-promotion, is specific to some oligomeric hydrolyzable tannins. Agrimoniin, a dimer isolated from *Agrimonia pilosa* which is used as a herbal medicine and vegetable, was the first oligomer of this type. The macro cyclic ellagitannin dimer, such as oenothein B, often showed this activity, which was also found for some trimers and tetramers. These tannins stimulated macrophage *in vitro* and *in vivo*, resulting in inducation of interleukin-1 and increase of cytotoxic immunocytes. The anticancer activity is attributable to potential of the host-mediated defence *via* activation of macrophage.

In East Asia, the concept that "*medicine and foods are fundamentally the same*" is closely related to the *Kampo* medicine and has been widely accepted from old times. This means that we unconsciously take medicine from foods, and that many kinds of effective components are present in the plant foods and herbs. Among them, polyphenolic compounds are widely, and in large amounts, distributed in the plants. The low-molecular-weight polyphenols such as flavonoids, coumarins, lignans, etc. have been extensively studied both chemically and pharmacologically. However, little information was available on the biological activities of tannins which consist of a major part of plant polyphenols until various chemical structures of tannins isolated from plants were elucidated (*1*). It is also known that activities of tannins are due to their binding to proteins (*2*), metals (*3*), alkaloids (*4*) and polymers through the polyhydroxy groups in the molecule. Tannins have thus been used to convert hide to leather, to give astringent taste, and to treat diarrhea.

The Okuda's group has developed new methods of tannin analysis and has isolated numerous tannins from medicinal plants (*5*). Studies on biological activity of purified tannins has made rapid progress in recent years. Among the major activities of tannins found are antioxidant and radical scavenging activities (*6,7*). These are the

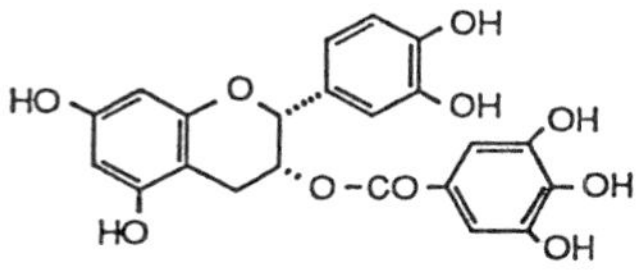

(-)-Epicatechin gallate (ECG)

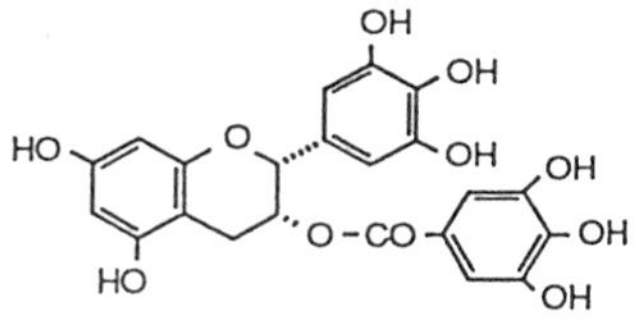

(-)-Epigallocatechin gallate (EGCG)

Figure 1. Chemical structures of tea polyphenols which are structural units of condensed tannins.

basic activities underlying the effects of tannin-rich medicinal plants which are effective in preventing and treating many diseases such as arteriosclerosis, heart dysfunction and liver injury, as well as inhibiting lipid-peroxidation (*8,9*). The inhibition of hepatotoxins (*10*) and mutagens (*11*) and the antitumor-promoter action of polyphenols are also correlated with their antioxidant activity (*12,13*). Most of these actions have been studied using low-molecular-weight polyphenols of less than 1000 Da, including epigallocatechin gallate (EGCG) (Figure 1). Our collaborators have demonstrated that hydrolyzable tannin monomers, oligomers and galloylated condensed tannins inhibit the replication of herpes simplex virus by blocking virus adsorption to the target cells (*14*), replication of human immunodeficiency virus by blocking the virus adsorption and inhibition of reverse transcriptase activity of the virus (*15,16*). The host-mediated anticancer activity of the tannins, as presented here, is the most chemical-structure specific anticancer activity. There are very few studies on anticancer activity of purified tannins, although the experience accumulated in East Asia has shown that many tannin-rich medicinal herbs may exhibit anticancer effects.

Anticancer Activity of Tannins on Mouse Sarcoma-180

We have developed a simple screening system in order to examine the anticancer activity of a large number of tannins and related polyphenols, using a small amount of compounds isolated and purified from many medicinal plants (*17*). Each compound was intraperitioneally (i.p.) injected into mice at 5 or 10 mg/kg 4 days before the i.p. innoculation of sarcoma-180 (S-180) cancer cells (1 x 10^5 cells/head), and the survival time was observed. Table I summarizes the anticancer activities of 108 compounds

Table I. Anticancer Activity of Polyphenols against S-180 in Mice [a]

Polyphenol	Number of tannins tested	Anticancer activity[b]			
		−	+	++	+++
Condensed tannins and related polyphenols	8	8	0	0	0
Hydrolyzable tannins and related polyphenols	100	67	6	20	7
Caffeic acid derivatives	3	3	0	0	0
Bergenin derivatives	3	3	0	0	0
Gallotannins	10	9	1	0	0
Ellagitannins					
Monomers	24	22	0	1	1
Dimers	31	14	2	10	5
Other oligomers	5	2	0	3	0
Macro cyclic oligomers	8	0	1	6	1
Dehydroellagitannins	16	14	2	0	0

[a]Each compound (5 or 10 mg/kg) was i.p. injected once into six mice in a group 4 days before the i.p. inoculation of S-180 (10^5 cells/mouse).

[b]Evaluation: −, <100 percent increase in the life span (%ILS); +, <100 %ILS but one survivor; ++, >100 %ILS or survivors; +++, >3 survivors out of six 60 days after the cancer cell innoculation. %ILS = 100 x (T-C)/C, where T and C indicate the mean survival time of the treated group and the mean survival days of the vehicle control group, respectively.

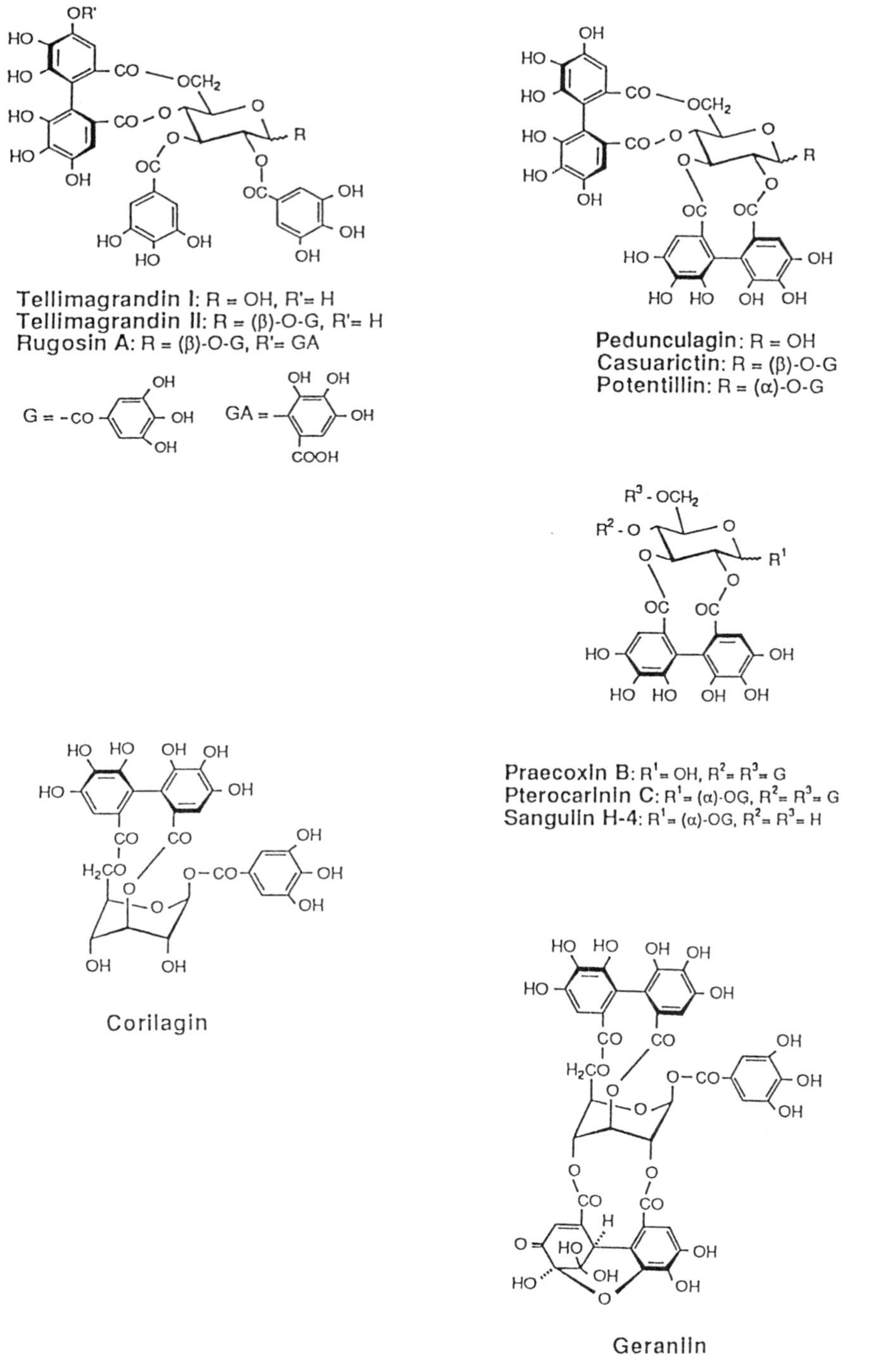

Tellimagrandin I: R = OH, R'= H
Tellimagrandin II: R = (β)-O-G, R'= H
Rugosin A: R = (β)-O-G, R'= GA

G = -CO-(3,4,5-trihydroxyphenyl)   GA = (2,3,4-trihydroxy-6-carboxyphenyl)

Pedunculagin: R = OH
Casuarictin: R = (β)-O-G
Potentillin: R = (α)-O-G

Praecoxin B: R^1 = OH, R^2 = R^3 = G
Pterocarinin C: R^1 = (α)-OG, R^2 = R^3 = G
Sanguiin H-4: R^1 = (α)-OG, R^2 = R^3 = H

Corilagin

Geraniin

Figure 2. Chemical structures of some monomeric ellagitannins.

tested (*17,18*). Condensed tannins, caffeic acid derivatives, bergenin derivatives, and gallotannins showed negligible or no anticancer activity. Monomeric ellagitannins, except for rugosin A and telllimagradin II, also showed very weak activities. Rugosin A and tellimagradin II, at 10 mg/kg, cured one and three out of six mice, respectively, but did not prolong the life-span of the other mice. On the other hand, nearly half of the oligomeric ellagitannins (20 out of 36) and all macrocyclic ellagitannin oligomers showed potent anticancer activities. It is clear that the presence of free phenolic hydroxyl groups is essential for the anticancer activity, has shown that results of nonacosa-*O*-methylcoriariin was inactive while the parent compound coriariin A showed a strong anticancer activity. In addition, the activity of dehydroellagitannins, either monomers or dimers, was very weak.

Oligomeric ellagitannins, such as coriariins A and C, cornusiins A and C, hirtellins A and B, isorugosin D, rugosins D and E, tamarixinin A, and woodfordins B and H, which consist of rugosin A and tellimagradins I and/or II as their monomer units exhibited a strong anticancer activity (see Figures 2 and 3). Agrimoniin, a dimer of potentillin, and gemin A, which consists of one potentillin unit and one tellimagrandin II unit, also showed potent anticancer activity. On the other hand, oligomeric tannins having a casuarictin (β-anomer of potentillin) unit, together with other monomeric units, had generally a very low anticancer activity (Figure 4.) All macrocyclic ellagitannins tested have tellimagradin I or II in their structures (Figure 5) and showed potent anticancer activity. This agrees with the results for open-chain ellagitannin oligomers. Thus, although it is not always necessary that the active oligomeric tannins consist of active monomer units, the structures of the monomer units play an important role in anticancer activity of the oligomers. The increment in the activity is not proportional to the degree of polymerization. For example, oenothein B, a macrocyclic dimer of tellimagradin I, exhibited a potent anticancer activity, but addition of the tellimagradin I or 2,3-digalloylglucose unit to the dimer molecule to form a trimer (oenothein A and woodfordin E) decreased the activity of the molecule. Furthermore, woodfordin F, a tetramer, did not cause regression of tumor growth (Figure 5). These results support the hypothesis that both the composition of the monomer units and their appropriate molecular size are important for the anticancer activity of oligomeric ellagitannins. Table II lists tannins with a potent anticancer activity as well as their distribution in different plant species and plant parts.

Anticancer Activity of Agrimoniin and Oenothein B

Agrimonia pilosa has been traditionally used as an intidiarrheic, a hemostatic, and an antiparasitic product in Japan and China (19,20). Its young leaves have also been used as a vegetable. Effectiveness of this plant on some carcinomas was described in Ben-cao-gang-mu (21), a classical encyclopedia of Chinese drugs. This plant is also used in cancer therapy in China today (22). In 1985, Koshiura *et al.* (23) experimentally confirmed that the methanol extract from root of *Agrimonia pilosa* Ledeb. exhibited activity against several rodent cancers (23) and that its active constituent was agrimoniin (24,25). This substance was already isolated from Rosaceous medicinal plants (26,27). Oenothein B was first isolated from *Oenothera erythrosepala* (28) and

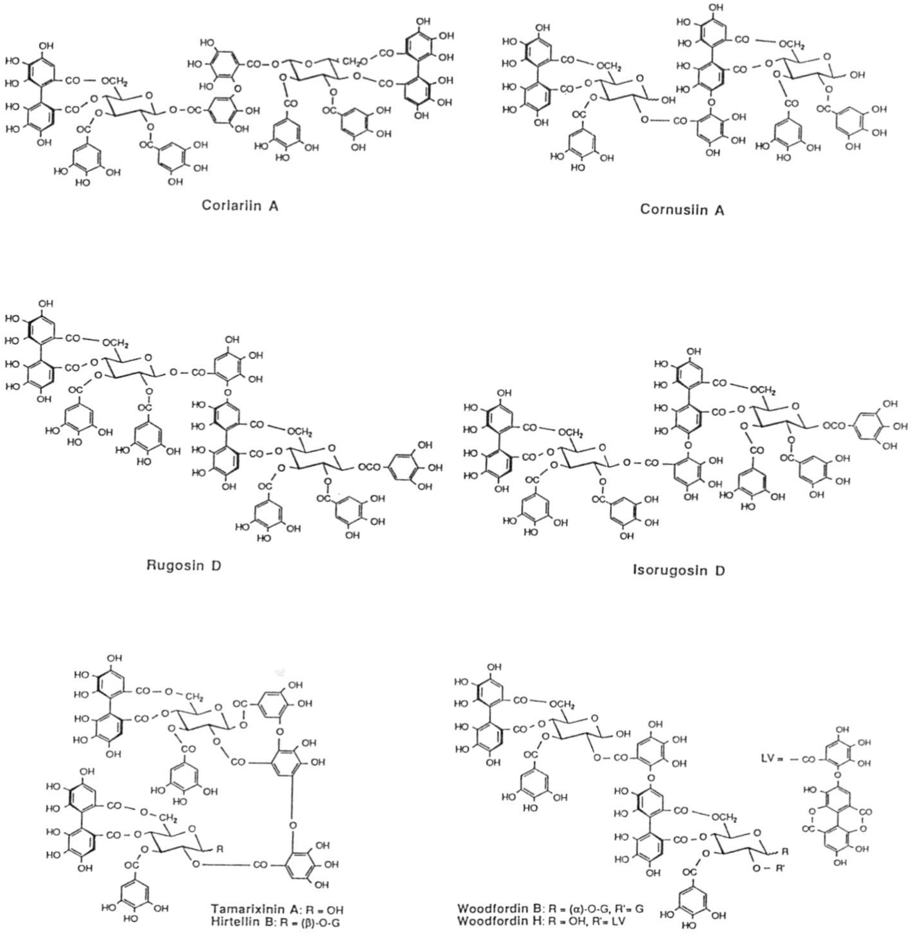

Corlariin A

Cornusiin A

Rugosin D

Isorugosin D

Tamarixinin A: R = OH
Hirtellin B: R = (β)-O-G

Woodfordin B: R = (α)-O-G, R' = G
Woodfordin H: R = OH, R' = LV

Figure 3. Chemical structures of some dimeric ellagitannins, which consist of the tellimagrandins monomer units.

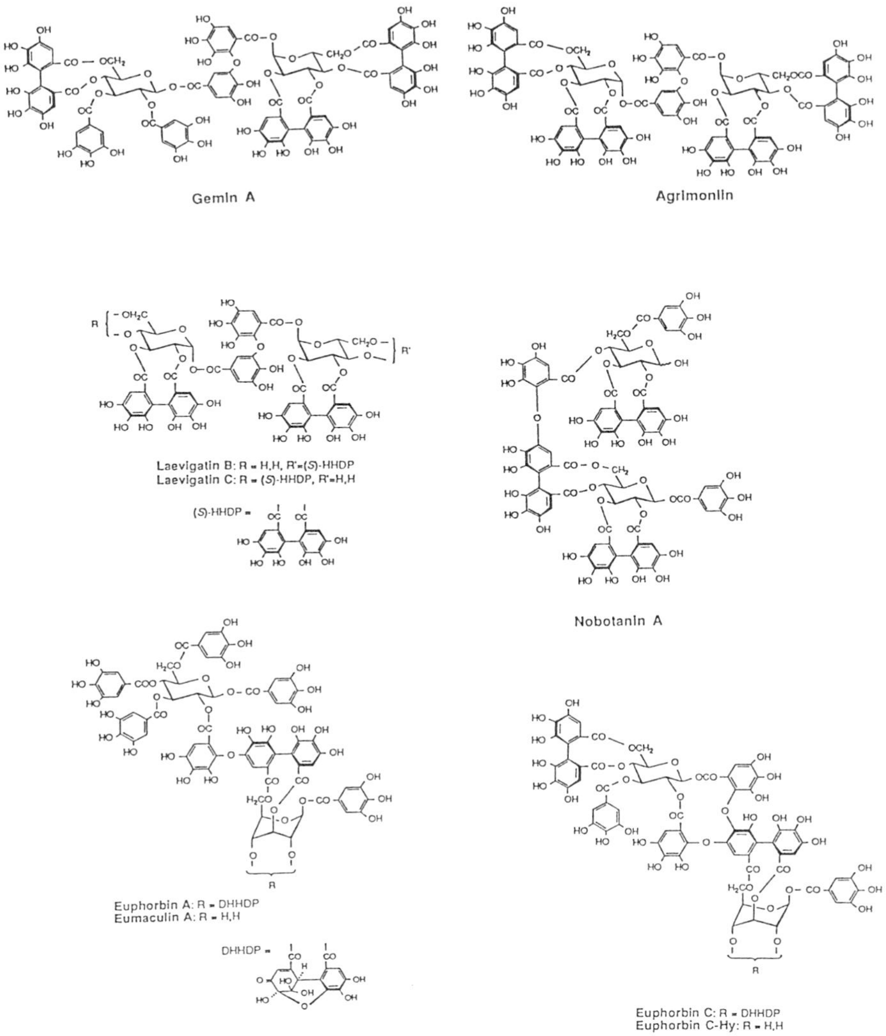

Gemin A

Agrimonlin

Laevigatin B: R = H,H, R'=(*S*)-HHDP
Laevigatin C: R = (*S*)-HHDP, R'=H,H

(*S*)-HHDP =

Nobotanin A

Euphorbin A: R = DHHDP
Eumaculin A: R = H,H

DHHDP =

Euphorbin C: R = DHHDP
Euphorbin C-Hy: R = H,H

Figure 4. Chemical structures of some oligomeric ellagitannins, which consist of the potentillin and monomer unit of related structures.

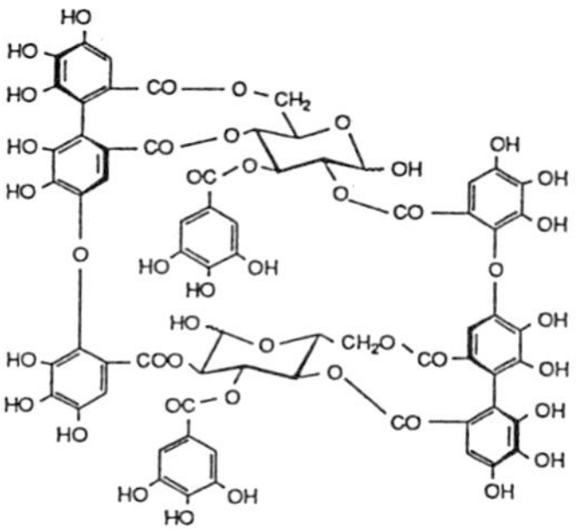

Oenothein B

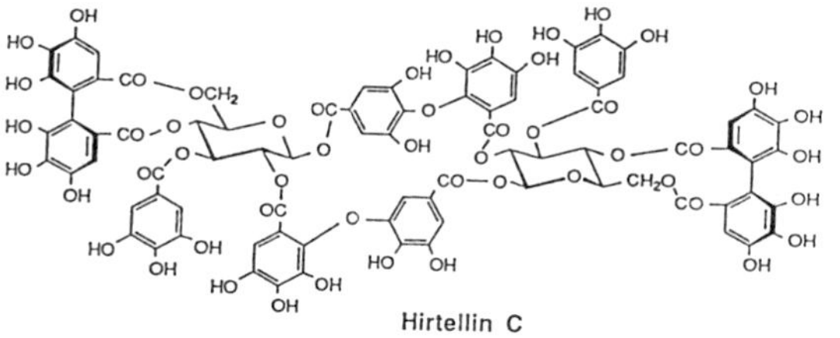

Hirtellin C

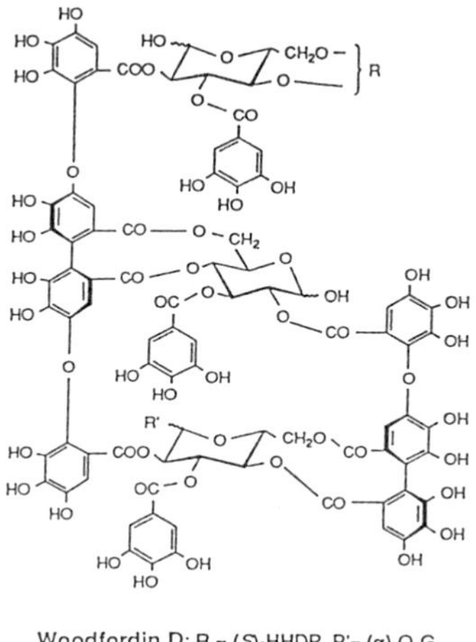

Woodfordin D: R = (*S*)-HHDP, R'= (α)-O-G
Woodfordin E: R = H,H, R'= OH
Oenothein A: R = (*S*)-HHDP, R'= OH

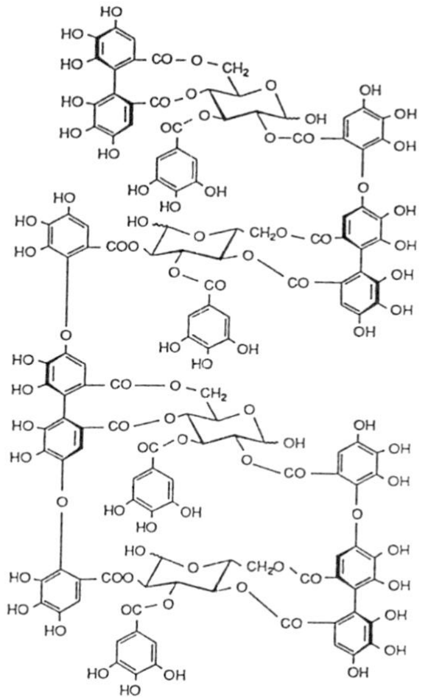

Woodfordin F

Figure 5. Chemical structures of some macrocyclic ellagitannins.

Table II. Sources of Tannins Showing Potent Anticancer Activity

Tannin	Main Species	Family	Part
Oligomeric ellagitannins			
Agrimoniin	*Agrimonia pilosa*	Rosaceae	roots
Coriariins A, C	*Coriaria japonica*	Coriariaceae	leaves
Cornusiin A	*Cornus officianalis*	Cornaceae	fruits
Hirtellins A, B	*Reaumuria hirtella*	Tamaricaceae	leaves
Isorugosin D	*Liquidambar formosana*	Hamamelidaceae	leaves
Rugosins D, E	*Rosa rugosa*	Rosaceae	petals
Tamarixinin A	*Tamarix pakistanica*	Tamaricaceae	leaves
Macro cyclic ellagitannins			
Camelliin B	*Camellia japonica*	Theaceae	flowers
Oenotheins A, B	*Oenothera erythrosepala*	Onagraceae	flowers
Woodfordins C, D, F	*Woodfordia fruticosa*	Lythraceae	flowers

then from the flower of *Woodfordia fruticosa* (29,30), a popular crude drug "sidowayah" in Indonesia and Malaysia for the treatment of dysentery, sprue, rheumatism, dysuria, and hematuria, when used in combination with other macrocyclic ellagitannin oligomers, such as oenothein A and woodfordins C and D. In spite of the unique macrocyclic structure of oenothein B, this dimer showed potent anticancer activity, similar to that of agrimoniin. Both agrimoniin and oenothein B exhibited a dose-dependent anticancer effect on mouse mammary cancer MM2 by a single i.p. administration 4 days before i.p. inoculation of the cancer cells, as did OK-432, a streptococcal preparation with a potent immune-stimulatory activity (31-33). The i.p., intravenous, and oral premedications of agrimoniin caused cancer regression even when the treatment began 14 days before the cancer inoculation. Agrimoniin was also effective in the postmedication by each administration route. However, oral administration of oenothein B did not show any anticancer activity. This may be attributed to the difficulty of absorption of the macrocyclic tannin through the intestine, and relatively easy absorption of agrimoniin. Agrimoniin and oenothein B were also effective on the solid type cancers, such as Meth-A fibrosarcoma and MH134 hepatoma (25,34).

Potentiation of Immune System

***In Vitro* Cytotoxicity.** Agrimoniin showed only a weak *in vitro* cytotoxicity against MM2 cells in the presence of calf serum (IC_{50}: 63 μg/ml), but was cytotoxic in the absence of the serum (IC_{50}:2.6 μg/ml)(25). The cytotoxicity of oenothein B was decreased, to a large extent, by the addition of serum (34). These data indicate that the above mentioned tannins easily bind to serum proteins, and their amount as unbound, i.e. in the free form, is very small when administered to the animals. It is therefore

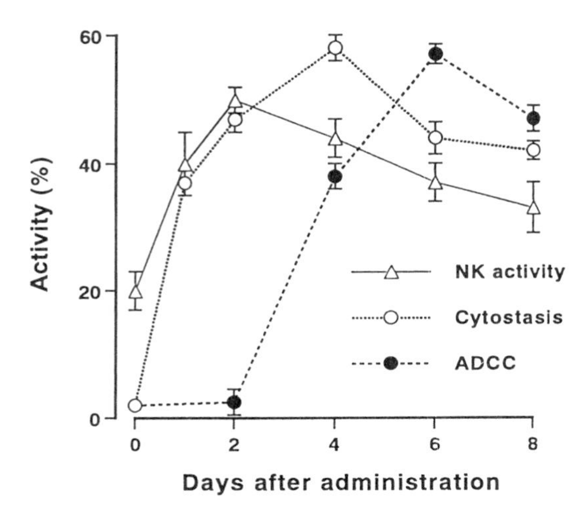

Figure 6. Changes in NK cell activity, cytostatic activity, and antibody-dependent cell-mediated cytotoxicity (ADCC) of peritoneal exudate cells from mice on the days indicated after an i.p. injection of agrimoniin (10 mg/kg).

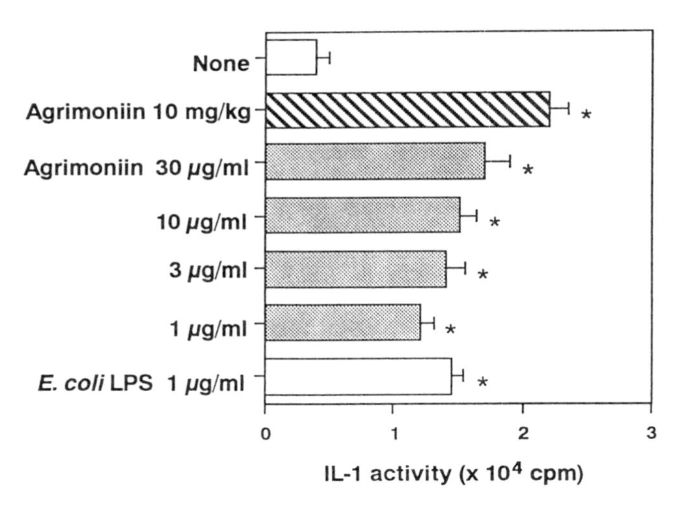

Figure 7. Secretion of IL-1 activity of macrophages treated with agrimoniin *in vitro* and *in vivo*. Culture supernatants of macrophages from mice 4 days after an i.p. injection of agrimoniin (10 mg/kg) or of macrophages from normal mice in the presence of the indicated concentrations of agrimoniin or *E. coli* LPS for 48 hr were used for the IL-1 assay. The IL-1 activity is defined as the radioactivity in thymocytes (1 x 10^6 cells/well) loaded 0.5 µCi of ^{3}H-TdR for 6 hr. Data are the means ± SE. $^*P < 0.01$, compared with the control by Student's *t*-test.

difficult to attribute the anticancer activity of the tannins only to their direct cytotoxicity on cancer cells.

Induction of Cytotoxic Cells. We found on 4 to 10 days after a single i.p. injection of agrimoniin into mice that peripheral white blood cells and peritoneal exudate cells, including monocytes and lymphocytes, increased significantly. The spleen weight also increased after the administration of the tannin. These cells exhibited a self-growing activity and cytotoxic activity against cancer cells (25,34,35). Agrimoniin enhanced the cytotoxic potential of several effector of peritoneal exudate cells via the natural killer (NK) cell activity as an earlier response, the cytostatic activity of macrophage, and the antibody-dependent macrophage-mediated cytotoxic activity as a later response (Figure 6). These results indicate that the tannin activates macrophage and lymphocytes, and exhibits anticancer activity through the cytocydal or cytostatic actions of the blood cells.

Stimulation of Interleukin-1 Induction. The potentiation of cell-mediated immunity by agrimoniin in mice suggests that the tannin stimulated some type of white blood cells and enhanced cytokine secretion. Agrimoniin never exhibited direct mitogenic activity on lymphocytes (*25*) and also did not stimulate interleukin-2 (IL-2) secretion from lymphocytes (Murayama *et al.*, unpublished data), but this compound enhanced the induction and secretion of IL-1 from macrophage (*36*). Figure 7 shows agrimoniin induced IL-1 secretion from macrophage in both the *in vitro* and *in vivo* treatments with a similar or higher potency as that of *in vitro* treatment with *E. coli* LPS. Secretion of IL-1β from agrimoniin-stimulated human macrophage was detected as early as 4 hr after stimulation and was maintained at very high levels, same as those after LPS-stimulation (*36*). These indicate that the targets of the tannin for anticancer activity are macrophage but not lymphocytes and cancer cells.

Table III shows anticancer activities and IL-1β-inducer activities of tannins in human macrophage. Mmonocytes-macrophage were isolated from human peripheral blood cells and treated with both tannin (10 μM) and *E. coli* LPS (10 μg/ml) for 4 hr and then cultured for 24 hr. Following this, IL-1β in the culture supernatant was measured (*37*). Monomeric ellagitannins increased IL-1β production by 2-fold over the non-stimulated basal production, and oligomeric ellagitannins with strong anticancer activity more potently stimulated the IL-1β. Figure 8 shows that agrimoniin induced the cell IL-1β (Table III and Figure 8). These results indicate that the primary mechanism for anticancer activity of tannins is perhaps direct stimulation of monocytes-macrophage and IL-1 induction. However, the correlation between their anticancer activity and IL-1β inducation of some tannins remains unclear. These tannins may be unstable in the host. It is unlikely that oligomeric tannins act on the cells after degradation to the monomers, because the IL-1 inducer activities of tellimagrandins I and II and potentillin were higher than those of laevigatins B and C and euophorbin C-Hy. Similarly, tellimagrandin II showed anticancer activity, but euphorbin C-Hy did not (Figures 2 and 4). In contrast, potentillin was ineffective, while agrimoniin, a dimer of potentillin unit, was a strong anticancer compound. This indicates that monomeric units and the structure and conformation of the oligomers are more important in the host-mediated anticancer activity.

Table III. Anticancer Activities and IL-1β Inducer Activities of Tannins

Compound	Molecular weight	Anticancer activity[a]		IL-1β production[b]
		%ILS	Regressors	% Increased
Condensed tannin				
ECG	442.4	38.7	0	35.1
EGCG	458.4	-1.9	0	44.3
Ellagitannin				
monomer				
Tellimagrandin I	786.6	35.2	0	162.3
Casuarictin	936.6	47.1	0	220.1
Potentillin	936.6	-1.9	0	212.0
Tellimagrandin II	938.7	18.1	3	198.4
Rugosin A	1106.8	25.2	1	209.2
dimer				
Laevigatin B	1569.1	-9.7	0	124.9
Laevigatin C	1569.1	8.4	0	129.5
Euphorbin C-Hy	1569.1	-6.5	0	108.2
Eumaculin A	1573.1	8.4	0	132.8
Agrimoniin	1871.3	136.2	3	359.0
Hirtellin B	1873.3	114.2	3	301.6
Euphorbin A	1891.3	45.2	0	188.5
Oenothein B	1569.1	196.0	4	293.4
Oenothein A	2353.7	102.7	1	254.1
OK-432		79.2	2	-
E. coli LPS		-	-	277.0

[a]Each tannin (10mg/kg) or OK-432 (100 KE/kg) was i.p. injected into mice once 4 days before the i.p. innoculation of S-180 cells (1 x 10^5).

[b]Macrophages from humanperipheral blood were treated with 10 μM each tannin or 10 μg/ml *E. coli* LPS for 4 hr and cultured for 24 hr. The concentration of IL-1β in the culture supernatant represenets the increasedpercentageof the untreated control (305 ± 45 pg/ml).

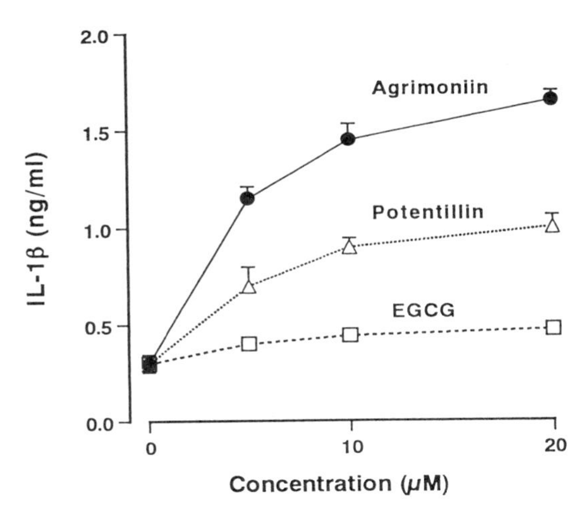

Figure 8. IL-1β secretion by human peripheral monocytes treated with tannins *in vitro*. Cells treated with 5, 10, and 20 μM of each tannin for 4 hr were cultured for 24 hr, and then IL-1β in the culture supernatant was measured. Data are the mean ± SE.

It is well documented that IL-1 causes macrophage and lymphocytes to induce IL-2, IL-2 receptors and other lymphokines and potentiates many functions of host defence against cancer (*38,39*). Moreover, there is evidence that IL-1 increases the binding of NK cells to tumor targets (*40*). Since IL-1 induces interferon, which synergizes with IL-1 with respect to its action on NK cells (*41*), one could view both mechanisms as being an efficient aspect of host defence against tumors. Additionally, agrimoniin has a potency to induce IL-8 production by macrophage, but not IL-2 secretion by lymphocytes (unpublished data), resulting in an increase in the number of white blood cells (*25*). Consequently, tannins such as agrimoniin initially stimulate monocytes-macrophage and act in host defence against cancer cells through complex action of activated immunocytes.

Conclusion

Anticancer activity is one of the important biological activities of tannins. The anticancer tannins are present in many medicinal plants and foods. Among them, oligomeric ellagitannins, which consist of monomer units, such as potentillin, tellimagrandin I and II, and related structures, showed a potent anticancer activity. The anticancer activity of tannins is mediated by potentiation of immune-defence activity of the host, and this is the concept underlying the Oriental medicine. The inhibitory activity on cancer initiation and promotion by tannins, and that the effects of the tannin-rich plant foods on diseases due to decreased immunity is documented. Development of tannin preparations for clinical application in cancer treatment is expected.

Literature Cited

1. Okuda, T.; Yoshida, T.; Hatano, T. In *Progress in the Chemistry of Organic Natural Products*; Herz, W., Ed.; Springer Verlag, Wein, 1995, Vol. 66, p. 1.

2. Haslam, E. In *Plant Polyphenols. Vegetable Tannins Revisited*; Cambridge: Cambridge University Press, 1989.
3. Okuda, T.; Mori, K.; Shiota, M.; Ida, K. *Yakugaku Zasshi* **1982**, *102*, 734.
4. Okuda, T.; Mori, K.; Shiota, M. *Yakugaku Zasshi* **1982**, *102*, 854.
5. Okuda, T.; Yoshida, T.: Hatano, T. *J. Nat. Prod.* **1989**, *52*, 1
6. Hatano, T.; Edamatsu, R.; Hiramatsu, m.; Mori, A.; Fujita, Y.; Yasuhara, T.; Yoshida, T.; Okuda, T. *Chem Pharm. Bull.* **1989**, *37*, 2016.
7. Okuda, T.; Yoshida, T.; Hatano, T. In *ACS Symposium Series no. 507*, **1992**, p. 160, American Chemical Society, Washington.
8. Hong, C.-Y.; Wang, C.-P.; Huang, S.-S.; Hsu, F.-L. *J. Pharm. Pharmacol.* **1995**, *47*, 138.
9. Kimura, Y.; Okuda, H.; Mori, K.; Okuda, T.; Arichi, S. *Chem. Pharm. Bull.* **1984**, *32*, 1866.
10. Hikino, H.; Kiso, Y.; Hatano, T.; Yoshida, T.; Okuda, T. *Ethnopharmacology* **1985**, *14*, 19.
11. Okuda, T.; Mori, K.; Hayatsu, H. *Chem. Pharm. Bull.* **1984**, *32*, 3755.
12. Yoshizawa, S.; Horiuchi, T.; Fujiki, H.; Yoshida, T.; Okuda, T. *Phytotherapy Res.* **1987**, *1*, 44.
13. Nishida, H.; Omori, M.; Fukutomi, Y.; Ninomiya, M.;Nishiwaki, S.; Suganuma, M.; Moriwaki, H.; Muto, Y. *Jpn. J. Cancer Res.* **1994**, *85*, 221.
14. Fukuchi, K.; Sakagami, H.; Okuda, T.; Hatano, T.; Tanuma, S.; Kitajima, K.; Inoue, Y.; Inoue, S.; Ichikawa, S.; Nonoyama, M.; Konno, K. *Antiviral Res.* **1989**, *11*, 285.
15. Asanaka, M.; Kurimura, T.; Koshiura, R.; Okuda, T.; Mori, M.; Yokoi, H. *4th International Conference on Immunopharmacology* **1988**, p. 35 (Abstr.).
16. Nakashima, H.; Murakami, T.; Yamamoto, N.; Sakagami, H.; Tanuma, S.; Hatano, T.; Yoshida, T.; Okuda, T. *Antiviral Res.* **1992**, 18, 91.
17. Miyamoto, K.; Kishi, N.; Koshiua, R.; Yoshida, T.; Hatano, T.; Okuda, T. *Chem. Pharm. Bull.* **1987**, *35*, 814.
18. Miyamoto, K.; Nomura, M.; Murayama, T.; Furukawa, T.; Hatano, T.; Yoshida, T.; Koshiura, R.; Okuda, T. *Biol. Pharm. Bull.* **1993**, *16*, 379.
19. Akamatsu K. In *Shintei Wakanyaku*. **1970**, p. 378, Ishiyakushuppan, Tokyo.
20. Chiang Su New Medical College, In *Dictionary of Chinese Crude Drugs*. **1977**, p. 665, Shanghai Scientific Technologic Publisher, Shanghai.
21. Li, S.Z. In *Shinchukotei Kokuyaku-Honzokomoku*. (Kimura K. ed.) **1974**, p. 389, Shunyodo-shoten, Tokyo.
22. Sugi, M. In *Cancer Therapy in China Today*. (Kondo, K. ed.) **1977**, p. 95, Shizensha, Tokyo.
23. Koshiura, R.; Miyamoto, K.; Ikeya, Y.; Taguchi, H. *Jpn. J. Pharmacol.* **1985**, *38*, 9.
24. Miyamoto, K.; Koshiura, R.; Ikeya, Y.; Taguchi, H. *Chem. Pharm. Bull.* **1985**, *33*, 3977.
25. Miyamoto, K.; Kishi, N.; Koshiura, R. *Jpn. J. Pharmacol.* **1987**, *43*, 187.
26. Okuda, T.; Yoshida, T.; Kuwahara, M.; Memon, M.U.; Shingu, T. *J. Chem. Soc., Chem. Commun.* **1982**, 163.
27. Okuda, T.; Yoshida, T.; Kuwahara, M.; Memon, M.U.; Shingu, T. *Chem. Pharm. Bull.* **1984**, *32*, 2165.

28. Hatano, T.; Yasuhara, T.; Matsuda, M.; Yazaki, K.; Yoshida, T.; Okuda, T. *J. Chem. Soc., Perkin Trans. 1*, **1990**, 2735.
29. Yoshida, T.; Chou, T.; Nitta, A.; Miyamoto, K.; Koshiura, R.; Okuda, T. *Chem. Pharm. Bull.* **1990**, *38*, 1211.
30. Yoshida, T.; Chou, T.;Matsuda, M.; Yasuhara, T.; Yazaki, K.; Hatano, T.; Okuda, T. *Chem. Pharm. Bull.* **1991**, *39*, 1157.
31. Okamoto, H.; Shoin, S.; Koshimura, S. In *Bacterial Toxins and Cell Membrane*. (Jeljaszewicz, J.; Wadstrom, T. eds.) **1978**, p. 259, Academic Press, New York.
32. Kai, S.; Tanaka, J.; Nomoto, K.; Torisu, M. *Clin. Exp. Immunol.* **1979**, *37*, 98.
33. Murayama, T.; Natsuume-Sakai, S.; Ryoyama, K.; Koshimura, S. *Cancer Immunol. Immunother.* **1982**, *12*, 141.
34. Miyamoto, K.; Nomura, M.; Sasakura, M.; Matsui, E.; Koshiura, R.; Murayama, T.; Furukawa, T.; Hatano, T.; Yoshida, T.; Okuda, T. *Jpn. J. Cancer Res.* **1993**, *84*, 99.
35. Miyamoto, K.; Kishi, N.; Murayama, T.; Furukawa, T.; Koshiura, R. *Cancer Immunol. Immunother.* **1988**, *27*, 59.
36. Murayama, T.; Kishi, N.; Koshiura, R.; Takagi, K.; Furukawa, T.; Miyamoto, K. *Anticancer Res.* **1992**, *12*, 1417.
37. Miyamoto, K.; Murayama, T.; Nomura, M.; Hatano, T.; Yoshida, T.; Furukawa, T.; Koshiura, R.; Okuda, T. *Anticancer Res.* **1993**, *13*, 37.
38. Durum, S.K.; Schmidt, J.A.; Oppenheim, J.J. *Annu. Rev. Immunol.* **1985**, *3*, 263.
39. Dinarello, C.A. *Fed. Am. Soc. Exp. Biol. J.* **1988**, *2*, 108.
40. Herman, J.; Dinarello, C.A.; Kew, M.C.; Rabson, A.R. *J. Immunol.* **1985**, *135*, 2882.
41. Dempsey, R.A.; Dinarello, C.A.; Mier, J.W.; Rosenwasser, L.J. *J. Immunol.* **1982**, *129*, 2504.

Chapter 15

Chemiluminescence of Catechins and Soybean Saponins in the Presence of Active Oxygen Species

Kazuyoshi Okubo[1], Yumiko Yoshiki[1], Kiharu Igarashi[2], and Kazuhiko Yotsuhashi[3]

[1]Department of Applied Biological Chemistry, Faculty of Agriculture, Tohoku University, 1–1 Amamiyamachi, Tsutsumitori, Aoba-ku Sendai 981, Japan
[2]Department of Bioproduction, Faculty of Agriculture, Yamagata University, Tsuruoka, Yamagata 997, Japan
[3]Chugai Pharmaceutical Company, Limited, Chuo-ku, Tokyo 104, Japan

The photon emission (chemiluminescence; CL) of catechin in the presence of active oxygen species and acetaldehyde was corroborated to occur non-enzymatically at room temperature and neutral condition. In the presence of 2% acetaldehyde (Z), photon emission [P] increased linearly depending upon the concentrations of active oxygen species [X] and of catechins or gallic acid [Y]; $[P] = k\,[X]\,[Y]\oint(Z)$ (k: photon constant).
Soyasaponins βg and Ab are present at a high level in the hypocotyl of soybean seeds, soyasaponin Ab is found only in the hypocotyl, but soyasaponin βg is distributed widely in the fibrovascular bundle of the leguminous plant. The CL derived from soyasaponins Ab and βg was compared with (-)-epigallocatechin (EGC) as the most effective Y and acetaldehyde (MeCHO) as the most effective Z in the presence of H_2O_2 as X. We have found that soyasaponin Ab and soyasaponin βg are equivalent to EGC(Y) and MeCHO(Z), respectively. In the presence of soyasaponin Ab as Y, CL of soyasaponin βg showed about 100 fold than that of MeCHO(Z). These results suggest that the combination of group A and DDMP saponins is important in the scavenging function of active oxygen species.

Chemiluminescence Measurement of Catechins and Gallic Acid in the Presence of Aldehyde and Active Oxygen Species

Many reports on carcinogenesis, DNA damage and lipid peroxidation emphasize the significant role of active oxygen species and alkyl peroxide-derived radicals ($R^{\bullet}$, $RO^{\bullet}$, $ROO^{\bullet}$) (*1-5*). Virus infection also causes activation of oxygen radical generation with deleterious consequences as a result of the host's over-reactive immunoresponse (*6, 7*). The scavenging and dismutation of active oxygen by enzymes such as superoxide dismutase and catalase, and by low molecular weight compounds such as ascorbic acid and glutathione have also been a subject of extensive studies from a pharmacological point of view. Flavan-3-ol catechins, natural products occurring in green tea which possess a gallyl moiety in their chemical structure, show inhibitory effects against H_2O_2-induced DNA strand breakage (*8*), lipid peroxidation (*9*) and oxidative cell damage by H_2O_2 (*10*). However, the differential activity and/or contradictive effect of catechins have also been revealed (*11*).

Recently, a radical scavenger was shown to exhibit a very weak light emission (chemiluminescence; CL) in the presence of acetaldehyde and active oxygen (*12, 13*). In the present study, the active oxygen- or radical-scavenging activity of 8 types of catechins and gallic acid were examined, and their photon constants determined by measuring their CL intensity in the presence of acetaldehyde and hydrogen peroxide (H_2O_2), hydroxyl radical ($HO^{\bullet}$), *tert*-butyl hydroperoxide (*tert*-BuOOH)or its derived radical (*tert*-$BuO^{\bullet}$). In addition, this contribution describes in detail the CL effect and the CL mode of these compounds in the presence of acetaldehyde and H_2O_2.

Gallic acid was purchased from Nakarai Co., Ltd. (Kyoto, Japan), *tert*-BuOOH and H_2O_2 were purchased from Katayama Chem. (Tokyo, Japan) and Santoku Chem. (Tokyo, Japan), respectively. (-)-Catechin (CA), (-)-epicatechin (EC), (-)-catechin gallate (CG), (-)-epigallo-catechin (EGC), (-)-gallocatechin gallate (GCG) and (-)-epigallocatechin gallate (EGCG) of *Camellia sinensis* were purchased from Kurita Co., Ltd. (Tokyo, Japan).

Gallic acid and catechins were solubilized in 50% MeOH by saponification prior to use. The chemiluminescence (CL) of these compounds was measured using a filter-equipped photon counting-type spectrophotometer (CLD-110, Tohoku Electronic Ind.), connected to a Waters Model 510 pump and U6K injector. The dispersed light at the grating was simultaneously detected on the photocathode with the image sensor. The photons counted in the wavelength between 300 and 650 nm were computed as total spectral intensities.

To determine the effect of the buffer pH on the CL intensity of catechins, a 50% MeOH solution was diluted with 50 mM citric acid-HCl buffer (pH 2.6), 25 mM phosphate buffer (pH 7.0) and 25 mM $Na_2B_4O_7$-HCl buffer (pH 9.0) was used for reactions under acidic, neutral and basic conditions, respectively. It was found that the CL intensity of individual catechins changed according to the pH of the reaction mixture. At neutral pH, the CL of the catechins was stronger than that observed at either acidic or basic conditions; with the exception of EGC and EGCG, all catechins failed to show the CL in acidic and basic conditions. Therefore, the CL of 8 types of catechins was measured in the presence of acetaldehyde and active oxygen (H_2O_2, $HO^{\bullet}$, *tert*-BuOOH, *tert*-$BuO^{\bullet}$) at pH 7.0 and 23°C. Although all catechins exhibited CL under the above neutral conditions, such a phenomenon did not occur if acetaldehyde or active oxygen was omitted from the reaction mixture. On their own, catechins also failed to exhibit CL; this result suggests that the CL of catechins in the presence of both acetaldehyde and active oxygens was a result of the formation of an excited form of catechin following the action of active oxygen.

The enhancement effect of various aldehydes for the catechin CL was compared using formaldehyde, acetaldehyde and chloral in the presence of H_2O_2. Because acetaldehyde specifically stimulated CL of catechin, this aldehyde was used throughout this experiment. Hydrogen peroxide was most effective n increasing the CL of EGC,

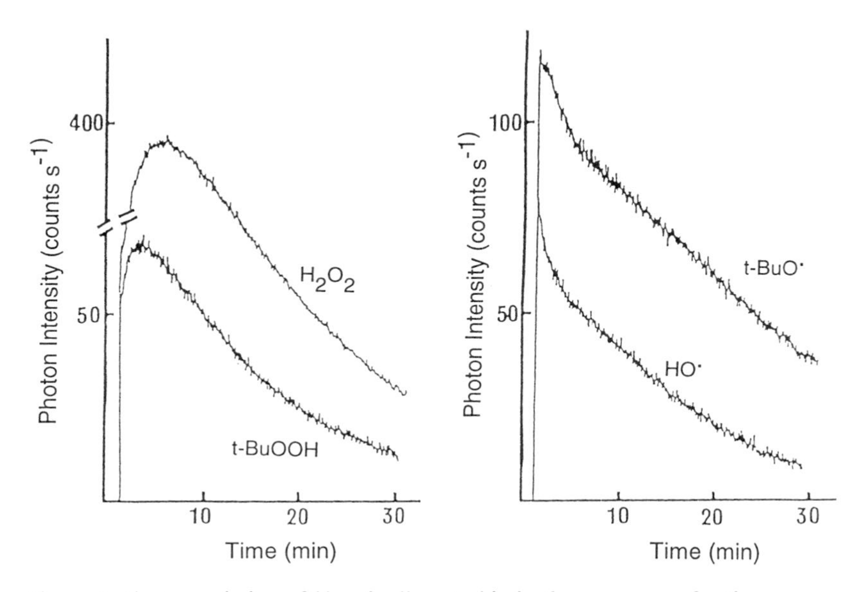

Figure 1. Photon emission of (-)-epigallocatechin in the presence of active oxygen species and acetaldehyde.
H_2O_2: 2.35 mM hydrogen peroxide (H_2O_2), 10.63 mM (-)-epigallocatechin (EGC) and 0.11 mM acetaldehyde (MeCHO);
t-BuOOH: 4.44 mM *t*-butyl hydroperoxide (*t*-BuOOH), 10.63 mM EGC and 0.11 mM MeCHO;
$HO^{\bullet}$: 1.18 mM H_2O_2, 10.63 mM EGC, 0.11 mM MeCHO and 9.5 mM $FeCl_2$;
t-$BuO^{\bullet}$: 4.44 mM *tert*-BuOOH, 10.63 mM EGC, 0.11 mM MeCHO and 9.5 mM $FeCl_2$.

followed by *tert*-BuOOH > *tert*-BuO˙ > HO˙. The time course study of EGC CL was conducted by adding acetaldehyde and active oxygen, and subsequently stopping the reaction-mixture in the flow cell (Figure 1). Two different types of time course were observed; one was for hydroperoxide (H_2O_2, *tert*-BuOOH), the second for peroxide-derived radical (HO˙, *tert*-BuO˙). In the presence of either H_2O_2 or *tert*-BuOOH, EGC CL showed a slow initial increase; with highest intensity being observed after 4.7 and 3.0 min, respectively. Thereafter, EGC CL decreased linearly to about 1.1 and 1.7 photon counts sec^{-1}, respectively. The CL of EGC in the presence of HO˙ or *tert*-BuO˙, showed an initial rapid increase, and a subsequent decay with two phases. The first phase continued until 3.5 and 5.8 min, respectively, to reach intensities of 7.5 (HO˙) and 12.8 (*tert*-BuO˙) photon counts sec^{-1}, respectively. Thereafter, the second phase continued with intensities of 4.5 and 4.4 photon counts sec^{-1} were reached.

The CL of catechins, including gallocatechins which produce protocatechuic acid or gallic acid following acid or alkaline hydrolysis, was measured in the presence of 2% acetaldehyde and 2% H_2O_2. The intensity of the CL of catechins was found to decrease in the order of EGC > EGCG > GC > GCG > EC > CA > ECG > CG > gallic acid (data not shown). The efficacy of these catechins to emit CL appears to be closely related to the number of hydroxyl groups on the B-ring of the molecule, in good agreement with the results of previous work (*14, 15*). In the presence of excess H_2O_2 and acetaldehyde, the CL of catechin was enhanced by an increase in catechin concentration. When a reaction mixture composed of EGC, acetaldehyde and H_2O_2 was subjected to HPLC (Superox ODS, 4.6 x 150 mM, S-5 µm, Shiseido Co.) with acetonitrile-water-0.05% phosphoric acid (1000:5:9000, v/v/v) as mobile phase (flow rate; 0.9ml/min), it was confirmed the EGC was almost unconsumed during the reaction; however, EGC itself did not show CL. These findings suggest that his reaction may proceed via a route similar to that described by Gotoh and Niki (*16*).

$$H_2O_2 + \text{catechin} \longrightarrow \text{catechinO}^{\bullet} + H_2O \quad (1)$$
$$\text{catechinO}^{\bullet} + \text{acetaldehyde} \longrightarrow \text{light} + \text{catechin} + \text{products} \quad (2)$$

Under the measurement conditions employed in this experiment, the CL intensity of the reaction mixture was dependent on the concentrations of H_2O_2 (X; active oxygen or radical species), catechin (Y; catalytic species) and acetaldehyde (Z; receptive species). Figure 2 shows the results obtained for H_2O_2 and EGC. The CL intensity, P, in the presence of H_2O_2 and acetaldehyde is given by equation. 3.

$$P = k\,[X]\,[Y]\oint(z) \quad (3)$$

Hence, the plot of P/[X] should give a straight line with a slope corresponding to Y concentration (Figure 2). A similar result was obtained when 7 other catechins and gallic acid were employed as Y. For EGC, the P value against concentration of X (HO˙, *tert*-BuOO˙, *tert*-BuO˙) increased linearly. The photon constants (i) of catechins and gallic acid against X, except H_2O_2, were therefore, calculated from a knowledge of [P], [X] and [Y]. The value of $\oint$ (z) is the concentration of acetaldehyde used for the reaction. The calculated photon constants are summarized in Table I. The photon constants of catechins in the presence of HO˙ and *tert*-BuO˙ were calculated using H_2O_2

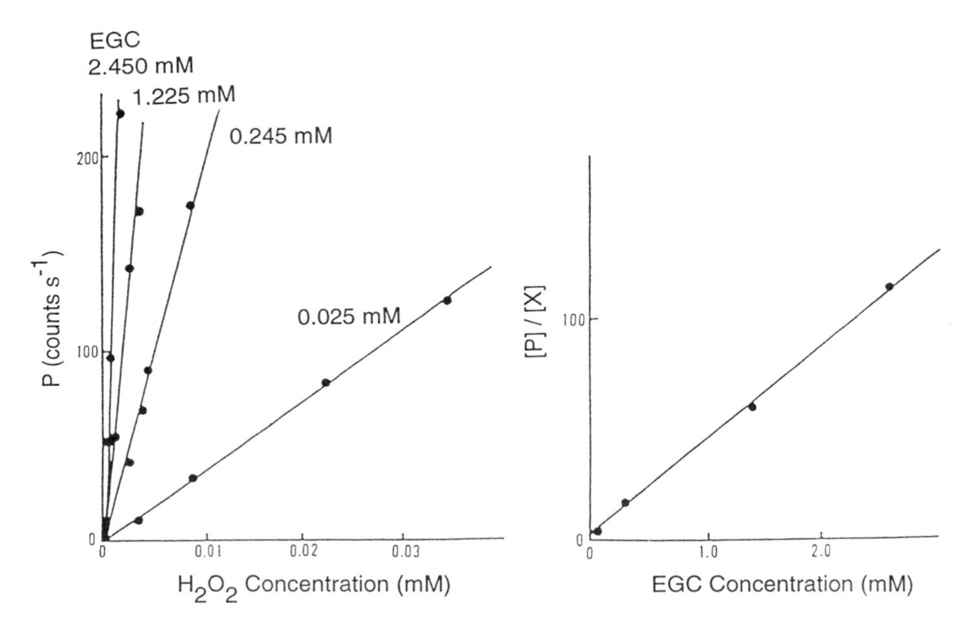

Figure 2. Effect of (-)-epigallocatechin (Y; catalytic species) and hydrogen peroxide (X; active oxygen species) concentrations on photon intensity (P) in the presence of acetaldehyde (Z; receptive species).

and *tert*-BuOOH concentrations since an excessive amount of ferric ion (Fe^{+2}) was added to the reaction mixture to produce $HO^{\bullet}$ and *tert*-$BuO^{\bullet}$.

Table I. Photon constants of catechins and gallic acid (Y; catalytic species) in the presence of acetaldehyde (Z; receptive species) and active oxygen species (X), in $M^{-2}S^{-1}$ counts at pH 7.0 and 23°C

Name	Substituent at		log k (k; photon constant)			
	C-3	C-5	HOOH	$HO^{\bullet}$	t-BuOOH	t-$BuO^{\bullet}$
(-)-Catechin	OH	H	6.52	6.38	5.98	5.72
(-)-Epicatechin	OH	H	7.47	6.50	6.25	6.20
(-)-Gallocatechin	OH	OH	7.99	6.39	6.93	7.46
(-)-Epigallocatechin	OH	OH	10.64	7.12	6.90	7.73
(-)-Catechin gallate	O-Ga	H	6.46	5.85	5.80	6.23
(-)-Epicatechin gallate	O-Ga	H	6.85	5.69	5.79	6.26
(-)-Gallocatechin gallate	O-Ga	OH	7.39	5.98	6.72	6.27
(-)-Epigallocatechin gallate	O-Ga	OH	9.41	5.51	6.42	6.77
Gallic acid	O-Ga		6.38	6.84	5.46	6.71

Ga - galloyl group; C-3 and C-5′ are substituted as follows.

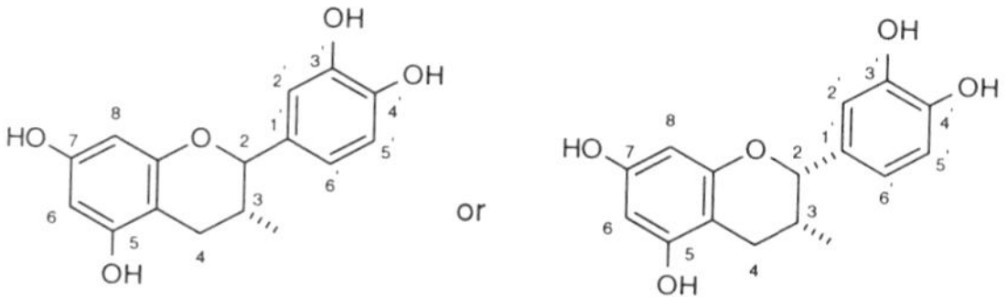

In the presence of H_2O_2 as X, the k of catechins was closely related to their chemical characteristics. The k was increased by the presence of following structural features: (a) the pyrogallol structure rather than catechol structure in the B-ring as described by Bors *et al.* (*13*), (b) a free hydroxyl group at C-3, rather than its esterified form, (c) the stereoscopic structure between C-3 hydroxyl group and B-ring, which is epimeric with respect to carbon atom 2. When CA and GC, which respectively possess catechol and pyrogallol moieties in the B-ring, were compared, the k was increased approximately 10-fold by the presence of C-5′ hydroxyl group in the B-ring [CA vs. GC; log 6.2 vs. log 7.99]. The significance of epimers and the C-3 hydroxyl group is evident from the following data: CA vs. EC, log 6.52 vs. log 7.45 for the epimer and GC vs. GCG, log 7.99 vs. log 7.39; EGC vs. EGCG, log 10.69 vs. log 9.41 for C-3 hydroxyl group (Table I).

EGC which fulfills all three requirements showed the highest increase in k, 10000-fold in comparison with that of CA. On the other hand, substitution of the C-3 hydroxyl group with gallic acid resulted in a decrease of the k of unsubstituted EGC by about

10-fold. The k of EGCG, which is the epimer of GCG, was 100-fold higher than that of GCG. The reason for this higher activity may be the increased reactivity between the C-5′ hydroxyl group of EGCG and H_2O, caused by the preferential steroscopic structure at C-2 position, CG and ECG, which both lack a hydroxyl group at C-5′, exhibited similar k. These results also suggested that hydroxyl groups at C-3 and/or C-5′ contributed to active oxygen-scavenging activity, in addition to the catechol structure with o-dihydroxyl groups. It is known that the o-dihydroxyl structure in the B-ring of flavonoids is the radical target site for all flavonoids (*2, 11, 14, 15*). Other structural moieties, such as hydroxyl groups at C-3 and/or C-5′ may be involved in radical-scavenging in flavonoids which possess there partial structures.

In the presence of *tert*-BuOOH or *tert*-BuO$^•$, the k of catechins had similar values, but these were lower than that measured in the presence of H_2O_2. In the presence of HO$^•$, gallic acid showed the highest k among catechins, except for that produced by EGC.

Although it is generally reported that HO$^•$ or *tert*-BuO$^•$ are both highly reactive and may attack biological membranes and tissues directly, the k of catechin was higher in H_2O_2 than in HO$^•$, suggesting that H_2O_2 is also an active oxygen species which may cause severe damage in biological systems. The radical scavenging activity of antioxidants is generally addressed from the standpoint of how rapidly they react with active oxygen species; however, only these oxidative substances or radicals which have been formed by secondary chemical reactions have been examined in detail with respect to their cytotoxic and mutagenic/antimutagenic activities.

From the results presented in this chapter, it may now be necessary to investigate the precise mechanism by which the high potential energy within the cell, which may be produced in the presence of active oxygens and antioxidants in the biological systems, is dissipated. For the CL of catechin, Y may depend on the rate of release of energy from catechins, which reflect their radical scavenging activities, can be calculated from the concentration of active oxygen or radical species, [X], the concentration of the catalytic species, [Y], and the concentration of the receptive species, [Z]. In addition, it has been shown that the method employed in this study is both sensitive and rapid, and can be applied for measurement of the reactivities of various natural and synthetic antioxidants toward active oxygen.

Chemiluminescence of Soybean Saponins in the Presence of Active Oxygen Species

Many types of saponins have been isolated from soybean seeds, and these may be divided into two groups, group A (*17*) and DDMP (2,3-dihydro-2,5-dihydroxy-6-methyl-4H-pyran-4-one) saponins (*18*) (Figure 3). Group A saponins are present only in the soybean hypocotyl, while DDMP saponins are widely distributed in leguminous seeds and are present in both hypocotyl and cotyledon, especially in the case of soybean. From pharmacological or health point of view, soybean saponins have been reported to show various effects such as hypolipidemic, antioxidative activity (*19*) and HIV infection inhibitory properties (*20*). Although both group A and DDMP saponins are present in the hypocotyl of soybean seeds at high concentrations, the interrelationship concerning their physiological role is not clear.

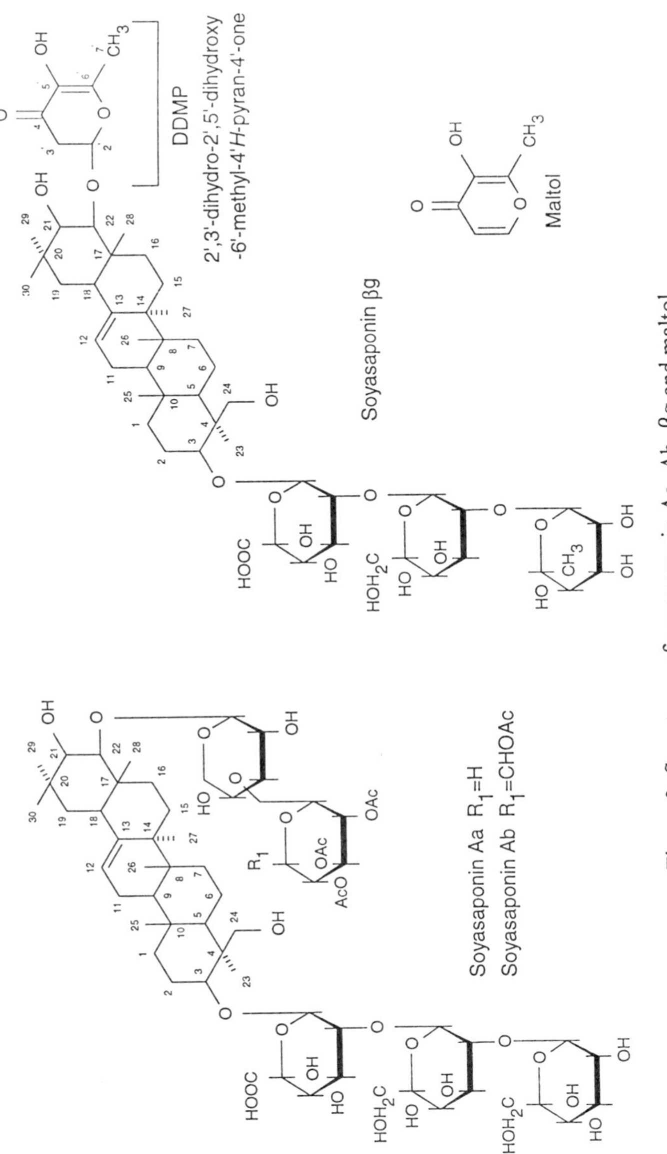

Figure 3. Structures of soyasaponin Aa, Ab, βg and maltol.

In this communication, we further describe interrelationship between group A and DDMP saponins based on the CL exhibited in the presence of active oxygen species.

Soyasaponin Ab and soyasaponin βg were isolated from hypocotyls (500 g) of soybean by extraction with 70% ethanol containing 0.01% EDTA and centrifuged at 8,000 rpm for 15 min. The extract was evaporated and dissolved in H_2O:n-BuOH (1:1, v/v); the n-BuOH layer was evaporated and the soyasaponin Ab and DDMP saponin fraction were isolated as described previously (*17, 21*). Soyasaponin Ab and soyasaponin βg were solubilized in 50% MeOH by saponification just before use. The chemiluminescence (CL) of soyasaponin Ab and soyasaponin βg were measured as described earlier.

Hydrogen peroxide, soyasaponin Ab and soyasaponin βg in 50 mM phosphate buffer (pH 7.0) did not exhibit CL. The latter two compounds as a mixture also did not exhibit CL. To study the role of group A and DDM0 saponin on CL, the CL of soyasaponin Ab and soyasaponin βg was measured in the presence of H_2O_2 (X) and acetaldehyde (Z), conditions which showed the strongest CL with phenolic compounds such as flavonoids, anthocyanins, catechins as factor Y. Although soyasaponin βg did not exhibit significant CL, soyasaponin Ab showed CL of about one tenth when compared with that of (-)-epigallocatechin (EGC) which exhibited the highest CL among phenolic compounds tested. The CL of soyasaponin Ab and soyasaponin βg was also stronger in comparison with that of MeCHO; soyasaponin Ab exhibited only a minor CL. These results indicate that soyasaponin Ab and soyasaponin βg act as factors Y and Z, respectively. Soyasaponin βg is thermally unstable and in basic conditions and is easily converted to soyasaponin I, identical to soybean saponin Bb, and maltol (*22*). Maltol exhibited CL as factor Z in the presence of H_2O_2 and EGC, however the CL intensity of 1 mM maltol (0.5 counts s^{-1}) was low when compared with that of 1 mM soyasaponin βg (1.1 x 103 counts s^{-1}) (Table II).

Table II. Photon counts (s^{-1}) as Z (receptive species) in the presence of X (active oxygen species) and Y (catalytic species)

	Photon counts (s^{-1})		
	pH 2.6	pH 7.0	pH 9.0
X ($HO^{\bullet}$), Y (GA)			
acetaldehyde	1.3	1.5	2.3
soyasaponin βg	1.5 x 1.0^3	1.0 x 10^3	8.9 x 10^2
maltol		0.1	
X (H_2O_2), Y (EGC)			
acetaldehyde	1.1	4.1	3.6
soyasaponin βg	1.3 x 10^3	1.1 x 10^3	2.1 x 10^3
maltol		0.5	

0.24M H_2O_2, 1.48 mM (-)-epigallocatechin (EGC), 1.16 mM gallic acid (GA) and 5 mM $FeCl_2$ for the Fenton reaction. Photon counts calculated to 1 mM Z.

For determining the effect of different pH buffers on the CL intensity of soyasaponin βg, 50% MeOH diluted with 50 mM citric acid-HCl buffer (pH 2.6), 25 mM phosphate buffer (pH 7.0) and 25 mM $Na_2B_4O_7$-HCl (pH 9.0) was used for reactions under acidic, neutral and basic conditions, respectively. It was found that the CL intensity of soyasaponin βg as factor Z in the presence of H_2O_2 (X) and EGC (Y), was only slightly dependent on the pH value of the reaction mixture (Table II). Photon intensity of soyasaponin βg against H_2O_2 was effected by the concentration of Y and Z in the presence of H_2O_2. The photon counts (s^{-1}) of acetaldehyde and soyasaponin βg as factor Z were measured under the same condition (0.24M H_2O_2 and 1.48 mM EGC or 0.24M H_2O_2, 1.16 mM gallic acid and 5 mM $FeCl_2$) and calculated. Under the acidic and neutral conditions, the photon counts of soyasaponin βg were stronger (1.5×10^3 and 1.0×10^3) than in acidic (8.9×10^3). The CL intensity of MeCHO as factor Z against H_2O_2 in the presence of EGC was 1.3, 1.5 and 2.3 at pH 2.6, pH 7.0 and pH 9.0, respectively (Table II). Therefore, a 50% MeOH solution containing 25 mM phosphate buffer (pH 7.0) was used in this experiment.

The CL of soyasaponin βg was higher than that of MeCHO (Z) by 10 fold in the presence of the same concentration of EGC and H_2O_2. This result suggests that the combination of Y and Z factors is important for CL. The CL of soyasaponin Ab and soyasaponin βg in the presence of H_2O_2 was measured to identify the interrelationship between group A and DDMP saponins (Figure 4). When CL of interrelationship Ab [Y; catalytic species] was measured in the presence of H_2O_2 [X; active oxygen species] and soyasaponin βg [Z; receptive species], the photon intensity [P] increased linearly depending on the concentrations of H_2O_2 and of soyasaponin Ab. This result indicates that the CL of soybean saponins is emitted by XYZ mechanism, $[P] = k\,[X]\,[Y] \oint (Z)$ similar to that found for phenolic compounds. CL of soyasaponin βg in the presence of soyasaponin Ab against H_2O_2 was stronger than that of MeCHO(Z) by about 100 fold.

The polymorphisms of group A saponin composition that exist in soybean seeds are controlled by the same gene (*23*) and the presence of both soyasaponin Aa and Ab as the main saponin of group A can be explained by co-dominant acting genes. Soybean hypocotyl contains soyasaponin Aa or Ab and a relatively high concentration of 1 to 5% soyasaponin βg (*24*). Therefore, some possibilities can be discussed by using the main soybean saponin present in the hypocotyl. The CL occurs in the presence of X (active oxygen species), Y (catalytic species), and Z (receptive species). We found that the soybean saponin played the role as Y or Z. It has been reported that group A saponins of soybean have antioxidative activity (*19*) and may act as a Y catalyst, because phenolic compounds such as flavonoids, anthocyanins and catechins which function as catalytic species, Y, in the presence of X and Z, also have antioxidative activity (*14*). Since a mixture of H_2O_2 and soyasaponin Ab does not exhibit CL, it is thought that CL occurs in two stages: (1) electron exchange to excite the state of soyasaponin Ab ($Y''^{\bullet}$), since H_2O_2 is not a radical, and CL is intense and rapid in the presence of the hydroxyl radical ($HO^{\bullet}$) generated by the Fenton reaction, and (2) electron exchange to ground sate ($Y' \longrightarrow Y$) in the presence of Z.

A clear difference in the role for CL reaction against H_2O_2 between soyasaponin Ab and soyasaponin βg from soybean seed hypocotyls seems to be explained by structural differences; the sugar chain attached to C-22 of the aglycone occurs in soyasaponin Ab,

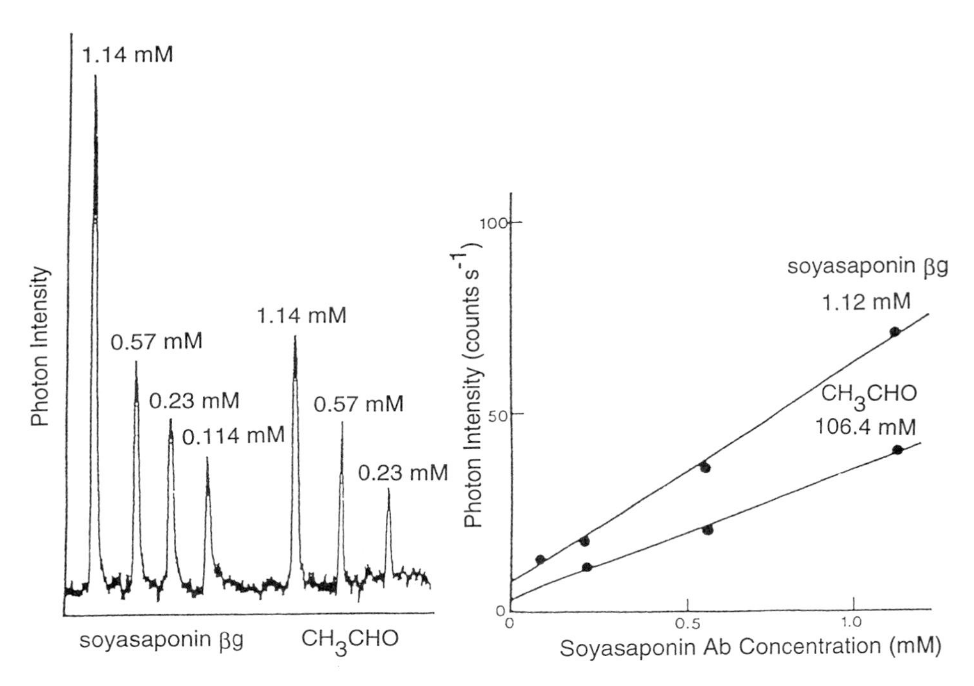

Figure 4. Chemiluminescence of soyasaponin Ab (Y; catalytic species) in the presence of hydrogen peroxide (X; active oxygen species) and soyasaponin βg or acetaldehyde (Z; receptive species).

while it is absent in soyasaponin βg (Figure 3). Nishida *et al.* (*25*) have shown that group A saponins strongly inhibit the CCl_4-dependent lipid peroxidation of mouse liver microsomes, and suggested that the sugar chain attached to C-22 of the sapongenol in the group A saponins may play a similar role to the phytyl side chain of α-tocopherol which is known to react directly with peroxy radicals (*25*). Although another saponin, glycyrrhizin, which has a similar chemical structure except for substitution at C-10, C-26 and C-30, and a sugar chain at C-3 also exhibited CL as Y in the presence of H_2O_2 and MeCHO, glycyrrhetinic acid (the aglycone)did not (data not shown). These results also suggest that the presence of a sugar chain is related to the occurrence of CL. The sugar chain may also play an important role in making the structure suitable for the electron exchange; this may be due to the structural complexity of saponins, because CL was affected by species and combination of Y and Z. Since maltol, which is a derivative of soyasaponin βg, formed after heating or under basic conditions, exhibited a weak CL in comparison with that of soyasaponin βg as factor Z, the DDMP moiety of soyasaponin βg might be slightly effected by the reaction with H_2O_2. The C-22 of the saponin aglycone and C′-2 of DDMP moiety conjugated at C-22 of the glycone of soyasaponin βg may play an important role as receptive species because these signals were found to decrease with time according to a ^{13}C-NMR spectral study of soyasaponin βg. These findings are the subject of further studies.

Chemiluminescence resulting from a soybean seedling extract in the presence of H_2O_2 and MeCHO continued for 30 min (data not shown) and may be due to more than one compound because two different types of CL were observed in the presence of H_2O_2 and MeCHO, namely an initial rapid increase and fast decay type, and a slow decay type (data not shown). In conclusion, the present results indicate that soyasaponin Ab and soyasaponin βg (obtained from soybean seed hypocotyls are able to act as factors Y and Z, respectively, on the CL reaction in the presence of H_2O_2. Soyasaponin Ab and soyasaponin βg, in combination, show strong scavenging of active oxygen and, additionally, there may be other compounds acting as Y and Z factors, because the autoclaved extracts of soybean seedling have also been found to exhibit strong CL.

Acknowledgments

This work was supported in part by a grant from the Bio Renaissance Program (BRP 93- VI-B-1) of the Ministry of Agriculture, Forestry, and Fisheries (MAFF), Japan, and by a grant in aid for Scientific Research (Project No. 05660128) from the Ministry of Education, Science, and Culture of Japan. We thank Dr. G.R. Fenwick of the Institute of Food Research, UK, for his assistance in the preparation of this manuscript.

Literature Cited

1. Basaga, H.S. *Biochem. Cell Biol.* **1990**, *68*, 989.
2. Audic, A.; Giacomoni, P.U. *Photochem. Photobiol.* **1993**, *57*, 508.
3. Aruoma, O.I.; Halliwell, B.; Gajewski, E.; Dizdaroglu, M. *J. Biol. Chem.* **1989**, *264*, 20509.
4. Witz, G. *Proc. Soc. Exp. Biol. Med.* **1991**, *198*, 675.
5. Slater, T.F. *Biochem. J.* **1984**, *222*, 1.

6. Oda, T.; Akaike, T.; Hamamoto, T.; Suzuki, F.; Hirano, T.; Maeda, H. *Science* **1989**, *244*, 974.
7. Maeda, H.; Akaike, T. *Proc. Soc. Exp. Biol. Mec.* **1991**, *198*, 721.
8. Nakayama, T.; Kuno, T.; Hiramitsu, M.; Osawa, T.; Kawakishi, S. *Biosci. Biotech. Biochem.* **1993**, *57*, 174.
9. Ariga, T.; Hamano, M. *Agric. Biol. Chem.* **1990**, *54*, 2499.
10. Nakayama, T.; Yamada, M.; Osawa, T.; Kawakishi, S. *Biochem. Pharmacol.* **1993**, *45*, 265.
11. Stich, F.H. *Mutat. Res.* **1991**, *259*, 307.
12. Yoshiki, Y.; Okubo, K.; Onuma, M.; Igarashi, K. *Phytochemistry* **1995**, *39*, 225.
13. Yoshiki, Y.; Okubo, K.; Igarashi, K. *J. Biolumi. Chemilumi.* **1995**, in press.
14. Bors, W.; Heller, W.; Michel, C.; Saran, M. *Methods in Enzymol.* **1990**, *186*, 343.
15. Puppo, A. *Phytochemistry* **1992**, *31*, 85.
16. Gotoh, N.; Niki, E. *Biochim. Biophys. Acta* **1992**, *1115*, 201.
17. Shiraiwa, M.; Kudou, S.; Shimoyamada, M.; Harada, K.; Okubo, K. *Agric. Biol. Chem.* **1991**, *55*, 315.
18. Kudou, S.; Tonomura, M.; Tsukamoti, C.; Uchida, T.; Sakabe, T.; Tamura, N.; Okubo, K. *Biosci. Biotech. Biochem.* **1993**, *57*, 546.
19. Ohminami, H.; Kimura, Y.; Okuda, H.; Arich, S.; Yoshikawa, M.; Kitagawa, I. *Planta Medica* **1984**, *46*, 440.
20. Nakashima, H.; Okubo, K.; Honda, Y.; Tamura, T.; Matsuda, S.; Yamamoto, N. *AIDS* **1989**, *3*, 655.
21. Yoshiki, Y.; Kim, J.H.; Okubo, K.; Nagoya, I.; Sakabe, T.; Tamura, N. *Phytochemistry* **1995**, *38*, 229.
22. Okubo, K.; Yoshiki, Y.; Okuda, K.; Sugihara, T.; Tsukamoto, C.; Hoshikawa, K. *Biosci. Biotech. Biochem.* **1994**, *58*, 2248.
23. Tsukamoto, C.; Kikuchi, A.; Harada, K.; Kitamura, K.; Okubo, K. *Phytochemistry* **1993**, *34*, 1351.
24. Shiraiwa, M.; Harada, K.; Okubo, K. *Agric. Biol. Chem.* **1991**, *55*, 323.
25. Nishida, K.; Araki, Y.; Nagamura, Y.; Ohta, Y.; Ito, M.; Ishiguro, I. *Med. Biol.* **1993**, *126*, 87.

Chapter 16

Phytoestrogens and Lignans: Effects on Reproduction and Chronic Disease

Sharon E. Rickard and Lilian U. Thompson

Department of Nutritional Sciences, Faculty of Medicine, University of Toronto, 150 College Street, Toronto, Ontario M5S 3E2, Canada

Phytoestrogens and lignans are components of a variety of plant foods. The adverse estrogenic effects of phytoestrogens on reproductive development have been observed in domestic and experimental animals. The type of estrogenic effect seen depends on both the relative potency of the compound and the time of exposure. Human consumption of foods rich in isoflavones or lignans has led to menstrual cycle changes, alleviation of menopausal symptoms as well as anti-osteoporosis effects. Human correlation and animal studies suggest that consumption of foods rich in phytoestrogens and lignans may result in reduced risk of chronic disease such as cancer and cardiovascular disease. Anticancer mechanisms include antagonism of estrogen metabolism, antioxidant activity, and modulatory effects on key control points of the cell cycle. Hypocholesterolemic effects as well as inhibition of platelet activation are postulated antiatherogenic mechanisms. The relative adverse and beneficial effects of each compound should be considered separately.

It is widely accepted that diet plays an important role in the development of many chronic diseases. Government agencies in both the United States and Canada recommend increased consumption of fruits, vegetables and grains in light of the growing body of evidence supporting their ability to reduce the risk of chronic disease, especially cancer (*1*, *2*). Phytoestrogens and lignans, non-nutrient components present in fruits, vegetables and grains (*3-5*), may be partly responsible for their observed protective effects. However, adverse effects have been observed for some phytoestrogens and lignans, particularly on reproductive development. In this chapter, some of the recent work on the effects of phytoestrogens and lignans will be reviewed. First, a description of typical compounds and sources will be presented in addition to their biological mode of action. Second, the estrogenic effects of phytoestrogens and lignans observed in exposed domestic and experimental animals will be discussed. Third, human exposure

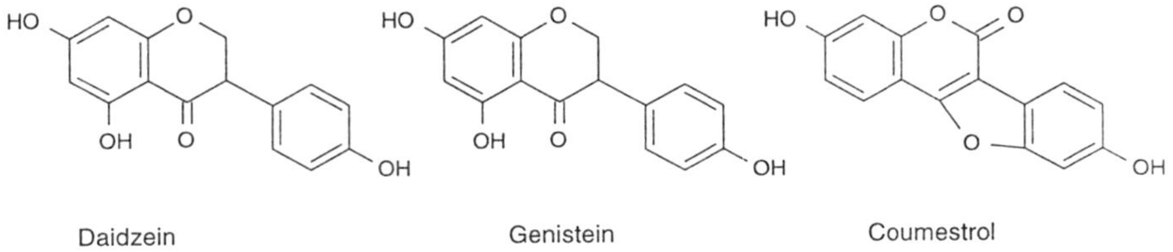
HO
O
OH
O
OH
Daidzein
HO
O
OH
O
OH
Genistein
HO
O
O
O
OH
Coumestrol

Figure 1. Typical structures of the isoflavones and coumestans.

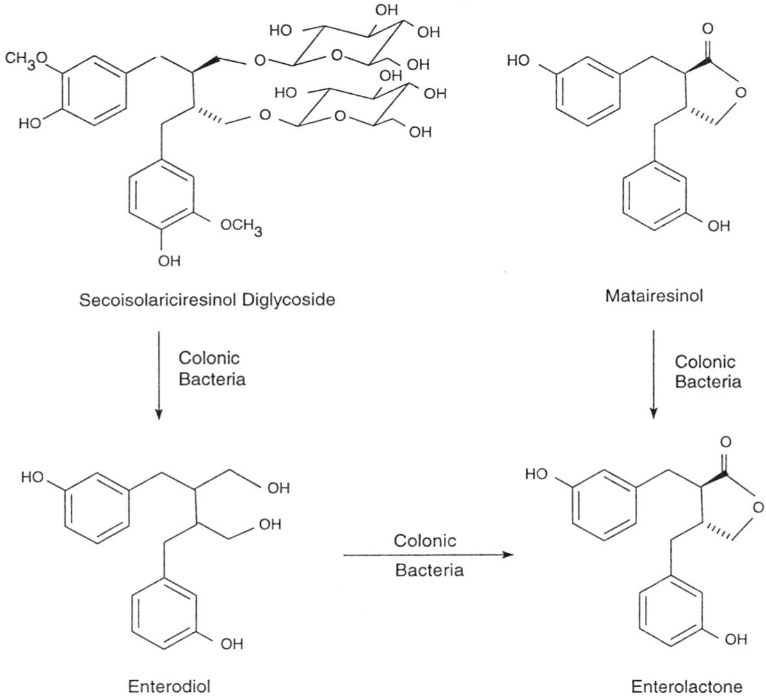
CH3O
HO
OH
OCH3
HO
OH
OH
O
O
OH
HO
OH
O
O
OH
Secoisolariciresinol Diglycoside
HO
O
O
OH
Matairesinol
Colonic
Bacteria
Colonic
Bacteria
HO
OH
OH
OH
Enterodiol
Colonic
Bacteria
HO
O
O
OH
Enterolactone

Figure 2. Mammalian lignan production from plant precursors.

to these compounds will be examined with specific effects in premenopausal and postmenopausal women. Finally, studies demonstrating the beneficial effects of phytoestrogens and lignans in chronic diseases, such as cancer and cardiovascular disease, will be reviewed with emphasis on their possible mechanisms of action.

Phytoestrogens and Lignans: Types of Compounds, Sources and Mode of Action

Types of Compounds. Phytoestrogens may be subdivided into two main classes, isoflavones and coumestans with typical structures shown in Figure 1. The most abundant isoflavones are glycosides of genistein (4′,5,7-trihydroxyisoflavone) and daidzein (4′,7-dihydroxyisoflavone) and their respective precursors biochanin A (5,7-dihydroxy-4′-methoxyisoflavanone) and formononetin (7-hydroxy-4′-methoxy-isoflavone). Of the coumestans, coumestrol, 4′-methoxycoumestrol, sativol, trifoliol and repensol are the most important (*6*). Lignans are a class of diphenolic compounds generally containing a dibenzylbutane skeleton structure with characteristics similar to that of the phytoestrogens (*7*). Although many different plant lignans have been identified and/or isolated (*8*), attention has recently focused on the mammalian lignans enterolactone [*trans*-2.3-bis-(3-hydroxybenzyl) butyrolactone] and enterodiol [2,3-bis(3-hydroxybenzyl)butane-1,4-diol] discovered in the biological fluids of humans and animals (*9-15*). Enterodiol and enterolactone are directly produced by colonic bacterial action on the plant lignans secoisolariciresinol and matairesinol, respectively (*16*) (Figure 2). Similarly, formononetin and daidzein can be converted to equol and *O*-desmethylangolensin by intestinal bacteria (*17*). Phytoestrogens, plant lignans and their bacterial products undergo enterohepatic circulation, a major proportion being detected in the urine (*7*).

Sources. Phytoestrogens and lignans are widely distributed in the plant kingdom. There are at least 220 species of plants containing isoflavonoids and 58 species with coumestan (*18*). Of these plants, soybean contains one of the highest levels of phytoestrogens (*5*) (Table I). Flaxseed has been shown, both in vivo (*3*) and in vitro (*4*), to be the richest source of mammalian lignan precursors, particularly secoisolariciresinol, compared to a variety of whole grains, seeds, legumes, fruits and vegetables tested (Table II). Despite their relatively lower concentrations of lignans, whole grains, seeds, legumes, fruits and vegetables may make significant contributions especially in vegetarians (*19*). Urinary excretion levels of phytoestrogens and lignans to phytoestrogen and lignan intake because they are consumed in larger quantities, are commonly used to assess the intake and metabolism of precursor compounds in foods. Studies comparing levels of phytoestrogens and lignans in the urine of subjects fed specific diets showed higher urinary phytoestrogen excretion with legume/allium or soy-based diets and higher urinary lignan excretion with high fruit and vegetable diets (*20*, *21*).

Mode of Action. Phytoestrogens and lignans are thought to exert their estrogenic or antiestrogenic effects by binding to the estrogen receptor. Endogenous steroidal estrogens, such as estrone, estradiol and estriol, bind to the estrogen receptor in the cell nucleus. The transformed estrogen receptor complex then binds to specific DNA sites for transcription of specific genes and induction of the known estrogenic responses in the reproductive tract, such as increased growth rate, hypertrophy and hyperplasia (*22*). The ability of non-steroidal compounds such as phytoestrogens, lignans or the potent synthetic estrogen diethylstilbestrol (DES) to bind to the estrogen receptor is partly determined by the distance between the two extreme hydroxyl groups (*23*). Phytoestrogens and lignans lacking this requirement have no estrogen binding activity (*24*). Because phytoestrogens and lignans are weak

Table I. Phytoestrogen Levels in Various Foods[1]

Food	*Isoflavones[2]*	*Coumestrol*
	(μg/g sample)	
Soy bean seeds, dry	1,953.0	nd
Soy flour	1,777.3	nd
Black soybean seeds, dry	1,310.7	nd
Tofu	278.8	nd
Soy bean seeds, fresh, raw	181.7	nd
Green split peas, dry	72.6	nd
Clover sprouts	30.7	280.6
Kala chana seeds, dry	19.0	61.3
Soy bean hulls	18.4	nd
Black eyed bean seeds, dry	17.3	nd
Small white bean seeds, dry	15.6	nd
Garbanzo bean seeds, dry	15.2	nd
Pink bean seeds, dry	10.5	nd
Small lima bean seeds, dry	9.2	nd
Yellow split peas, dry	8.6	nd
Mung bean seeds, dry	6.1	nd
Great northern bean seeds, dry	6.0	nd
Pinto bean seeds, dry	5.6	36.1
Red bean seeds, dry	3.1	trace
Green beans, fresh, raw	1.5	nd
Round split peas, dry	nd	81.1
Alfalfa sprouts	trace	46.8
Large lima bean seeds, dry	trace	14.8

[1]Determined by HPLC analysis (Adapted from ref. 5)
[2]Isoflavones = daidzein + genistein + formononetin + biochanin A

estrogens, their lower affinity for the estrogen receptor results in a less stable complex and a reduced ability for the complex to be transformed (*25-27*). Whether phytoestrogens act as estrogens or antiestrogens depends on the presence and relative concentrations of stronger steroidal estrogens (*28*). Stimulatory effects seen for phytoestrogens and lignans alone are inhibited by stronger antiestrogens (*29*, *30*).

Exposure of Domestic Animals to Estrogenic Compounds in Plants

Subterranean clover, found in some grazing pastures, may contain up to 5% dry weight of estrogenic isoflavones, including genistein, formononetin and biochanin A (*31*). Because genistein and biochanin A may be broken down into non-estrogenic products in the rumen, only formononetin is considered to be important in ruminants. The estrogenicity of formononetin is enhanced by its extensive metabolism by ruminal microflora to the more estrogenic compound equol (4',7-dihydroxyisoflavan), the alleged infertility agent (*32*). The estrogenic effects of phytoestrogens described for sheep and cattle below may be partially explained by their relatively low concentrations of circulating estradiol (*33*).

Table II. Mammalian Lignan Production from Various Plant Foods[1]

Food Group	*Mammalian Lignans[2] (µg/100 g sample[3])*	*Range*
Oilseeds	20,461 ± 12,685 (n = 6)	161 (peanuts) - 67,541 (flaxseed)
Dried Seaweeds	900 ± 247 (n = 2)	653 (hijiki) - 1,147 (mekuba)
Whole Legumes	562 ± 211 (n = 7)	201 (pinto bean) - 1,787 (lentil)
Cereal Brans	486 ± 90 (n = 5)	181 (rice bran) - 651 (oat bran)
Legume Hulls	371 ± 52 (n = 5)	222 (fava bean) - 535 (navy bean)
Cereals	359 ± 81 (n = 9)	115 (barley) - 924 (triticale)
Vegetables	144 ± 23 (n = 27)	21 (fiddle head, tomato) - 407 (garlic)
Fruits	84 ± 22 (n = 7)	35 (apple) - 181 (pear)

[1] Summary of results from in vitro fermentation of food with human fecal microbiota for 48 h (Adapted from ref. 4)
[2] Enterodiol and enterolactone
[3] As-is (wet) basis

Effects in Sheep. The adverse effects of dietary phytoestrogens were first documented by Bennetts and colleagues in the 1940s with reports of severe clinical abnormalities in ewes grazing pastures with a high content of the clover *Trifolium subterraneum* in Western Australia (*34*). This condition, known as "clover disease", is characterized by low lambing rates, prolapse of the uterus and difficulty in birth. Severe uterine inflammation with pus and fluid accumulation have also been observed (*31*).

With the use of low-formononetin cultivars of subterranean clover, severe cases of the classical clover disease are now uncommon. However, attention has now been focused on the subclinical infertility of female sheep (males are unaffected) exposed to low concentrations of phytoestrogens. Temporary infertility results from activational effects of phytoestrogens on sheep mated on estrogenic pastures. Reduced ovulation and possibly decreased conception rates are observed, with some ewes developing swelling and reddening of the vulva and enlargement and hypertrophy of the mammary glands. Morphological defects of the central nervous system include microgliosis and neuronal damage to the hypothalamus although few functional defects are evident. These effects disappear within 3 to 5 weeks after removal of the ewes to a non-estrogenic pasture (*33*). Permanent infertility is due to organizational effects of phytoestrogens from prolonged exposure to estrogenic pastures. The primary symptom is failure to conceive due to low viscoelasticity of cervical mucus and hence impaired transport of spermatozoa. This cervical dysfunction results in an irreversible redifferentiation of the cervix to uterine-like tissue both histologically and functionally (*33*). Similarly, loss of cervical structure has also been seen in mice treated with estrogen during neonatal life (*35*) and in women exposed to DES in utero (*36*). Interestingly, a recent study by Croker and co-workers in Australia observed that Merino ewes developed a resistance to the infertility effects of grazing on the estrogenic clover pastures within three generations (*37*), suggesting that phytoestrogen-induced infertility may be eradicated in the future through a natural selection process. However, the significance of this finding requires further exploration.

Effects in Cattle. Although reports of infertility of cattle on subterranean clover are uncommon, infertility effects have been observed after consumption of red clover (*Trifolium pratense*) silage containing isoflavones and alfalfa (*Medicago sativa*) containing coumestans. In addition to the effects seen for sheep, cystic ovarian disease was common in cows consuming estrogenic feed (*31*). Daily consumption of red clover isoflavones by dairy cows may reach 50-100 g, but permanent infertility does not occur (*38*). However, ovarian function may take several weeks or months to recover after removal of cattle from estrogenic feed (*31*). The differences in effects of the phytoestrogens in sheep and cattle may be due to differences in metabolism. Contrary to what might be expected, sheep have greater clearance rates of equol and a higher capacity to conjugate phytoestrogens than cows, indicating a greater ability to detoxify and eliminate these compounds (*38*). However, estrogen receptor levels in the uteri of sheep are up to four times higher than that in cattle, suggesting that sheep may be more sensitive to the lower plasma levels (*38*).

Effects in Pigs. Swine feed may also contain significant amounts of phytoestrogens from red clover and soy products. However, reports of hyper-estrogenicity of isoflavones in pigs have not been documented even though they are known to be highly sensitive to the mycoestrogen zearalenone produced by mold fungus *Fusarium sp.* contaminating feed (*38*). In pigs fed a 20% red clover diet, the formononetin plasma levels were 10-fold greater than the highest levels found in bovine plasma. Total plasma levels of conjugated equol are 10-15 times lower in the pig than the sheep or cow. However, plasma levels of the biologically active equol are similar in these three species (*38*). The possible estrogenic effects of phytoestrogens in pigs should be explored.

Effects in Poultry. Flaxseed, due to its high content of alpha-linolenic acid, has been used in poultry diets to increase yolk n-3 fatty acids for designer egg production (*39*). It was recently determined that avian cecal microflora have the ability to produce lignans from flaxseed (*40*). In addition, enhanced follicle uniformity, increased incidence of double-yolk eggs and potential oviduct regression were found in laying hens consuming flaxseed during early reproduction (*40*). However, these effects, attributed to decreased circulating serum estradiol, were overcome by the onset of peak production. Although the long-term consequences of flaxseed feeding during early reproduction require further study, these preliminary results suggest that hens fed flaxseed during peak production exhibit no adverse effects.

Exposure of Experimental Animals to Phytoestrogens and Lignans: Effects at Different Stages of Development

The effects phytoestrogens on reproduction in rats have been well documented. In contrast, the effects of lignan exposure have only been studied after weaning (21 days of age). Most work has centered around the most potent estrogenic phytoestrogen coumestrol which has a relative activity of 10^{-3} of estradiol whereas isoflavones have relative estrogenic activities of 10^{-5} (*22*). Exposure to phytoestrogens gestationally or during the period of lactation may have marked effects on sexual differentiation or on normal reproductive development, with some changes manifesting later in puberty. As seen below, the type of effect observed with phytoestrogen exposure is dependent not only on the relative potency of the phytoestrogen but on the timing of administration as well.

In Utero. Injection of pregnant dams with a low dose (5 mg) genistein caused external genital feminization, identified by a reduction in the anogenital distance (AGD = length of tissue separating the anus from the genitalia papilla) in rats. Although AGD was not affected by high doses (25 mg) of genistein in utero, an antiestrogenic response was observed where vaginal opening (the onset of puberty) was delayed (*41*). A morphological marker of sexual differentiation, the sexually dimorphic nucleus in the preoptic area of the hypothalamus (SDN-POA), increases with exposure to estrogens (*42*). In utero injections of either high or low doses of genistein did not have any effect on SDN-POA volume (*41*). However, postnatal exposure to high dose genistein was found to cause a significant increase in SDN-POA volume (*43*, *44*), indicating an estrogenic effect. The differential effect of genistein gestationally vs. neonatally is due to the variation in hormone sensitivities at different critical stages of development.

Lactation/Neonatal. Persistent cornification of vaginal epithelia (permanent estrus state) and an inability to increase luteinizing hormone after estradiol exposure was observed in female rats exposed to coumestrol at the 0.01% level throughout the 21-day lactation period (*45*). No effects were seen on the onset of puberty in these rats although vaginal opening did occur at a lighter body weight. Precocious vaginal opening, persistent vaginal cornification and morphological alterations in the ovary, uterus and cervicovaginal tissue have been observed in mice treated with similar doses (up to 100 μg) of coumestrol (*46*). A subsequent study by Whitten and colleagues (*47*) found that female rats had normal cycling patterns if exposure was restricted to the first 10 days of life. However, male offspring exposed during this 10 day period exhibited significant deficits in adult sexual behavior measured by frequency and prolongation of latencies in mounting and ejaculation. Although gonadal function was not impaired, the males produced female cyclic patterned gonadotropin releasing hormone secretions, indicating demasculization of the males with coumestrol exposure (*47*). Medlock and colleagues (*48*) found that initial increases in uterine weight and premature gland development in rats exposed to 100 μg coumestrol during the first 5 days of lactation resulted in severe depression of both uterine weight and estrogen receptor levels at later ages. Although the same amount of equol also lowered uterine weight, it had no effect on estrogen receptor levels (*48*). When given during the period of rapid uterine development (postnatal days 10-14), the dose-dependent inhibition of gland genesis by coumestrol (100 μg) and equol (100 μg) was not as severe as that seen for DES or tamoxifen, an antiestrogenic drug commonly used in adjuvant breast cancer therapy (*48*). Contrary to the studies by Whitten and colleagues (*45, 47*), Medlock and colleagues (*48*) found no long-term adverse effects following neonatal exposure to similar levels of coumestrol nor to equol.

Weaning. Short-term exposure of immature rats to human dietary concentrations of coumestrol (0.005%-0.1%) results in significant increases in uterine weight accompanied by induction of cytosolic progestin receptors and increased nuclear estrogen binding (*49*). A later study showed that orally administered coumestrol had additive effects on the uterotrophic response to estradiol, but cytosolic levels of the estrogen receptor were reduced due to elevated and extended estrogen exposure (*50*). Precocious vaginal opening occurred in weanling rats chronically exposed to coumestrol (0.01%-0.1%) for four weeks, and initial regular cycling at 48-60 days of age became increasingly irregular at 116-131 days of age (*51*).

Because lignans are structurally similar to isoflavones, they are also thought to have similar estrogenic and antifertility effects. However, studies on cycling in rats suggest that the effects of lignans are antiestrogenic. Dietary supplementation with 10% flaxseed or 1.5 or 3.0 mg/day of its purified precursor secoisolariciresinol diglycoside (SDG) in 43 day old rats has caused either disruption, evident by a permanent diestrus state (as opposed to permanent estrus with coumestrol), or lengthening of the overall estrous cycle (*52*). Similar effects were seen for the antiestrogen tamoxifen (1 mg/kg BW/d) (*52*). Weanling rats given the same daily doses of SDG for four weeks also exhibited significant increases in acyclicity of the permanent diestrus state (Rickard, S. E., University of Toronto, unpublished data).

Estrogenic effects have been seen in rodents and other animals consuming isoflavones from dietary sources. Uterotrophic effects have been described in rats fed soy (*53*) or commercially prepared rat cake (*54*). Commercial chow diets contain significant amounts of soy meal, exposing rats to isoflavonoid doses of at least 400 μg/day (*55*). Rat chow is also a significant source of lignans (*56*). Infertility and liver disease in captive cheetahs were linked to the intake of commercial feline diet containing soy meal (*57*). Chow exposure during lactation may also have some effect on sexual function later in life (*47*). It should be noted that isoflavonoid concentrations in commercial chow may fluctuate depending upon ingredient availability and price and environmental stresses on plant sources (*31*).

Contrary to the estrogenic effects seen for purified phytoestrogens and commercial chow, antiestrogenic effects have been observed in male mice exposed to soy gestationally. In neonatally estrogenized male mice, dietary soybean from gestation to adulthood significantly reduced prostatic growth inhibition and the development of prostatic dysplasia induced by DES. Exposure to dietary soybean from fertilization to 12 months of age resulted in a significant increase in the growth of accessory sex glands. However, no effect on accessory gland weight, besides a slight increase in overall prostatic lobe weight, was observed in mice exposed to soybean for two months after weaning (*53*), further demonstrating the significance of the timing of administration of isoflavones on reproductive development.

Other Potential Adverse Effects of Phytoestrogens and Lignans

Although phytoestrogens and lignans are weaker estrogens in comparison to steroidal estrogens like estradiol, there are concerns regarding their potential to enhance tumor growth (*58*). These compounds are aromatic, have molecular sizes similar to that of many carcinogenic compounds and are formed by bacterial action like many carcinogens (*13*). In the estrogen-dependent breast cell lines T47D and MCF-7, both enterolactone and equol stimulated growth and progesterone receptor induction in the absence of estradiol (*30*). Mousavi and Adlercreutz (*59*) found that both enterolactone and estradiol stimulated proliferation of MCF-7 cells separately, but in combination, the stimulatory effect decreased. However, binding inhibition studies using ^{3}H-12-*O*-tetradecanoylphorbol-13-acetate (TPA) with a mouse skin particulate fraction found no tumor-promoting effects of mammalian lignans or phytoestrogens at levels as high as 10^{-4} M (*7*). Formononetin was found to stimulate mammary gland proliferation in mice, but its stimulatory effects were 15,000 times less than that observed for the endogenous estrogen estradiol (*60*). In addition, no evidence of mutagenicity was found for the phytoestrogens daidzein, genistein, formononetin and biochanin A using the Ames *Salmonella*/mammalian microsome assay (*61*).

In addition to some promoting effects on tumor growth, some purified plant lignans are toxic at relatively low levels. Ingestion of the plant lignan podophyllin has been shown to cause adverse gastrointestinal effects, such as nausea, vomiting and diarrhea, which may end in delirium, stupor, coma and even death. Renal impairment and immunosuppression are other side effects (*8*). Nordihydroguaiaretic acid (NDGA; 2,3-dimethyl-1,4-bis(3,4-dihydroxyphenyl) butane), previously used as an antioxidant in foods, have been shown after long-term, low dose (0.5%-1%) treatment in rats, to induce cystic enlargement of the mesenteric lymph nodes at the ileocecal junction and vacuolization of kidney tubular epithelium (*62*).

Effects of Human Female Exposure to Phytoestrogens and Lignans

High urinary levels of phytoestrogens and/or mammalian lignans have been observed in vegetarian women (*63-65*) and in Japanese women consuming their traditional diet (*66, 67*), indicating high consumption of precursor compounds from foods. High plasma levels of mammalian lignans have also been observed in Finnish vegetarian women (*68*). Dietary supplementation with rich sources of phytoestrogens (red clover or soy flour) or lignans (flaxseed) has resulted in plasma isoflavone levels as high as 150 ng/mL and mammalian lignan levels as high as 500 ng/mL (*69*). These concentrations are approximately 10,000 times greater than normal circulating levels of the steroidal estrogen estradiol and thus may have the ability to exert significant biological effects.

In contrast to the above mentioned animal studies, there is no epidemiological evidence of abnormalities in the reproductive tracts of humans consuming high levels of phytoestrogens or lignans. However, one study in premenarcheal girls (9-10 years of age) found delayed breast development and menarche as well as lower plasma levels of follicular stimulating hormone (FSH), luteinizing hormone (LH) and estradiol by age 12 in girls consuming diets with relatively higher dietary fiber levels (*70*). This result was taken as protective against future breast cancer development in light of the risk factors for this disease described below. Possibly phytoestrogens or lignans derived from high fiber foods may have played a role in these effects. Interestingly, both premenopausal and postmenopausal women with breast cancer have significantly lower urinary excretion levels of phytoestrogens and lignans (*64, 65*).

Alterations in Menstrual Cycling. Breast cancer risk is positively related to early menarche, late menopause, low parity and delayed first pregnancy. This increased risk appears to be related to the regularity and length of ovulatory cycles and hence overall exposure level to estrogens (*71*, *72*). Reduced risks of breast cancer have been observed in women with irregular or long cycles in several studies (*73-76*). Lengthening of the luteal phase or follicular phase of the menstrual cycle have been shown in women consuming 10 g of flaxseed/day (*77*) and 60 g of textured soybean protein/day (*78*), respectively, but without a significant increase in menstrual cycle length. These effects could be attributed in part to hormonal changes although the types of hormonal changes seen in each study were different. Consumption of 60 g/day of Arcon F, a soybean product where the isoflavones have been chemically removed, had no effect phase length or hormonal levels (*79*). It should be noted that these feeding trials were only for short periods of time (one to three months), so long-term consumption of foods rich in phytoestrogens or lignans may eventually lead to lengthening of the menstrual cycle. In fact, there is a trend for longer menstrual cycles in Japanese women, shown to excrete high levels of phytoestrogens (*66*), compared to women in the U. S. (*72*).

Estrogenic Effects in Postmenopausal Women. In premenopausal women, phytoestrogens and lignans act as antiestrogens because of high circulating levels of the more potent steroidal estrogens (*28*). Postmenopausal women, however, have low circulating levels of estrogens due to the nearly negligible estrogen production by the ovary and adrenal cortex and increased reliance on the estrogen synthesis from androgens in peripheral tissues (*80*). Thus, one would expect phytoestrogens and lignans to act as estrogens in postmenopausal women due to the low levels of steroidal estrogens.

For this reason, it was hypothesized that dietary phytoestrogens and lignans may alleviate the symptoms of menopause (*81*). Menopausal symptoms include vaginal dryness, hot flashes, night sweats, mood changes, insomnia and urinary dysfunction (*82*). Estrogenic effects on vaginal epithelium were observed in Australian postmenopausal women consuming soy for two weeks (*83*). Twelve week supplementation of soy (high in isoflavones) or wheat (high in lignans) flour resulted in significant reductions in the frequency of hot flashes by 40% and 25% respectively. The response was quicker in the soy-supplemented group with significant reductions seen in six weeks (*84*). However, no effects were seen on vaginal cell maturation, plasma lipids or urinary calcium. In contrast, Baird and colleagues (*85*) did not find any estrogenic response measured by changes in follicular stimulating hormone (FSH), luteinizing hormone (LH), and sex hormone binding globulin (SHBG), except for a small effect on vaginal cytology, in postmenopausal women supplemented with soy (average intake of 165 mg isoflavones/day) for a four week period. As previously mentioned for changes in menstrual cycling, demonstrable effects on menopausal symptoms may only be evident after long-term supplementation of foods rich in phytoestrogen and lignans which may not be reflected by changes in hormonal status.

Low levels of estrogen also put postmenopausal women at increased risk of osteoporosis. Ipriflavone, a synthetic isoflavone derivative, has been shown to be effective in improving bone density in animal models (*86*, *87*) as well as in elderly women without (*88*) or with (*89*) osteoporosis. In an ovariectomized rat model of osteoporosis, dietary soybean protein isolate inhibited bone mineral loss. However, it could not be concluded whether the effect was due to isoflavones present in the protein isolate or to the protein itself (*90*). Recent work by Anderson and colleagues showed that low dose, but not high dose, genistein was as effective as the estrogenic compound Premarin, which has established bone-retaining properties, in preventing the loss of trabecular bone tissue in ovariectomized rats (*91*). Thus isoflavones appear to play a protective role in bone metabolism.

Beneficial Effects of Phytoestrogens and Lignans in Chronic Disease

Although phytoestrogens and perhaps lignans may have potentially adverse effects on reproductive development , it is becoming increasingly evident that these compounds have protective effects against a variety of chronic diseases, particularly cancer. Possible anticancer mechanisms include antagonism of estrogen metabolism, antioxidant activity, and modulatory effects on key control points of the cell cycle. Isoflavones and, to a lesser extent, lignans have also been implicated as possible anti-atherogenic factors by exhibiting hypocholesterolemic effects and inhibiting platelet activation. Epidemiological and experimental data on these effects are presented below.

Cancer-Protective Effects of Phytoestrogens and Lignans. Preliminary evidence for the link between cancer and these weak estrogens came from epidemiological studies monitoring urinary excretion levels of these compounds in women with different habitual diets. Vegetarian women were found to excrete much higher levels of isoflavones (equol, daidzein, *O*-desmethylangolensin) and mammalian lignans (enterodiol and enterolactone) than women consuming omnivorous diets (*63-65*). The urinary excretion levels of these compounds were even lower in women with breast cancer (*64, 65*), which suggested that phytoestrogens and lignans may be protective against the disease. Japanese men and women consuming their traditional diets have high concentrations of phytoestrogens in their urine (*66, 67*) and plasma (*92*) and are at lower risk for breast and prostate cancer than Finnish and North American populations (*93*). In fact, phytoestrogens have been detected in breast aspirate and prostatic fluid (*94*). Chimpanzees, found to have a great resistance to the carcinogenic effect of estrogens, excrete large amounts of lignans and phytoestrogens when consuming their normal diet (*95*).

Soy and Cancer Risk. Because soybean is a rich source of phytoestrogens such as isoflavones, many experimental feeding studies have been done with soy to determine the effects of isoflavones on carcinogenesis. Mammary tumor induction by *N*-methyl-*N*-nitrosourea (MNU) or 7,12-dimethylbenz(*a*)anthracene (DMBA) was inhibited by feeding of soya bean either as chips or protein isolates (*94*). To provide further evidence that the isoflavones in soy were responsible for this protective effect and not some other component (e.g. protease inhibitors), the soya was also autoclaved (inactivation of protease inhibitors) or the isoflavones chemically extracted (Arcon F). The autoclaved soya produced a similar tumor inhibitory effect whereas the soya devoid of isoflavones had no effect on mammary carcinogenesis (*96*). A recent review on the intake of soy products and risk of cancer in a variety of sites found that 17/26 animal experiments reported a protective effect with no studies showing an increased risk (*97*). However, epidemiological studies on the consumption of soy products and cancer risk were inconsistent (*97*).

Flaxseed and Cancer Risk. Similar to the use of soy in testing the effects of isoflavones on cancer risk, our laboratory has used flaxseed, a rich source of mammalian lignan precursors, to test the effects of lignans on carcinogenesis. Flaxseed supplementation at the 5% level significantly lowered epithelial cell proliferation and aberrant crypt foci formation in the colon (*98*) and epithelial cell proliferation and nuclear aberration in the mammary gland of rats (*99*). Significant reductions in mammary tumor size were also observed in rats fed at the promotion stage of tumorigenesis (*100*). Isolation of the major mammalian lignan precursor in flaxseed, secoisolariciresinol diglycoside (SDG), allowed us to determine that the protective effect of flaxseed on mammary carcinogenesis was due in part to its precursor-derived mammalian lignans. SDG feeding at levels equivalent to that found in 5% flaxseed significantly reduced mammary tumor number at early promotion phase (Thompson, L. U. et al. *Nutr. Cancer*, in press) and mammary tumor size at the late promotion phase (Thompson, L. U. et al. *Carcinogenesis*, in press) of carcinogenesis in DMBA-treated rats. SDG has also been shown to decrease the size and multiplicity of aberrant crypt foci in rats treated with the colon carcinogen azoxymethane (Jenab, M. and Thompson, L. U., *Carcinogenesis*, in press).

Potential Mechanisms of Phytoestrogens and Lignans in Carcinogenesis. In view of their weak estrogenic/antiestrogenic effects, one mechanism proposed for the action of phytoestrogens and lignans is antagonism of estrogen metabolism. Many studies have demonstrated their ability to bind to the estrogen receptor and prevent the growth-promoting actions of steroidal estrogens (*24*, *101*). Isoflavones (daidzein, equol and formononetin) and lignans (enterolactone, enterodiol, matairesinol and isolariciresinol) compete with estradiol for binding to the rat uterine type II estrogen binding site (the bioflavonoid receptor) (*102*) believed to play a role in estrogen-stimulated uterine growth (*103*, *104*). DMBA-treated rats given 5 mg genistein on days 2, 4 and 6 post-partum had increased latency and reduced incidence and multiplicity of mammary tumors (*105*). Exposure to genistein during the neonatal period induced cell differentiation in the mammary gland and subsequently decreased the number of the highly proliferative terminal end buds (*105*), the proposed site for carcinogenic attack (*106*). Isoflavones, and to a lesser extent, lignans inhibit the enzyme aromatase, denying the tumor an endogenous source of estrogens from its action on androgens (*107-110*). Epidemiological studies have found positive correlations between urinary excretion of total lignans and total phytoestrogens and plasma SHBG (*111*, *112*) with direct stimulatory effects seen for enterolactone and genistein on SHBG synthesis from HepG2 liver cancer cells in vitro (*102*, *113*). Thus more free estradiol and testosterone, the biologically active forms, would be bound to SHBG and inactivated since phytoestrogens have a low affinity for this binding protein (*25*). However, secoisolariciresinol, the precursor to mammalian lignans, has been shown to reduce binding activity of SHBG towards 5-α-dihydrotestosterone (DHT) in vitro by competition with DHT for the binding site (*114*).

A second mechanism for the protective effect of phytoestrogens and lignans are their ability to act as an antioxidant. Free radicals formed endogenously through cellular respiration or obtained externally from food, polluted air or cigarette smoke can attack DNA, proteins and lipids in cellular membranes causing cell and tissue damages. The body normally has defenses against free radical damage, but an imbalance in oxidative events and endogenous antioxidant activity plays a role in the aging process as well as disease development such as cancer or heart disease (*19*). Antioxidant effects have been demonstrated for the lignans sesaminol and sesamin (*115-118*), NDGA (*119*) and dibenzocyclooctene lignans (*120*) in vitro and for genistein both in vitro and in vivo (*121*, *122*). The antioxidant activity of NDGA is thought to be partly responsible for its protective effects against TPA-induced skin tumors (*123*). However, as mentioned previously, the antioxidant effect of NDGA is counteracted by its toxic effects observed in rats (*62*).

Some isoflavones and lignans have inhibitory effects on enzymes involved in cell proliferation, can affect programmed cell death (apoptosis) or even induce differentiation of cancer cells. Genistein is a potent inhibitor of tyrosine-specific protein kinase (*124-127*) which is associated with the cellular receptor of several growth factors as well as oncogene products of retroviral src family, suggesting that genistein also plays a role in neoplastic transformation (*6*). Other inhibitory effects on cellular enzymes involved in proliferation include genistein against DNA topoisomerase (*124*) and NDGA against ornithine decarboxylase (*128*) and protein kinase C (*129*). Apoptosis induction in human gastrointestinal and breast cancer cell lines has been found for daidzein and biochanin A (*130*) and genistein (*131*), respectively. Induction of differentiation has also been observed with genistein (*132-137*) and daidzein (*138*).

Many other anti-cancer effects have been reported in the literature for lignans and isoflavones. Cytostatic or cytotoxic effects on human breast cancer cells (*131, 139, 140*), prostatic cancer cells (*141*), gastric cancer cells (*142*), liver cancer cells (*113*), promyelocytic leukemic cell lines (*143-146*) and peripheral blood lymphocytes (*147*). Synthesized or extracted genistein has also been shown to inhibit the growth of cells derived from solid pediatric tumors (*148*). Antimitotic effects of lignans on a variety of tumors have also been observed (*149*). NDGA, equol, enterolactone and enterodiol inhibit the binding of estrone and estradiol to rat and human alpha-fetoprotein, a protein characteristic of neoplastic development (*150*). Both phytoestrogens and lignans have been shown to inhibit in vitro angiogenesis, the generation of capillaries, thus denying tumors access to nutrients from the blood supply (*108, 151*). NDGA also inhibits cytochrome P-450 monooxygenase activity in epidermal and hepatic microsomal preparations, suggesting an inhibitory effect on the metabolic activation of promutagens and procarcinogens (*119*). Invasion, a characteristic of metastatic cancer, was inhibited by genistein at normal physiological concentrations in vitro (*152*). The protective role of the mammalian lignans enterolactone and enterodiol in colon cancer may involve their ability to inhibit activity of cholesterol-7-alpha-hydroxylase, the rate limiting enzyme in formation of primary bile acids from cholesterol (*153*). Secondary bile acids (e.g. deoxycholic acid) are formed by bacterial action on primary bile acids in the colon and are thought to increase colon cancer risk (*19*). Anti-inflammatory effects have also been noted (*154, 155*).

Anti-Atherogenic Mechanisms of Phytoestrogens and Lignans. In addition to the beneficial effects observed for cancer, isoflavonoid phytoestrogens and lignans may also protect against heart disease. Soy protein has been associated with lower total or low-density lipoprotein (LDL) cholesterol levels compared to casein diets in animal models (*156, 157*). Similar reductions in cholesterol have been seen in clinical studies comparing animal and soy protein (*158, 159*). This hypocholesterolemic effect can be partly attributed to the phytoestrogen content of soy protein since rats fed isoflavones either in a purified form (*160*) or as part of a crude extract (*161*) had similar reductions in plasma cholesterol. In addition, humans fed textured soy protein containing 45 mg isoflavones had reductions in total cholesterol but not with consumption of an isoflavone-free product (79). Significant reductions in plasma cholesterol levels have also been observed in premenopausal and surgically postmenopausal monkeys fed phytoestrogen-enriched soy-protein vs. low-phytoestrogen soy protein (*162, 163*). Genistein appears to exhibit anti-thrombotic activity with the prevention of thromboxane A2- and collagen-induced platelet activation (*164*) and as well as thrombin-induced platelet activation (*165, 166*) and aggregation (*167*) in vitro. This is believed to be mediated though its anti-tyrosine kinase activity. Genistein may have an indirect blocking effect of platelet-derived growth factor (PDGF) activity due to its inhibitory effect on platelet activation (*168*). The potential effects of mammalian lignans on blood lipid levels and coronary heart disease are only circumstantial due to their positive relationship with SHBG and the positive association between SHBG with high density lipoprotein (HDL) cholesterol and apolipoprotein A1 in epidemiological studies (*169*). However, lignans may still have protective effects in heart disease due to observed digitalis-like action (*170, 171*), antagonistic action of platelet-activating factor (PAF) receptor (*172*), and diuretic effects (*173, 174*).

Conclusions

The health benefits attributed to soybean and flaxseed, both rich sources of isoflavonoids and lignans respectively, has led to the increased incorporation of these foods in a variety of food products (*175, 176*). Concern has been raised regarding the potential adverse effects of the current exposure levels to these phytochemicals (*177*). Animal studies suggest that phytoestrogens may be harmful during critical, hormone-sensitive phases of neuroendocrine development, suggesting that human infants exposed to these compounds may be more susceptible to their possible harmful effects. Since large quantities of phytoestrogens have been detected in the urine of infants consuming soy-based formula (*178*) and cow's milk contains appreciable amounts of equol (*179*), investigations into the effects of early exposure to phytoestrogens and lignans in humans is warranted.

The impact of the intake of high levels of phytoestrogens and lignans on health, whether adverse or beneficial, is difficult to assess for many reasons. Firstly, concentrations of phytoestrogens and lignans in plants varies depending on environmental stressors such as mechanical injury, temperature, humidity or fungal infestation (*8, 31*). Studies in our laboratory have indicated that secoisolarciresinol levels in flaxseed and subsequent mammalian lignan levels vary with the variety, harvest location and harvest year of flaxseed (Thompson, L. U., University of Toronto, unpublished data). Similar variability in soybean isoflavones have been observed (*180*). Secondly, large inter-individual variations in plasma (*69*) and urinary levels of phytoestrogens and lignans (*20, 21, 181-183*) in subjects fed controlled diets indicate variations in metabolism and bioavailability of these compounds. Thirdly, not all phytoestrogens are created equal. For example, coumestrol is more potent than other phytoestrogens in its estrogenic effects (*22*). Fourthly, some individuals may be more susceptible to the effects of phytoestrogens than others. Sex differences in mammalian lignan production have been observed in humans (*21*) and in rats (Thompson, L. U., University of Toronto, unpublished data) with males excreting more of the biologically active enterolactone vs. enterodiol compared to females. In addition, other studies have found that not all individuals are able to produce equol from daidzein (*20, 21, 179, 181, 183*).

In conclusion, evidence exists for the protective effects of phytoestrogens and lignans against chronic disease. Both in vitro and in animal models demonstrate the beneficial effects of these compounds against diseases such as cancer, heart disease and osteoporosis. The adverse effects on reproduction observed for phytoestrogens in young animals have not been reported for humans in the literature. Although it does caution against over-consumption of phytoestrogen- or lignan-rich foods in susceptible individuals like infants or young children, moderate consumption of foods such soy and flaxseed in adults may have long-term benefits.

Literature Cited

1. Health and Welfare Canada. *Nutrition Recommendations. The Report of the Scientific Review Committee;* Health Protection Branch: Ottawa, ON, 1990; pp 5-6.
2. National Research Council, Food and Nutrition Board, and Commission on Life Sciences. *Diet and Health: Implications for Reducing Chronic Disease Risk.* National Academy Press: Washington, D. C., 1990.
3. Axelson, M.; Sjovall, J.; Gustafsson, B. E.; Setchell, K. D. R. *Nature* **1982**, *298*, 659-660.
4. Thompson, L. U.; Robb, P.; Serraino, M.; Cheung, F. *Nutr. Cancer* **1991**, *16*, 43-52.
5. Franke, A. A.; Custer, L. J.; Cerna, C. M.; Narala, K. *Proc. Soc. Exper. Biol. Med.* **1995**, *208*, 18-26.
6. Thompson, L. U. *Food Res. Int.* **1993**, *26*, 131-150.
7. Setchell, K. D. R.; Adlercreutz, H. In *Role of the Gut Flora, Toxicity and Cancer*; Rowland, I. R., Ed.; Academic Press: London, 1988; pp 315-345.
8. *Lignans: Chemical, Biological and Clinical Properties*; Ayres, D. C.; Loike, J. D., Eds.; Chemistry and Pharmacology of Natural Products; Cambridge University Press: Cambridge, 1990.
9. Stich, S. R.; Toumba, J. K.; Groen, M. B.; Funke, C. W.; Leemhuis, J.; Vink, J.; Woods, G. F. *Nature* **1980**, *287*, 738-740.
10. Setchell, K. D. R.; Lawson, A. M.; Mitchell, F. L.; Adlercreutz, H.; Kirk, D. N.; Axelson, M. *Nature* **1980**, *287*, 740-742.
11. Setchell, K. D. R.; Bull, R.; Adlercreutz, H. *J. Steroid Biochem.* **1980**, *12*, 375-384.
12. Setchell, K. D. R.; Lawson, A. M.; Conway, E.; Taylor, N. F.; Kirk, D. N.; Cooley, G.; Farrant, R. D.; Wynn, S.; Axelson, M. *Biochem. J.* **1981**, *197*, 447-458.
13. Setchell, K. D. R.; Lawson, A. M.; Borriello, S. P.; Harkness, R.; Gordon, H.; Morgan, D. M.; Kirk, D. N.; Adlercreutz, H.; Anderson, L. C.; Axelson, M. *Lancet* **1981**, *2*, 4-7.
14. Axelson, M.; Setchell, K. D. R. *F. E. B. S. Lett.* **1981**, *123*, 337-342.
15. Dehennin, L.; Reiffsteck, A.; Jondet, M.; Thibier, M. *J. Reprod. Fert.* **1982**, *66*, 305-309.
16. Borriello, S. P.; Setchell, K. D. R.; Axelson, M.; Lawson, A. M. *J. Applied Bacteriol.* **1985**, *58*, 37-43.
17. Adlercreutz, C. H. T.; Goldin, B. R.; Gorbach, S. L.; Hockerstedt, K. A. V.; Watanabe, S.; Hamalainen, E. K.; Markkanen, M. H.; Makela, T. H.; Wahala, K. T.; Hase, T. A.; Fotsis, T. *J. Nutr.* **1995**, *125*, 757S-770S.
18. Reddy, C. S.; Hayes, A. W. In *Principles and Methods of Toxicology*; Hayes, A. W., Ed.; Raven Press Ltd.: New York, NY, 1994; pp 67-110.
19. Thompson, L. U. *Crit. Rev. Food Sci. Nutr.* **1994**, *34*, 473-497.
20. Hutchins, A. M.; Lampe, J. W.; Martini, M. C.; Campbell, D. R.; Slavin, J. L. *J. Am. Diet. Assoc.* **1995**, *95*, 769-774.

21. Kirkman, L. M.; Lampe, J. W.; Campbell, D. R.; Martini, M. C.; Slavin, J. L. *Nutr. Cancer* **1995**, *24*, 1-12.
22. Adams, N. R. In *Phenolics*; Cheeke, P. R., Ed.; *Toxicants of Plant Origin*; CRC Press: Boca Raton, FL, 1989, Vol. 4; pp 24-51.
23. Jordan, V. C.; Mittal, S.; Gosden, B.; Koch, R.; Lieberman, M. E. *Environ. Health Perspect.* **1985**, *61*, 97-110.
24. Molteni, A.; Brizio-Molteni, L.; Persky, V. *J. Nutr.* **1995**, *125*, 751S-756S.
25. Martin, P. M.; Horwitz, K. B.; Ryan, D.; McGuire, W. L. *Endocrinology* **1978**, *103*, 1860-1867.
26. Tang, B. Y.; Adams, N. R. *J. Endocrinol.* **1980**, *85*, 1203-1210.
27. Kitts, W. D.; Newsome, F.; Runeckles, V. *Can. J. Anim. Sci.* **1984**, *63*, 823-830.
28. Folman, Y.; Pope, G. S. *J. Endocrinol.* **1966**, *34*, 215-222.
29. Markiewicz, L.; Garey, J.; Adlercreutz, H.; Gurpide, E. *J. Steroid Biochem. Molec. Biol.* **1993**, *45*, 399-405.
30. Welshons, W. V.; Murphy, C. S.; Koch, R.; Calaf, G.; Jordan, V. C. *Breast Cancer Res. Treat.* **1987**, *10*, 169-175.
31. Adams, N. R. *J. Anim. Sci.* **1995**, *73*, 1509-1515.
32. Cox, R. I.; Braden, A. W. H. *Proc. Aust. Soc. Anim. Prod.* **1974**, *10*, 122.
33. Adams, N. R. *Proc. Soc. Exper. Biol. Med..* **1995**, *208*, 87-91.
34. Bennetts, H. W.; Underwood, E. J.; Shier, F. L. *Aust. Vet. J.* **1946**, *22*, 2-12.
35. Plapinger, L.; Bern, H. A. *J. Natl. Cancer Inst.* **1979**, *63*, 507-518.
36. Herbst, A. L.; Kurman, R. J.; Scully, R. E. *Obstet. Gynecol.* **1972**, *40*, 287-298.
37. Croker, K. P.; Lightfoot, R. J.; Johnson, T. J.; Adams, N. R.; Carrick, M. J. *Aust. J. Agri. Res.* **1989**, *40*, 165-176.
38. Lundh, T. *Proc. Soc. Exper. Biol. Med.* **1995**, *208*, 33-39.
39. Hargis, P. S.; Van Elswyk, M. E. *World's Poul. Sci. J.* **1993**, *49*, 251-264.
40. Kennedy, A. K.; Dean, C. E.; Thompson, L. U.; Van Elswyk, M. E. *Proceedings of the 56th Flax Institute of the United States of America*; Fargo, ND, 1996.
41. Levy, J. R.; Faber, K. A.; Ayyash, L.; Hughes, C. L. J. *Proc. Soc. Exper. Biol. Med.* **1995**, *208*, 60-66.
42. Dohler, K. D.; Coquelin, A.; Davis, F.; Hines, M.; Shryne, J. E.; Gorski, R. A. *Brain Res.* **1984**, *302*, 291-295.
43. Faber, K. A.; Hughes, C. L. *Biol. Reprod.* **1991**, *45*, 649-653.
44. Faber, K. A.; Hughes, C. L. *Reprod. Toxicol.* **1993**, *7*, 35-39.
45. Whitten, P. L.; Lewis, C.; Naftolin, F. *Biol. Reprod.* **1993**, *49*, 1117-1121.
46. Burroughs, C. D. *Proc. Soc. Exper. Biol. Med.* **1995**, *208*, 78-81.
47. Whitten, P. L.; Lewis, C.; Russell, E.; Naftolin, F. *Proc. Soc. Exper. Biol. Med.* **1995**, *208*, 82-86.
48. Medlock, K. L.; Branham, W. S.; Sheehan, D. M. *Proc. Soc. Exper. Biol. Med.* **1995**, *208*, 67-71.
49. Whitten, P. L.; Russell, E.; Naftolin, F. *Steroids* **1992**, *57*, 98-106.
50. Whitten, P. L.; Russell, E.; Naftolin, F. *Steroids* **1994**, *59*, 443-449.
51. Whitten, P. L.; Naftolin, F. *Steroids* **1992**, *57*, 56-61.
52. Orcheson, L.; Rickard, S.; Seidl, M.; Cheung, F.; Luyengi, L.; Fong, H.; Thompson, L. U. *Fed. Am. Soc. Exp. Biol. J.* **1993**, *7*, A291.

53. Makela, S. I.; Pylkkanen, L. H.; Santti, R. S.; Adlercreutz, H. *J. Nutr.* **1995**, *125*, 437-445.
54. Drane, H.; Patterson, D. S.; Roberts, B. A.; Saba, N. *Food Cosmet. Toxicol.* **1975**, *13*, 491-492.
55. Axelson, M.; Sjovall, J.; Gustafsson, B. E.; Setchell, K. D. R. *J. Endocrinol.* **1984**, *102*, 49-56.
56. Coert, A.; Vonk Noordegraaf, C. A.; Groen, M. B.; Van der Vies, J. *Experientia* **1982**, *38*, 904-905.
57. Setchell, K. D. R.; Gosselin, S. J.; Welsh, M. B.; Johnston, J. O.; Balistreri, W. F.; Kramer, L. W.; Dresser, B. L.; Tarr, M. J. *Gastroenterology* **1987**, *93*, 225-233.
58. Miller, W. R. *J. Steroid Biochem. Molec. Biol.* **1990**, *37*, 467-480.
59. Mousavi, Y.; Adlercreutz, H. *J. Steroid Biochem. Molec. Biol.* **1992**, *41*, 615-619.
60. Wang, W.; Tanaka, Y.; Han, Z.; Higuchi, C. M. *Nutr. Cancer* **1995**, *23*, 131-140.
61. Bartholomew, R. M.; Ryan, D. A. *Mutat. Res.* **1980**, *78*, 317-321.
62. Grice, H. C. *Food. Cosmet. Toxicol.* **1968**, *6*, 155-161.
63. Adlercreutz, H.; Fotsis, T.; Heikkinen, R.; Dwyer, J. T.; Goldin, B. R.; Gorbach, S. L.; Lawson, A. M.; Setchell, K. D. R. *Med. Biol.* **1981**, *59*, 259-261.
64. Adlercreutz, H.; Fotsis, T.; Heikkinen, R.; Dwyer, J. T.; Woods, M.; Goldin, B. R.; Gorbach, S. L. *Lancet* **1982**, *2*, 1295-1299.
65. Adlercreutz, H.; Fotsis, T.; Bannwart, C.; Wahala, K.; Makela, T.; Brunow, G.; Hase, T. *J. Steroid Biochem.* **1986**, *25*, 791-797.
66. Adlercreutz, H. In *Frontiers in Gastrointestinal Research*; Rozen, P., Kargar, S., Eds.; Basel, Switzerland, 1988; pp 165-176.
67. Adlercreutz, H.; Honjo, H.; Higashi, A.; Fotsis, T.; Hamalainen, E.; Hasegawa, T.; Okada, H. *Am. J. Clin. Nutr.* **1991**, *54*, 1093-1100.
68. Adlercreutz, H.; Fotsis, T.; Watanabe, S.; Lampe, J.; Wahala, K.; Makela, T.; Hase, T. *Cancer Detect. Prev.* **1994**, *18*, 259-271.
69. Morton, M. S.; Wilcox, G.; Wahlqvist, M. L.; Griffiths, K. *J. Endocrinol.* **1994**, *142*, 251-259.
70. de Ridder, C. M.; Thijjssen, J. H. H.; Van't Veer, P.; van Duuren, R.; Bruning, P. F.; Zonderland, M. L.; Erich, W. B. M. *Am. J. Clin. Nutr.* **1991**, *54*, 805-813.
71. Kvale, G. *Acta Oncol.* **1992**, *31*, 187-194.
72. Bernstein, L.; Ross, R. K.; Henderson, B. E. *Am. J. Epidemiol.* **1992**, *135*, 142-152.
73. Kvale, G.; Heuch, I. *Cancer Lett.* **1988**, *62*, 1625-1631.
74. La Vecchia, C.; Decarli, A.; Di Pietro, S.; Franceschi, S.; Negri, E.; Parassini, F. *Eur. J. Cancer Clin. Oncol.* **1985**, *21*, 417-422.
75. Frisch, R. E.; Wyshak, G.; Albright, N. L.; Schiff, I.; Jones, K. P.; Witschi, J.; Shiang, E.; Koff, E.; Marguglio, M. *Br. J. Cancer* **1985**, *52*, 885-891.
76. Pederson, A. B.; Bartholomew, M. J.; Dolence, L. A.; Aljadir, L. P.; Netteburg, K. L.; Lloyd, T. *Am. J. Clin. Nutr.* **1991**, *53*, 879-884.
77. Phipps, W. R.; Martini, M. C.; Lampe, J. W.; Slavin, J. L.; Kurzer, M. S. *J. Clin. Endocrinol. Metab.* **1993**, *77*, 1215-1219.
78. Cassidy, A.; Bingham, S.; Setchell, K. D. R. *Am. J. Clin. Nutr.* **1994**, *60*, 333-340.
79. Cassidy, A.; Bingham, S.; Setchell, K. D. R. *Br. J. Nutr.* **1995**, *74*, 587-601.
80. Judd, H. L.; Shamonki, I. M.; Frumar, A. M.; Lagasse, L. D. *Obstet. Gynecol.* **1982**, *59*, 680-686.

81. Adlercreutz, H.; Hamalainen, E.; Gorbach, S.; Golden, B. *Lancet*, **1992**, *339*, 1233.
82. Ansbacher, R. *Compr. Ther.* **1995**, *21*, 242-244.
83. Wilcox, G.; Wahlqvist, M. L.; Burger, H. G.; Medley, G. *Br. Med. J.* **1990**, *301*, 905-906.
84. Murkies, A. L.; Lombard, C.; Strauss, B. J.; Wilcox, G.; Burger, H. G.; Morton, M. S. *Maturitas* **1995**, *21*, 189-195.
85. Baird, D. D.; Umbach, D. M.; Lansdell, L.; Hughes, C. L.; Setchell, K. D. R.; Weinberg, C. R.; Haney, A. F.; Wilcox, A. J.; McLachlan, J. A. *J. Clin. Endocrinol. Metab.* **1995**, *80*, 1685-1690.
86. Brandi, M. L. *Bone Miner.* **1992**, *19 (Suppl.)*, S3-S14.
87. Civitelli, R.; Abbasi-Jarhomi, S. H.; Halstead, L. R.; Dimarogonas, A. *Calcif. Tissue Int.* **1995**, *56*, 215-219.
88. Nakamura, S.; Morimoto, S.; Takamoto, S.; Onishi, T.; Fukuo, K.; Koh, E.; Kitano, S.; Miyashita, Y.; Yasuda, O.; Tamatani, M. *Calcif. Tissue Int.* **1992**, *51 (Suppl.)*, S30-S34.
89. Passeri, M.; Biondi, M.; Costi, D.; Bufalino, L.; Castiglione, G. N.; Di Peppe, C.; Abate, G. *Bone Miner.* **1992**, *19 (Suppl.)*, S57-S62.
90. Arjmandi, B. H.; Alekel, L.; Hollis, B. W.; Amin, D.; Stacewicz-Sapuntzakis, M.; Guo, P.; Kukreja, S. C. *J. Nutr.* **1994**, *126*, 161-167.
91. Anderson, J. J.; Ambrose, W. W.; Garner, S. C. *J. Nutr.* **1995**, *125*, 799S.
92. Adlercreutz, H.; Markkanen, H.; Watanabe, S. *Lancet* **1993**, *342*, 1209-1210.
93. Rose, D. P.; Boyar, A. P.; Wynder, E. L. *Cancer* **1986**, *58*, 2363-2371.
94. Finlay, E. M. H.; Wilson, D. W.; Adlercreutz, H.; Griffiths, K. *J. Endocrinol.* **1991**, *129 (Suppl.)*, 49.
95. Adlercreutz, H.; Musey, P. I.; Fotsis, T.; Bannwart, C.; Wahala, K.; Makela, T.; Brunow, G.; Hase, T. *Clin. Chim. Acta* **1986**, *158*, 147-154.
96. Barnes, S.; Grubbs, C.; Setchell, K. D. R.; Carlson, J. *Prog. Clin. Biol. Res.* **1990**, *347*, 239-353.
97. Messina, M. J.; Persky, V.; Setchell, K. D. R.; Barnes, S. *Nutr. Cancer* **1994**, *21*, 113-131.
98. Serraino, M.; Thompson, L. U. *Cancer Lett.* **1992**, *63*, 159-165.
99. Serraino, M.; Thompson, L. U. *Cancer Lett.* **1991**, *60*, 135-142.
100. Serraino, M.; Thompson, L. U. *Nutr. Cancer* **1992**, *17*, 153-159.
101. Waters, A. P.; Knowler, J. T. *J. Reprod. Fert.* **1982**, *66*, 379-381.
102. Adlercreutz, H.; Mousavi, Y.; Clark, J.; Hockerstedt, K.; Hamalainen, E.; Wahala, K.; Makela, T.; Hase, T. *J. Steroid Biochem. Molec. Biol.* **1992**, *41*, 331-337.
103. Markaverich, B. M.; Clark, J. H. *Endocrinology*, **1979**, *105*, 1458-1462.
104. Markaverich, B. M.; Upchurch, S.; Clark, J. H. *J. Steroid Biochem.* **1981**, *14*, 125-132.
105. Lamartiniere, C. A.; Moore, J. B.; Brown, N. M.; Thompson, R.; Hardin, M. J.; Barnes, S. *Carcinogenesis* **1995**, *16*, 2833-2840.
106. Russo, I. H.; Russo, J. *J. Natl. Cancer Inst.* **1978**, *61*, 1439-1449.
107. Adlercreutz, H.; Bannwart, C.; Wahala, K.; Makela, T.; Brunow, G.; Hase, T.; Arosemena, P. J.; Kellis, J. T.; Vickery, L. E. *J. Steroid Biochem. Molec. Biol.* **1993**, *44*, 147-153.
108. Fotsis, T.; Pepper, M.; Adlercreutz, H.; Fleischmann, G.; Hase, T.; Montesano, R.; Schweigerer, L. *Proc. Natl. Acad. Sci. U. S. A.* **1993**, *90*, 2690-2694.

109. Wang, C.; Makela, T.; Hase, T.; Adlercreutz, H.; Kurzer, M. S. *J. Steroid Biochem. Molec. Biol.* **1994**, *50*, 205-212.
110. Kellis, J. T.; Vickery, L. E. *Science* **1984**, *225*, 1032-1034.
111. Adlercreutz, H.; Hockerstedt, K.; Bannwart, C.; Bloigu, S.; Hamalainen, E.; Fotsis, T.; Ollus, A. *J. Steroid Biochem.* **1987**, *27*, 1135-1144.
112. Adlercreutz, H.; Hockerstedt, K.; Bannwart, C.; Hamalainen, E.; Fotsis, T.; Bloigu, S. In *Hormones and Cancer 3*; Bresciani, F., King, R. J. B., Lippman, M. E., Raynaud, J.-P., Eds.; Progress in Cancer Research and Therapy; Raven Press: New York, 1988, Vol. 35; pp 409-412.
113. Mousavi, Y.; Adlercreutz, H. *Steroids* **1993**, *58*, 301-304.
114. Ganber, D.; Spiteller, G. *Z. Naturforsch* **1995**, *50c*, 98-104.
115. Yamashita, K. E.; Nohara, Y.; Katayama, K.; Namiki, M. *J. Nutr.* **1992**, *122*, 2440-2446.
116. Fukuda, Y.; Osawa, T.; Namiki, M. *Agric. Biol. Chem.* **1985**, *49*, 301-306.
117. Fukuda, Y.; Nagata, M.; Osawa, T.; Namiki, M. *J. Am. Oil Chem. Soc.* **1986**, *63*, 1027-1031.
118. Osawa, T.; Nagata, M.; Namiki, M.; Fukuda, Y. *Agric. Biol. Chem.* **1985**, *49*, 3351-3352.
119. Agarwal, R.; Wang, Z. Y.; Bik, D. P.; Mukhtar, H. *Drug Metab. Disp.* **1991**, *19*, 620-624.
120. Lu, H.; Liu, G. T. *Plant Med.* **1992**, *58*, 311-313.
121. Wei, H. C.; Wei, L. H.; Frenkel, K.; Bowen, R.; Barnes, S. *Nutr. Cancer* **1993**, *20*, 1-12.
122. Wei, H.; Bowen, R.; Cai, Q.; Barnes, S.; Wang, Y. *Proc. Soc. Exper. Biol. Med.* **1995**, *208*, 124-130.
123. Nakadate, T.; Yamamoto, S.; Iseki, H.; Sonoda, S.; Takemura, S.; Ura, A.; Hosoda, Y.; Kato, R. *Gann* **1982**, *73*, 841-843.
124. Markovits, J.; Linassier, C.; Fosse, P.; Couprie, J.; Pierre, J.; Jacquemin-Sablon, A.; Saucier, J. M.; Le Pecq, J. B.; Larsen, A. K. *Cancer Res.* **1989**, *49*, 5111-5117.
125. Akiyama, T.; Ishida, J.; Nakagawa, S.; Ogawara, H.; Watanabe, S.; Itoh, N.; Shibuya, M.; Fukami, Y. *J. Biol. Chem.* **1987**, *262*, 5592-5595.
126. Ogawara, H.; Akiyama, T.; Watanabe, S.; Ito, N.; Kobori, M.; Seoda, Y. *J. Antibiotics* **1989**, *42*, 340-343.
127. Teraoka, H.; Ohmura, Y.; Tsukada, K. *Biochem. Int.* **1989**, *18*, 1203-1210.
128. Nakadate, T.; Yamamoto, S.; Ishii, M.; Kato, R. *Cancer Res.* **1982**, *42*, 2841-2845.
129. Nakadate, T. *Jap. J. Pharmacol.* **1989**, *49*, 1-9.
130. Yanagihara, K.; Ito, A.; Toge, T.; Numoto, M. *Cancer Res.* **1993**, *53*, 5815-5821.
131. Pagliacci, M. C.; Smacchia, M.; Migliorati, G.; Grignani, F.; Riccardi, C.; Nicoletti, I. *Eur. J. Cancer* **1994**, *30A*, 1675-1682.
132. Kiguchi, K.; Constantinou, A. I.; Huberman, E. *Cancer Commun.* **1990**, *2*, 271-277.
133. Honma, Y.; Okabe-Kado, J.; Kasukabe, T.; Hozumi, M.; Umezawa, K. *Jap. J. Cancer Res.* **1990**, *81*, 1132-1136.
134. Constantinou, A.; Kiguchi, K.; Huberman, E. *Cancer Res.* **1990**, *50*, 2618-2624.
135. Makishima, M.; Honma, Y.; Hozumi, M.; Sampi, K.; Hattori, M.; Umezawa, K.; Motoyoshi, K. *Leukemia Res.* **1991**, *15*, 701-708.

136. Makishima, M.; Honma, Y.; Hozumi, M.; Nagata, N.; Motoyoshi, K. *Biochim. Biophys. Acta* **1993**, *1176*, 245-249.
137. Kondo, K.; Tsuneizumi, K.; Watanabe, T.; Oishi, M. *Cancer Res.* **1991**, *51*, 5398-5404.
138. Jing, Y.; Nakaya, K.; Han, R. *Anticancer Res.* **1993**, *13*, 1049-1054.
139. Hirano, T.; Oka, K.; Akiba, M. *Res. Comm. Pathol. Pharmacol.* **1989**, *64*, 69-78.
140. Hirano, T.; Fukuoka, K.; Oka, K.; Naito, T.; Hosaka, K.; Mitsuhashi, H.; Matsumoto, Y. *Cancer Invest.* **1990**, *8*, 595-602.
141. Peterson, G.; Barnes, S. *Biochem. Biophys. Res. Commun.* **1991**, *179*, 661-667.
142. Matsukawa, Y.; Marui, N.; Sakai, T.; Satomi, Y.; Yoshida, M.; Matsumoto, K.; Nishino, H.; Aoike, A. *Cancer Res.* **1993**, *53*, 1328-1331.
143. Kuriu, A.; Ikeda, H.; Kanakura, Y.; Griffin, J. D.; Druker, B.; Yagura, H.; Kitayama, H.; Ishikawa, J.; Nishiura, T.; Kanayama, Y.; Yonezawa, T.; Tarui, S. *Blood* **1991**, *78*, 2834-2840.
144. Hirano, T.; Fukuoka, K.; Oka, K.; Matsumoto, Y. *Cancer Invest.* **1991**, *9*, 145-150.
145. Traganos, F.; Ardelt, B.; Halko, N.; Bruno, S.; Darzynkiewicz, Z. *Cancer Res.* **1992**, *52*, 6200-6208.
146. Hirano, T.; Gotoh, M.; Oka, K. *Life Sci.* **1994**, *55*, 1061-1069.
147. Hirano, T.; Oka, K.; Kawashima, E.; Akiba, M. *Life Sci.* **1989**, *45*, 1407-1411.
148. Schweigerer, L.; Christeleit, K.; Fleischmann, G.; Adlercreutz, H.; Wahala, K.; Hase, T.; Schwab, M.; Ludwig, R.; Fotsis, T. *Eur. J. Clin. Invest.* **1992**, *22*, 260-264.
149. Hartwell, J. L. *Cancer Treat. Rep.* **1976**, *60*, 1031-1067.
150. Garreau, B.; Vallette, G.; Adlercreutz, H.; Wahala, K.; Makela, T.; Benassayag, C.; Nunez, E. A. *Biochim. Biophys. Acta* **1991**, *1094*, 339-345.
151. Fotsis, T.; Pepper, M.; Adlercreutz, H.; Hase, T.; Montesano, R.; Schweigerer, L. *J. Nutr.* **1995**, *125*, 790S-797S.
152. Scholar, E. M.; Toews, M. L. *Cancer Lett.* **1994**, *87*, 159-162.
153. Sanghvi, A.; Diven, W. F.; Seltman, H.; Warty, V.; Rizk, M.; Kritchevsky, D.; Setchell, K. D. R. In *Proceedings of the 8th International Symposium on Drugs Affecting Lipid Metabolism*; Kritchevsky, D., Paoletti, R., Holmes, W. L., Eds.; Symposia of the Giovanni Lorenzini Foundation; Plenum Press: New York, NY, 1984, Vol. 8; pp 311-322.
154. Young, J. M.; Wagner, B. M.; Spires, D. A. *J. Invest. Dermatol.* **1983**, *80*, 48-52.
155. Coyne, D. W.; Morrison, A. R. *Biochem. Biophys. Res. Commun.* **1990**, *173*, 718-724.
156. Kritchevsky, D. *J. Nutr.* **1995**, *125*, 589S-593S.
157. Carroll, K. K.; Kurowska, E. M. *J. Nutr.* **1995**, *125*, 594S-597S.
158. Sirtori, C. R.; Lovati, M. R.; Manzoni, C.; Monetti, M.; Pazzucconi, F.; Gatti, E. *J. Nutr.* **1995**, *125*, 598S-605S.
159. Clarkson, T. B.; Anthony, M. S.; Hughes, C. L. *Trends Endocrinol. Metab.* **1995**, *6*, 11-16.
160. Sharma, R. D. *Lipids* **1979**, *14*, 535-539.
161. Potter, S. M. *J. Nutr.* **1995**, *125*, 606S-611S.
162. Anthony, M. S.; Clarkson, T. B.; Hughes, C. L. *Circulation* **1994**, *90*, I-235.
163. Anthony, M. S.; Clarkson, T. B.; Weddle, D. L.; Wolfe, M. S. In *First International Symposium on the Role of Soy in Preventing and Treating Chronic Disease;* Mesa, AZ, 1994; p 15.
164. Nakashima, S.; Koike, T.; Nozawa, Y. *Mol. Pharmacol.* **1991**, *39*, 475-480.

165. Ozaki, Y.; Yatomi, Y.; Jinnai, Y.; Kume, S. *Biochem. Pharmacol.* **1993**, *46*, 395-403.
166. Sargeant, P.; Farndale, R. W.; Sage, S. O. *F. E. B. S. Lett.* **1993**, *315*, 242-246.
167. Asahi, M.; Yanagi, S.; Ohta, S.; Inazu, T.; Sakai, K.; Takeuchi, F.; Taniguchi, T.; Yamamura, H. *F. E. B. S. Lett.* **1992**, *309*, 10-14.
168. Wilcox, J. N.; Blumenthal, B. F. *J. Nutr.* **1995**, *125*, 631S-638S.
169. Adlercreutz, H. *Scand. J. Clin. Lab. Invest. Suppl.* **1990**, *201*, 3-23.
170. Fagoo, M.; Braquet, P.; Robin, J. P.; Esanu, A.; Godfraind, T. *Biochem. Biophys. Res. Commun.* **1986**, *134*, 1064-1070.
171. Hirano, T.; Oka, K.; Naitoh, T.; Hosaka, K.; Mitsuhashi, H. *Res. Commun. Chem. Pathol. Pharmacol.* **1989**, *64*, 227-240.
172. Braquet, P.; Robin, J. P.; Esanu, A.; Landais, Y.; Vilain, B.; Baroggi, N.; Touvay, C.; Etienne, A. *Prostaglandins* **1985**, *30*, 692.
173. Plante, G. E.; Prevost, C.; Chainey, A.; Braquet, P.; Sirois, P. *Am. J. Physiol.* **1987**, *253*, R375-378.
174. Hirano, T.; Homma, M.; Oka, K.; Naito, T.; Hosaka, K.; Mitsuhashi, H. *Life Sci.* **1991**, *49*, 1871-1878.
175. Thompson, L. U. In *Flaxseed in Human Nutrition*; Cunnane, S. C., Thompson, L. U., Eds.; AOCS Press: Champaign, IL, 1995; pp 219-236.
176. Golbitz, P. *J. Nutr.* **1995**, *125*, 570S-572S.
177. Whitten, P. L.; Lewis, C.; Russell, E.; Naftolin, F. *J. Nutr.* **1995**, *125*, 771S-776S.
178. Cruz, M. L. A.; Wong, W. W.; Mimouni, F.; Hachey, D. L.; Setchell, K. D. R.; Klein, P. D.; Tsang, R. C. *Pediatr. Res.* **1994**, *35*, 135-140.
179. Setchell, K. D. R.; Borriello, S. P.; Hulme, P.; Kirk, D. N.; Axelson, M. *Am. J. Clin. Nutr.* **1984**, *40*, 569-578.
180. Eldridge, A. C.; Kwolek, W. F. *J. Agric. Food Chem.* **1983**, *31*, 394-396.
181. Kelly, G. E.; Joannou, G. E.; Reeder, A. Y.; Nelson, C.; Waring, M. A. *Proc. Soc. Exper. Biol. Med.* **1995**, *208*, 40-403.
182. Lampe, J. W.; Martini, M. C.; Kurzer, M. S.; Adlercreutz, H.; Slavin, J. L. *Am. J. Clin. Nutr.* **1994**, *60*, 122-128.
183. Xu, X.; Wang, W. J.; Murphy, P. A.; Cook, L.; Hendrich, S. *J. Nutr.* **1994**, *124*, 825-8321.

Chapter 17

Interactions and Biological Effects of Phytic Acid

Sharon E. Rickard and Lilian U. Thompson

Department of Nutritional Sciences, Faculty of Medicine, University of Toronto, 150 College Street, Toronto, Ontario M5S 3E2, Canada

Phytic acid is a compound found in cereal grains, legumes, nuts and oilseeds. It is considered to be an antinutrient due to its ability to bind minerals, proteins and starch at physiological pH. However, recent evidence indicates that phytic acid has healthful effects. Reductions in glycemic response to starchy foods as well as lower plasma cholesterol and triglyceride levels have been observed with endogenous phytate consumed in foods or with the addition of purified sodium phytate. In addition, phytic acid has anticancer effects in the colon and mammary gland in rodent models and in various tumor cell lines in vitro. In view of these beneficial effects, the term "antinutrient" used to describe food constituents like PA needs to be reevaluated.

Despite current recommendations to increase the intake of foods rich in dietary fiber (*1*, *2*), concerns have been raised regarding the potential adverse effects arising from higher consumption of antinutrients naturally present in these foods (*3*). However, these antinutrients have also been hypothesized to play a role in the beneficial effects observed with high fiber foods (*4-6*). In this chapter, the adverse effects of the antinutrient phytic acid (PA) will be reviewed with a subsequent discussion of its potential healthful effects.

Occurrence of Phytic Acid in Foods and its Interaction with Food Components

Phytic acid (*myo*-inositol 1,2,3,4,5,6-hexakis-dihydrogen phosphate) (Figure 1) is found in cereals and legumes at levels ranging from 0.4% to 6.4% by weight (*7*, *8*). In most seeds, PA is the primary phosphate reserve, accounting for 60-90% of the total phosphorus (*7*). Phytic acid is present in the germ of corn and in the aleurone or bran layer of monocotyledenous seeds such as wheat and rice (*7*). These discrete locations of PA allows easy separation through the milling process (*5*). However, in dicotyledenous seeds such as legumes, nuts and oilseeds, PA is found closely associated with proteins in crystalloid-type globoids (*7*). Consequently, PA is often isolated or concentrated with the protein fraction of these foods. (*5*).

PA has been termed an "antinutrient" due to its ability to bind minerals, proteins and starches, either directly or indirectly, and thus alter their solubility, functionality, digestibility and absorption. At normal pH ranges found in foods, the six phosphate groups on PA are negatively charged, making PA highly reactive with other positively charged particles such as minerals and proteins (*5*). This interaction is pH-dependent. Mineral ions may bind with one or more phosphate groups in one or more PA molecules, forming complexes of varying solubilities and stabilities (Figure 2A) (*5, 9*). Proteins, positively charged at a pH below their isoelectric pH, can bind directly to PA through electrostatic attraction (Figure 2B). At an intermediate pH above the isoelectric pH, binding between PA and proteins is mediated through multivalent cations such as calcium (Ca) since both particles are negatively charged (Figure 2C). At a high pH, the PA-cation-protein complex may dissociate with PA precipitating as cation-PA (*5, 9*). Binding to the PA molecule is not necessarily electrostatic since starch binding may also occur through the formation of hydrogen bonds (Figure 2D) or indirectly through the proteins with which the starch is associated. This binding with subsequent changes in nutrient digestibility and availability is thought to be responsible for both the adverse and beneficial effects of PA.

Bioavailability of Minerals

Many studies have demonstrated that PA reduces mineral bioavailability in both animals and humans (*9-16*). The degree to which a mineral is made unavailable for absorption depends on both the relative concentrations of PA and minerals as well as the strength of binding (*6*). Zinc (Zn), because it forms the most stable and insoluble complex with PA (*17*), appears to be one of the minerals most affected by high PA concentrations. Studies in rats (*18-20*) and humans (*21-23*) indicated that Zn utilization and growth was inhibited by a [PA]/[Zn] molar ratio greater than 10. Since Ca combined with Zn forms an even more insoluble complex with PA, later investigators have also used the [PA][Ca]/[Zn] molar ratio, with limits of 3.5 mol/kg BW in rats (*24*) or 100 mmol/1000 kcal (4.2 MJ) in humans (*25-27*) as an indicator of compromised Zn status. Of course, there are limitations to the use of these ratios since the effect of PA on mineral bioavailability is dependent on many other factors discussed below.

Food Processing. Studies have shown that food processing increases the insolubility of the mineral-PA complexes, subsequently decreasing mineral availability. Extrusion cooking of a starch-wheat bran-gluten mixture, rich in phytate, was found to decrease Zn balance in ileostomy patients (*28*). This

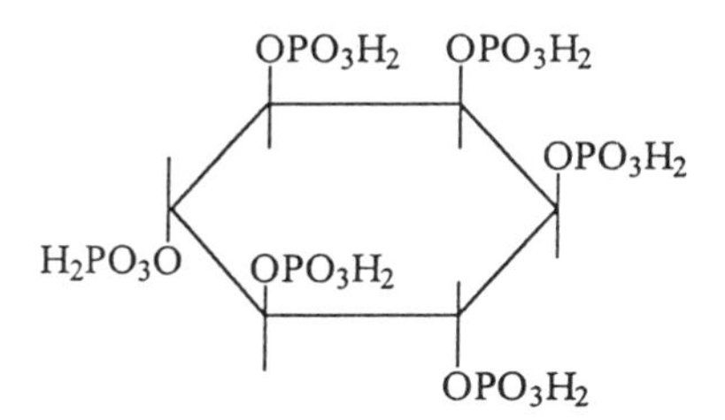

Figure 1. Basic structure of phytic acid.

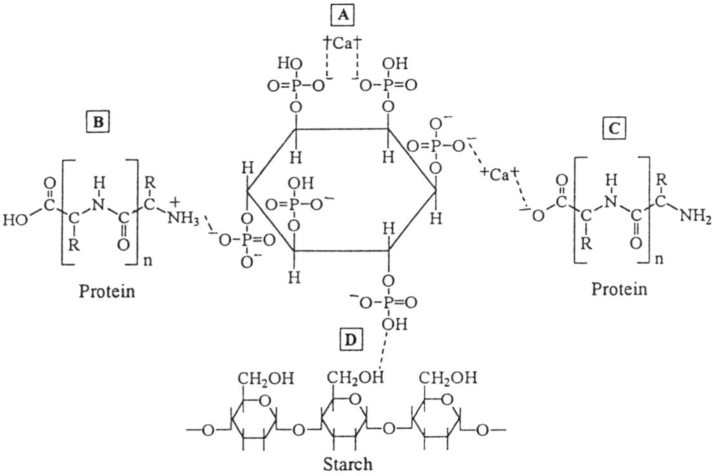

Figure 2. Possible interactions of phytic acid with minerals, protein and starch. Adapted from ref *4*.

reduction may be due to the inactivation of endogenous phytases naturally present in wheat bran or to the formation of indigestible PA complexes (*29*). Neutralized soy proteins form tightly bound complexes with PA and Zn which are resistant to digestion whereas acid-precipitated soy proteins, being protonated, are less likely to react with PA and minerals (*30*, *31*). Nevertheless, food processing can also enhance the degradation of PA through soaking (*32*, *33*), fermentation (*34-38*), germination or irradiation (*39*).

Added vs. Endogenous PA. Added PA, usually in the form of a sodium salt, is very reactive with other dietary components such as minerals due to its high solubility. Endogenous PA, however, may already be complexed with other food components and thus would be insoluble and unable to bind dietary minerals (*5*).

Presence of Dietary, Intestinal or Bacterial Phytase. Many legumes and cereals contain endogenous phytases which, if not destroyed by cooking or processing before intake, can hydrolyze PA in the GI tract (*5*, *40*). The addition of microbial (e.g. *Aspergillus niger*) phytases have been shown to enhance mineral availability in rats (*41*) and pigs (*42-46*). However, mineral complexes with PA have been shown to reduce alkaline phosphatase activity (*47-49*). Vucenik and colleagues showed that purified PA can be hydrolyzed to lower inositol phosphate products, i.e., inositol mono-, di-, tri-, tetra- and penta-phosphate (InsP1-InsP5), by murine and human cells in vitro (*50*). Therefore, dietary phytase may not be required for hydrolysis. The hydrolysis products of PA have lower mineral binding capability than PA (*51*, *52*). Although some studies suggest the less phosphorylated products (InsP1-InsP4) have no effect on mineral absorption (*13*), there are studies to the contrary (*12*).

Presence of Mineral Absorption Enhancers. Ascorbic acid (*15*, *53*, *54*), 1,25-dihydroxycholecalciferol (vitamin D3) (*55*, *56*) and meat protein (*57*) have been shown to be enhancers of mineral absorption. Frequently mineral deficiencies are seen in people of low socioeconomic status who consume foods rich in phytate in combination with marginal intakes of minerals and trace elements and low intakes of absorption enhancers such as ascorbic acid and meat protein (*53*, *58-60*).

Other Mineral Binding Constituents. It is still debated whether it is the fiber or the PA in high-fiber foods that is responsible for mineral binding (*9*, 61, *62*). Early studies by Andersson and co-workers (*63*) showed that varied wheat bran levels had no effect on mineral absorption. A later study by Fairweather-Tait and colleagues found that sugar beet fiber enhanced Zn and iron (Fe) absorption whereas wheat bran, high in phytate, caused a reduction (*64*). On the other hand, one study found no adverse effects on Zn or Ca balances with untreated or dephytinized wheat bran (*65*, *66*). Similarly, men fed wheat or corn bran for four weeks did not have adverse effects on Fe, Zn or copper (Cu) balances (*67*). In addition, removal of virtually all PA from a soy-protein meal still resulted in absorption of Fe that was 50% of an egg white protein control (*14*). The phytate-free conglycinin fraction of soy protein isolates were found to inhibit Fe absorption (*68*). Although isolated fiber fractions have also been shown to bind minerals in humans, the results are inconsistent (*61*). Oxalic acid (found at high levels in spinach and rhubarb) and tannins (found in beans) may also compete with PA for binding with minerals (*5*, 61).

Level and Type of Dietary Protein. Increased dietary protein levels can attenuate the effect of PA on Zn absorption (62, 69). Certain proteins provide amino acids which desorb Zn from the PA-Zn precipitate and improve its bioavailability (*70*). Protein and several amino acids were found to predict 50% of the variance in Zn availability in one study in rats (*71*). However, a recent study by Reddy et al. (*16*) did not find the inhibiting effect of phytate dependent on the protein composition of the meal when comparing egg white (neutral), meat (enhancer) and dephytinized soy (inhibitor).

Presence of Other Minerals. Mineral interaction through PA may affect absorption. Many studies have implicated Ca to interfere with the absorption of many minerals, especially Zn and Fe (*70*, *72*, *73*). In pigs, normal levels of Ca were found to reduce colonic degradation of phytate (*74*) as well as the efficacy of supplemental microbial phytase which was partially offset by vitamin D3 (*55*). Studies in rats found that dietary Ca inhibited phosphorus (P), magnesium (Mg) and other trace element bioavailability and reduced their internal deposition (*36*). High concentrations of Cu and Fe can also affect Zn absorption (*5*). Because PA forms strong complexes with Zn, PA has been shown to actually increase Cu absorption since Zn competes with Cu at the site of intestinal absorption (*75*). This may partially explain why high levels of PA had no effect on Cu absorption in young men in one study (*76*).

Metabolic Adaptation. Studies in the late 40's and early 50's have shown that negative mineral balance in humans at early weeks of feeding diets high in PA were followed by positive mineral balance after several weeks of treatment (*77*, *78*). Studies to the contrary (*21*, 79, *80*) may be attributable to the loss of adaptability with very high PA consumption (*21*, *81*).

Digestion and Absorption of Protein

Although the effect of PA on mineral bioavailability has been studied extensively, relatively fewer investigators have examined its effect on protein digestibility and amino acid availability. In the limited number of studies done to date, the evidence for an adverse effect is inconsistent. These discrepancies may be partly attributed to the differences in reactivity of added vs. endogenous PA towards substrates or digestive enzymes (*5*). The type and degree of processing of the protein source may also contribute to the conflicting findings. Results of in vitro, in vivo and ileostomate studies are described below.

In Vitro Studies. Enzymatic hydrolysis of gluten (*82*), casein (*83-86*), human or bovine serum albumin (*85-88*) and pea and fava bean protein (*89*) was lowered upon addition of purified PA in vitro. This reduction in protein hydrolysis may be due to PA forming a complex with either the protein substrate or the enzyme itself (*5*) since degradation of PA to lower inositol phosphates diminished this effect (*86*). However, removal of endogenous PA from soybean protein (*90*), northern bean protein (*91*), rapeseed flour (*92*) or lupine (*38*) had no effect on protein digestibility in vitro. Nevertheless, recent studies have found increased in vitro protein digestibility with removal of endogenous PA by soaking (*32*, *33*), fermentation (*34*, *35*, *37*, *93*), germination or irradiation (*39*). Although these processes also remove other antinutritional factors which can affect protein digestibility such as tannins (*6*), significant inverse correlations between PA and in vitro protein digestibility were reported in some studies (*34*, *93*). Similar correlations were also demonstrated by Chitra and colleagues when comparing unprocessed grain legumes with different endogenous PA levels (*94*).

Removal of PA has been shown to affect the pattern of amino acid release from proteins in vitro (*92*). A 51% reduction in rapeseed flour PA by dialysis with or without phytase treatment increased the availability of methionine, leucine and phenylalanine compared to untreated flour. Further reduction in the PA content (89% in total) did not enhance availability (*92*). Reduction in non-essential, polar, aromatic and acidic amino acids were seen with soaking of apricot kernals in a basic solution for 30 h at 47°C (*32*). However, the relative total essential amino acid availability was increased. Germination of soybean for 120 h also reduced non-essential and increased essential amino acid availability, but exposure of the seeds to irradiation (0.1 kGy) led to increases in both cases (*39*). It is not known if these alterations in amino acid release have any significant impact on their utilization in vivo.

In Vivo Studies. Even though in vitro digestibility data cannot account for the action of intestinal or bacterial phytase which play significant roles in the activity of PA in vivo (*5*), findings from in vivo studies on the effect of PA on protein digestibility and amino acid absorption are also inconsistent. Rapeseed protein quality (*95*), in vivo protein digestibility and amino acid absorption (*96*) were not affected by PA in rats. Addition of purified PA or its hydrolysis products at levels as high as 5% in casein diets did not alter protein digestibility, utilization or weight gain in rodent models (*5*, *86*, *97*). In addition, variations in the PA content of soy protein isolate had no effect on protein digestibility, protein efficiency ratio and biological value (*98*). In contrast, lower growth rates, food intakes and protein utilization efficiencies in rats fed rapeseed protein (1.24% PA) compared to those fed casein improved when the PA content of the rapeseed diet was decreased to 0.41% (*99*). Similar improvements in the nutritive value of proteins in a high protein wheat bran flour were also observed with PA reduction (*98*). In a human study, the nitrogen balance in humans was unaffected with the intake of 2.5 g of purified PA added to diets (*21*). However, consumption of PA-rich unleavened whole wheat bread (endogenous PA) resulted in a less positive and sometimes negative nitrogen balance. This adverse effect, however was attributed to higher dietary fiber intake during that phase of the study.

Ileostomate Studies. Ileostomates, healthy individuals with their colons removed, have been used to obtain more accurate information on protein digestibility. In vivo digestibility measurements are based on fecal nitrogen levels. However, unabsorbed nitrogen in the small intestines can be broken down by bacterial flora and eventually absorbed in the colon or used for bacterial synthesis (*5*). In ileostomates, the undigested products of digestion are collected in a bag attached to their terminal ileum, thus allowing direct quantification of nutrients that are not absorbed in the small intestines (*5*). Some preliminary work by Thompson and colleagues indicated a significant correlation between PA/protein and protein output in an ileostomate fed 21 different leguminous and cereal foods (*5*). Only a 3% rise in ileal protein recovery was found after PA supplementation of unleavened bread was increased almost two-fold, suggesting a small effect of PA on protein digestion and absorption (*5*). Although extrusion cooking significantly reduces PA degradation, Sandberg and co-workers found no significant difference in amount of nitrogen found in ileostomy contents when patients were fed a bran-gluten-starch mixture vs. the extruded product (*100*). Conversely, net nitrogen absorption from soy diets was found to be much less than that from meat protein in ileostomates (*101*). Thus, the potential adverse effect of PA on protein digestion and absorption is still unresolved. Consequently, the nutritional and physiological implications of possible alterations in protein digestion rate by PA remain to be determined.

Digestion and Absorption of Starch

One beneficial effect of high fiber foods has been the delay in digestion and absorption of starch, resulting in lower blood glucose, insulin and other endocrine responses (*102*). Large increases in blood glucose requires large amounts of insulin for its uptake in the liver. Continuous stimulation of the release of large amounts of insulin may lead to reduced sensitivity of insulin and possibly insulin resistance, a predisposing factor for the development of diabetes (*103*). Although soluble fiber is believed to be responsible for the slow digestion and absorption of starchy foods (*103*), other fiber-associated components such as PA also appear to be involved.

There is evidence supporting the role of antinutrients like PA in the inhibition of starch digestion and absorption. Legumes with the highest concentrations of PA were digested the slowest and produced the flattest blood glucose responses (*104*, *105*). To determine which component of legumes was responsible for the slow rate of starch digestion and absorption, navy beans were fractionated to starch, hull, protein, fiber and soluble whey fractions. The starch fraction was then digested alone or in combination with each fraction in the same proportions as they exist in the original whole bean. The fraction richest in antinutrients like PA, the soluble whey component, exhibited the greatest inhibitory effect (*4*). The reduction in starch digestion rate may be related to the inhibitory effect of PA on amylase, an enzyme essential to the breakdown of starch. Phytic acid has been shown to inhibit amylase activity in various legumes, the bacteria *Bacillus subtilis* (*106*, *107*), bovine pancreas (*108*), and human saliva (*109*, *110*). However, one study did not find an effect on amyloglucosidase and amylase activities and a significant effect on pancreatic amylase activity only at the early stage of digestion (*111*). Human feeding studies with healthy, normal volunteers showed reductions in starch digestion rate and glycemic response when PA was added (2%, starch basis) to unleavened wheat flour (*110*). In addition, increases in the in vitro rate of starch digestion and the blood glucose response with dephytinized navy bean flour were reversed by the readdition of PA to the dephytinized flour in normal subjects (*112*). Reductions in starch digestibility and glycemic response have also been observed in diabetic volunteers (*4*). Furthermore, the addition of Ca both in vivo and in vitro eliminates the inhibitory effect of PA on starch digestibility, presumably by forming an insoluble complex with PA and rendering it inactive toward digestive enzymes (*5*).

Possible mechanisms of the effect of PA on starch digestion and absorption include direct binding with starch, binding with proteins closely associated with starch, association with digestive enzymes, chelation of Ca required for amylase activity, effects on starch gelatinization during cooking or processing, and/or potential influences on gastric emptying (*5*). There is some evidence suggesting that PA's effect may not involve alterations in starch digestion since blood glucose response to a test meal of 50 g glucose was significantly reduced with the addition of PA at a 0.8% level (*113*).

A potential adverse effect for decreasing starch digestibility would be an increase in starch malabsorption. Excessive amounts of malabsorbed starch, fermented in the colon to short chain fatty acids (SCFA) such as acetate, proprionate and butyrate as well as hydrogen and methane gases (*114*), can result in flatulence, diarrhea, nausea and overall discomfort (*4*). However, relatively small amounts of carbohydrate are malabsorbed with PA as seen by breath hydrogen or ileostomate methods (*4*, *5*, *100*, *101*, *112*). Thus, most of the starch is digested and absorbed before it reaches the colon. Even so, some starch malabsorption may have beneficial effects which include (a) the use of SCFA, especially butyrate, as a fuel by colonic mucosal cells, (b) an increase in fecal bulk, inversely correlated with colon cancer risk (*105*), due to bacterial synthesis during fermentation and (c) a reduction in pH thereby minimizing the effects of the tumor promoter ammonia (*114*, *115*)

Reduction in Serum Lipids

Lower serum lipids have been associated with higher intakes of the PA-rich legumes (*30*, *116*, *117*). The specific role of PA in this hypocholesterolemic effect has been demonstrated in rats. Almost 20 years ago, Klevay and colleagues showed that addition of purified PA to the rat diet significantly reduced plasma lipid levels (*118*). A few years later, Sharma and co-workers demonstrated, in hypercholesterolemic rats, that the addition of 0.2% PA to a high cholesterol diet reduced both serum cholesterol and triacylglycerols (*119*). Similar results were obtained in a study with normocholesterolemic rats by Jariwalla and colleagues where 9% PA was added to a diet containing cholesterol (*120*). The hypolipidemic effect of PA is thought to be related to its higher affinity for binding zinc compared to copper since higher plasma Zn:Cu ratios predispose humans to cardiovascular disease (*118*).

The applicability of these results to humans may be questionable for two reasons. First, purified PA, not normally consumed by humans, may be more reactive than the endogenous form naturally present in foods (*5*). Wheat bran, which contains high levels of PA, does not lower blood lipid levels (*5*) although rat feeding studies with the legume Bengal gram containing 0.2% PA indicated hypolipidemic effects (*121*). Second, the study by Jariwalla and colleagues used fairly high levels of PA to observe an effect. Reported PA intake levels as high as 2,575 mg/day in vegetarians (*5*) translates to about 0.5% by weight assuming a daily food intake of 500 g. Nevertheless, the hypoglycemic effect of PA described above and seen in both rats and humans may be related to its hypolipidemic effects since reductions in plasma glucose and insulin may lead to decreased hepatic lipid synthesis (*4*, *102*). Recently, Katayama found that the rise in hepatic total lipids, hepatic and serum triacylglycerols, serum phospholipids and hepatic NADPH-generating enzymes by sucrose feeding in rats were all significantly depressed by dietary sodium phytate at the 0.5% level (*122*).

Anticancer Effects

It is well known that diets low in fat and high in fiber confer protective effects against chronic diseases such as cancer and cardiovascular disease. Recently, research has focused on fiber-associated substances, many of which are the "antinutrient" components of foods, that may be partly mediating the protective effects seen for dietary fiber (*6*). A limited number of studies, described below, have examined the role of PA in cancer of the colon, mammary gland and other organs or tissues at various stages of the carcinogenic process. Suggested mechanisms of action appear to be related to the binding ability of PA although there is recent evidence suggesting that the hydrolysis products of PA play a role in its anticancer effects.

Colon. Studies in ileostomates suggest that as much as 4 g PA may escape degradation in the small intestine and enter the colon (*5*). Thus, PA may have direct effects on the colonic mucosa. An in vitro study with the human colon cancer cell line HT-29 found a dose-dependent inhibition of cell proliferation with PA at levels of 0.66-10 mM as well as decreased expression of the tumor marker β-D-galactose-(1-3)-N-acetyl-galactosamine, indicating increased cell differentiation (*123*). Studies in rats have shown that PA reduces early risk markers for colon carcinogenesis such as epithelial cell proliferation expressed as either labeling index (*124*) or percentage mitotic rate (*125*, *126*) and multiplicity of aberrant crypt foci, putative precursor lesions to colonic tumor formation (*127*).

Much of the research examining the role of PA in colonic tumorigenesis has been done by Shamsuddin and colleagues in both rats and mice. The addition of 1% PA to the drinking water of rats one to two weeks prior to carcinogen (azoxymethane, AOM) administration ("initiation stage") and up to 12 months afterward significantly decreased tumor size and number (*128*, *129*). A significant decrease in tumor prevalence and frequency was also observed with 1% PA (*129*). Lower doses of PA (0.1%) in the drinking water given at the initiation stage inhibited tumor size to a greater extent (71% reduction) but had less of an effect on tumor prevalence (21% reduction) (*129*). Experiments where 2% PA was given in the drinking water either two weeks or five months after the last dose of AOM or 1,2-dimethylhydrazine (DMH) ("promotion stage") also exhibited significant reductions tumor incidence or prevalence (*126*, *128*), tumor size, number of tumors/rat and tumor load/unit area of the colon (*125*). Although the earlier studies by Shamsuddin and colleagues were at a high pH (10.8-11.4) (*125*, *126*, *128*), neutralization of the PA solution (pH 7.4) had no effect on tumor prevalence and frequency (*129*). Greater reductions in tumor size with the neutralized PA solution (65%) compared to the unadjusted PA solution (62.3%) may have been related to the greater water intake and hence higher PA intake by those animals (*129*).

Mammary Gland. Mammary carcinogenesis studies also indicate that PA is protective. Reductions in cell proliferation in MCF-7 human breast cancer cells were seen with PA treatment in vitro (*8*). Mammary epithelial cell proliferation (as labeling index) and nuclear aberration in carcinogen-treated mice were reduced by 30% when 1.2% PA was added to the diet (*130*). The reducing effect of PA was even greater (up to 53%) in mice fed diets high in iron (535 ppm) and calcium (1.5%) (*130*). The addition of 2% PA to the diet after administration of the carcinogen dimethylbenz(*a*)anthracence (DMBA) in rats significantly increased the number of survivors and decreased tumor size (*131*). No significant differences were found for tumor incidence and multiplicity. However, significant reductions in tumor incidence, number, multiplicity and burden were observed in DMBA-treated rats when PA was given in their drinking water at a concentration of 15 mM (*132*).

Other Tissues. Anticancer effects in tissues other than the colon and mammary gland have also been seen with PA. In vitro studies with the human K-562 erythroleukemia cell line (*133*) or PC-3 prostate cancer cell line (*134*) showed that PA inhibited growth and induced differentiation. Intraperitoneal injection of PA in mice with subcutaneously transplanted fibrosarcomas significantly reduced growth, prolonged survival and reduced the number of pulmonary metastases (*135*). In a similar tumor model, dietary treatment with PA (12%) as pentapotassium dimagnesium phytate, a form commonly found in grains (*7*), inhibited the tumor promoting effect of a diet containing 1.4% magnesium oxide (*136*).

However, the effect of PA has not always been protective in other cancer models tested. Hirose and co-workers used a wide-spectrum organ carcinogenesis model where tumors were initiated in rats by daily injections of 2,2-dihydroxy-di-n-propylnitrosamine followed by N-ethyl-N-hydroxy-ethyl-nitrosamine and then 3,2'-dimethyl-4-amino-biphenyl for three weeks (*137*). Starting one week after the last injection, PA was fed at the 2% level for 32 weeks. The incidence of liver hyperplastic nodules, hepatocellular carcinoma and eosinophilic focus in the pancreas tended to be lower in PA-treated animals. Yet, PA did not appear to have any effect on the lung, colon, esophagus, forestomach, small intestine, kidney or thyroid gland. In addition, urinary bladder papilloma incidence was significantly increased with PA treatment which was attributed to the higher urinary pH and sodium concentrations. Similarly, Takaba and colleagues found that preneoplastic and neoplastic lesions of the bladder induced in rats by N-butyl-N-(4-hydroxybutyl)nitrosamine were enhanced by feeding sodium or potassium salts of PA although PA itself and its magnesium salt had no effect (*138*).

Mechanisms of Action. The ability of PA to bind starch, proteins and minerals has been implicated as part of the mechanism whereby PA exerts its anticancer effects. Some recent studies indicate that PA's effect may be mediated through its hydrolysis products, particularly inositol triphosphate (InsP3) which acts in cellular signal transduction (*8*) and in the enhancement of natural killer (NK) cell activity (*139*, *140*). Phytic acid does not appear to act through a cytotoxic mechanism (*141*) but through promoting the differentiation of the malignant cells to a more normal phenotype (*8*).

Starch Malabsorption. As mentioned above, PA consumption can contribute to a small amount of starch reaching the colon. Colonic bacterial fermentation of the starch produces SCFA (*114*), possibly causing a decrease in pH. The reduction in pH is considered protective since tumor promotors such as bile acids and ammonia would be insolubilized or neutralized, respectively (*115*). Significant reductions in cecal pH have been observed in rats fed increasing levels of PA (*142*). In addition, butyric acid is a primary substrate used by colonic cells for differentiation (*114*).

Enzyme Inhibition. Increased activity of bacterial enzymes such as β-glucuronidase and mucinase have been associated with increased cancer risk (*143*, *144*). In rats, the activity of these enzymes decreased with increasing levels of PA (*4*, *142*). Although the exact mechanism remains to be determined, this reduction in activity may be through the direct binding of PA to the enzymes.

Mineral Binding. Many studies have demonstrated the ability of PA to reduce mineral bioavailability (*9*). The chelation of Zn, required for DNA synthesis, may reduce cell proliferation. The inhibition of colonic epithelial cell proliferation by PA was reversed by Zn supplementation in rats (*4*, *124*, *142*). PA may act as an antioxidant through its ability to chelate Fe^{3+}, a catalyst of hydroxy radical formation and subsequent lipid peroxidation and DNA damage (*8*, *145*). The promoting effect of high levels of dietary Fe (535-580 ppm) were reduced by PA treatment in colon and mammary tumorigenesis models (*130*, *146*). Although high dietary levels of Ca (>1%) have been shown to promote carcinogenesis (*130*, *147-150*), many studies have shown Ca to be cancer-protective by forming insoluble soaps with the tumor promoting bile and free fatty acids, by inducing cell differentiation and by reducing cell proliferation (*151-155*). Thus, the binding of Ca by PA may be adverse or beneficial depending on the experimental conditions.

Effects of Hydrolysis Products of PA. It has been suggested that the antineoplastic activity of PA may be mediated through its hydrolysis to lower phosphorylated compounds. Phytic acid has been shown to be taken up rapidly by cells from murine (YAC-1 T-cell leukemia) and human (K-562 erythroleukemia and HT-29 colon) tumor cell lines (*50*, *133*). Interestingly, only lower level inositol phosphates were found in the YAC-1 and K-562 cells whereas the HT-29 cells contained InsP1-InsP6 (PA), the latter of which grew the slowest (*50*). Thus PA may directly alter the cellular inositol phosphate pool and induce significant changes in cellular signaling. The lower inositol phosphates (InsP1-InsP4) act as intracellular messengers (*8*). Of these, InsP3 plays an important second messenger role, bringing about mitosis and mobilizing intracellular Ca. Even though both InsP4 and InsP5 can induce sequestration of Ca inside the cell (*8*), Shamsuddin et al. did find increased intracellular Ca to correspond with reduced cell division after PA treatment (*133*). Recent evidence in rodents also indicate that inositol (Ins) treatment has antitumorigenic effects in both the colon and mammary gland (*126*, *132*). Possibly, the free phosphate formed from the breakdown of PA to lower inositol phosphates may react with Ins to produce excess InsP3 which may in turn cause negative feedback and inhibition of cell proliferation (*126*). Inositol triphosphate is also generated by activated NK cells (*139*) and may be involved in attachment and fusion of NK cells with target cells, i.e., tumor cells (*140*). The enhanced NK cell activity observed both in mice and in vitro with PA treatment (*140*) may be through the formation of InsP3.

Conclusions

Phytic acid has both adverse and beneficial effects, both of which are attributed to its strong binding ability at physiological pH. The degree to which PA is adverse or beneficial in an individual is dependent upon many factors: (a) the relative concentrations of PA and other nutrients, (b) the conditions under which these constituents are consumed (i.e., effects of processing, presence of degrading enzymes, interactions with other food components), and (c) the nutritional and health status of the individual. For instance, those who consume foods rich in PA but poor in other nutrients such as minerals and proteins may be more prone to develop mineral deficiencies or symptoms of protein malnutrition. Similarly, infants and young children who have higher requirements for these nutrients may be more susceptible to the effects of PA. On the other hand, individuals consuming excess calories and nutrients, and thus at increased risk of chronic diseases such as cancer, cardiovascular disease or diabetes, may tolerate higher doses of PA and thereby benefit from its hypoglycemic, hypolipidemic and anticancer effects. Further studies are needed to determine proper dosing regimes to minimize possible adverse effects and yet maximize the potential health benefits of PA. In light of its healthful properties, the term "antinutrient" is an inappropriate and outdated label for food constituents like PA and thus requires reevaluation.

Literature Cited

1. National Research Council; Food and Nutrition Board; Commission on Life Sciences. *Diet and Health, Implications for Reducing Chronic Disease Risk*; National Academy Press: Washington, D.C. 1990,
2. Health and Welfare Canada. *Nutrition Recommendations. The Report of the Scientific Review Committee*; Health Protection Branch: Ottawa, ON, 1990, pp 5-6.
3. Morgan, M.; Fenwick, G. *Lancet*, **1990**, *336*, 1492-1495.
4. Thompson, L. U. *Food Technol.*, **1988**, *42*, 123-132.
5. Thompson, L. U. In *Food Proteins*; Kinsella, J. E., Soucie, W. G., Eds.; American Oil Chemical Society: Champaign, IL, 1989, pp 410-431.
6. Thompson, L. U. *Food Res. Int.*, **1993**, *26*, 131-149.
7. Reddy, N. R.; Sathe, S. K.; Salunkhe, D. K. *Adv. Food Res.*, **1982**, *28*, 1-92.
8. Shamsuddin, A. M. *J. Nutr.*, **1995**, *125*, 725S-732S.
9. Torre, M.; Rodriguez, A. R.; Saura-Calixto, F. *Crit. Rev. Food Sci. Nutr.*, **1991**, *30*, 1-22.
10. Zhou, J. R.; Fordyce, E. J.; Raboy, V.; Dickinson, D. B.; Wong, M. S.; Burns, R. A.; Erdman, J. W., Jr. *J. Nutr.*, **1992**, *122*, 2466-2473.
11. Heaney, R. P.; Weaver, C. M.; Fitzsimmons, M. L. *Am. J. Clin. Nutr.*, **1991**, *53*, 745-747.
12. Brune, M.; Rossander-Hulten, L.; Hallberg, L.; Gleerup, A.; Sandberg, A. S. *J. Nutr.*, **1992**, *122*, 442-449.
13. Lonnerdal, B.; Sandberg, A. S.; Sandstrom, B.; Kunz, C. *J. Nutr.*, **1989**, *119*, 211-214.
14. Hurrell, R. F.; Juillerat, M. A.; Reddy, M. B.; Lynch, S. R.; Dassenko, S. A.; Cook, J. D. *Am. J. Clin. Nutr.*, **1992**, *56*, 573-578.
15. Siegenberg, D.; Baynes, R. D.; Bothwell, T. H.; Macfarlane, B. J.; Lamparelli, R. D.; Car, N. G.; MacPhail, P.; Schmidt, U.; Tal, A.; Mayet, F. *Am. J. Clin. Nutr.*, **1991**, *53*, 537-541.
16. Reddy, M. B.; Hurrell, R. F.; Juillerat, M. A.; Cook, J. D. *Am. J. Clin. Nutr.*, **1996**, *63*, 203-207.
17. Evans, W. J.; Martin, C. J. *J. Inorg. Biochem.*, **1988**, *34*, 11-18.
18. Davies, N. T.; Olpin, S. E. *Br. J. Nutr.*, **1979**, *41*, 590-603.

19. Morris, E. R.; Ellis, R. *J. Nutr.*, **1980**, *110*, 1037-1045.
20. Lo, G. S.; Settle, S. L.; Steinke, F. H.; Hopkins, D. T. *J. Nutr.*, **1981**, *111*, 2223-2235.
21. Reinhold, J.; Nasr, K.; Lahimgarzadeh, A.; Hedayati, H. *Lancet*, **1973**, *i*, 283-288.
22. Morris, E. R.; Ellis, R. *Biol. Trace Elem. Res.*, **1989**, *19*, 107-117.
23. Turnlund, J. R.; King, J. C.; Keyes, W. R.; Gong, B.; Michel, M. C. *Am. J. Clin. Nutr.*, **1984**, *40*, 1071-1077.
24. Fordyce, E. J.; Forbes, R. M.; Robbins, K. R.; Erdman, J. W. *J. Food Sci.*, **1987**, *52*, 440-444.
25. Ellis, R.; Kelsay, J. L.; Reynolds, R. D.; Morris, E. R.; Moser, P. B.; Frazier, C. W. *J. Am. Diet. Assoc.*, **1987**, *87*, 1043-1047.
26. Ferguson, E. L.; Gibson, R. S.; Thompson, L. U.; Ounpuu, S. *Am. J. Clin. Nutr.*, **1989**, *50*, 1450-1456.
27. Gibson, R. S.; Smit Vanderkooy, P. D.; Thompson, L. *Biol. Trace Elem. Res.*, **1991**, *30*, 87-94.
28. Kivisto, B.; Andersson, H.; Cederblad, G.; Sandberg, A.-S.; Sandstrom, B. *Br. J. Nutr.*, **1986**, *55*, 255-?
29. Sandberg, A. S.; Andersson, H.; Carlsson, N. G.; Sandstrom, B. *J. Nutr.*, **1987**, *117*, 2061-2065.
30. Erdman, J. W., Jr.; Poneros-Schneier, A. *Adv. Exper. Med. Biol.*, **1989**, *249*, 161-171.
31. Ketelsen, S. M.; Stuart, M. A.; Weaver, C. M.; Forbes, R. M.; Erdman, J. W., Jr. *J. Nutr.*, **1984**, *114*, 536-542.
32. el-Adawy, T. A.; Rahma, E. H.; el-Badawey, A. A.; Gomaa, M. A.; Lasztity, R.; Sarkadi, L. *Nahrung*, **1994**, *38*, 12-20.
33. Estevez, A. M.; Castillo, E.; Figuerola, F.; Yanez, E. *Plant Foods Hum. Nutr.*, **1991**, *41*, 193-201.
34. Gupta, M.; Khetarpaul, N. *Nahrung*, **1993**, *37*, 141-146.
35. Khetarpaul, N.; Chauhan, B. M. *Plant Foods Hum. Nutr.*, **1991**, *41*, 321-327.
36. Larsen, T. *Biol. Trace Elem. Res.*, **1993**, *39*, 55-71.
37. Sharma, A.; Khetarpaul, N. *Nahrung*, **1995**, *39*, 282-287.
38. Camacho, L.; Sierra, C.; Marcus, D.; Guzman, E.; Campos, R.; von Baer, D.; Trugo, L. *Int. J. Food Microbiol.*, **1991**, *14*, 277-286.
39. Sattar, A.; Neelofar; Akhtar, M. A. *Plant Foods Hum. Nutr.*, **1990**, *40*, 185-194.
40. Pointillart, A. *J. Anim. Sci.*, **1991**, *69*, 1109-1115.
41. Rimbach, G.; Pallauf, J. *Z. Ernahrungswiss.*, **1993**, *32*, 308-315.
42. Lei, X. G.; Ku, P. K.; Miller, E. R.; Yokoyama, M. T.; Ullrey, D. E. *J. Anim. Sci.*, **1993**, *71*, 3368-3375.
43. Lei, X. G.; Ku, P. K.; Miller, E. R.; Yokoyama, M. T. *J. Anim. Sci.*, **1993**, *71*, 3359-3367.

44. Simons, P. C.; Versteegh, H. A.; Jongbloed, A. W.; Kemme, P. A.; Slump, P.; Bos, K. D.; Wolters, M. G.; Beudeker, R. F.; Verschoor, G. J. *Br. J. Nutr.*, **1990**, *64*, 525-540.
45. Jongbloed, A. W.; Mroz, Z.; Kemme, P. A. *J. Anim. Sci.*, **1992**, *70*, 1159-1168.
46. Mroz, Z.; Jongbloed, A. W.; Kemme, P. A. *J. Anim. Sci.*, **1994**, *72*, 126-132.
47. Martin, C. J.; Evans, W. J. *J. Inorg. Biochem.*, **1991**, *42*, 161-175.
48. Martin, C. J. *J. Inorg. Biochem.*, **1995**, *58*, 89-107.
49. McPherson, G. A. *Life Sci.*, **1990**, *47*, 1569-1577.
50. Vucenik, I.; Shamsuddin, A. M. *J. Nutr.*, **1994**, *124*, 861-868.
51. Tao, S. H.; Fox, M.; Phillippy, B.; Fry, B. E.; Johnson, M. L.; Johnston, M. R. *Fed. Proc.*, **1987**, *45*, 819.
52. Lonnerdal, B.; Kunz, C.; Sandberg, A.; Sandstrom, B. *Fed. Proc.*, **1987**, *46*, 599.
53. Taylor, P. G.; Mendez-Castellanos, H.; Martinez-Torres, C.; Jaffe, W.; Lopez de Blanco, M.; Landaeta-Jimenez, M.; Leets, I.; Tropper, E.; Ramirez, J.; Garcia Casal, M. N.; et al. *J. Nutr.*, **1995**, *125*, 1860-1868.
54. Davidsson, L.; Galan, P.; Kastenmayer, P.; Cherouvrier, F.; Juillerat, M. A.; Hercberg, S.; Hurrell, R. F. *Pediatr. Res.*, **1994**, *36*, 816-822.
55. Lei, X. G.; Ku, P. K.; Miller, E. R.; Yokoyama, M. T.; Ullrey, D. E. *J. Anim. Sci.*, **1994**, *72*, 139-143.
56. Edwards, H. M., Jr. *J. Nutr.*, **1993**, *123*, 567-577.
57. Fairweather-Tait, S. J.; Fox, T. E.; Wharf, S. G.; Eagles, J.; Kennedy, H. *Br. J. Nutr.*, **1992**, *67*, 411-419.
58. Brunvand, L.; Henriksen, C.; Larsson, M.; Sandberg, A. S. *Acta Obstet. Gynecol. Scand.*, **1995**, *74*, 520-525.
59. Fitzgerald, S. L.; Gibson, R. S.; Quan de Serrano, J.; Portocarrero, L.; Vasquez, A.; de Zepeda, E.; Lopez-Palacios, C. Y.; Thompson, L. U.; Stephen, A. M.; Solomons, N. W. *Am. J. Clin. Nutr.*, **1993**, *57*, 195-201.
60. Ferguson, E. L.; Gibson, R. S.; Opare-Obisaw, C.; Ounpuu, S.; Thompson, L. U.; Lehrfeld, J. *J. Nutr.*, **1993**, *123*, 1487-1496.
61. Kelsay, J. L. *Am. J. Gastroenterol.*, **1987**, *82*, 983-986.
62. Sandstrom, B. *Scand. J. Gastroenterol. Suppl.*, **1987**, *129*, 80-84.
63. Andersson, H.; Navert, B.; Bingham, S. A.; Englyst, H. N.; Cummings, J. H. *Br. J. Nutr.*, **1983**, *50*, 503-510.
64. Fairweather-Tait, S. J.; Wright, A. J. *Br. J. Nutr.*, **1990**, *64*, 547-552.
65. Morris, E. R.; Ellis, R. In *Nutritional Bioavailability of Zinc*; Englett, G. E., Ed.; ACS Symposium Series; American Chemical Society: Washington, DC, 1983, Vol. 210; pp 159-172.
66. Morris, E. R.; Ellis, R. In *Nutritional Bioavailability of Calcium*; Kies, C., Ed.; ACS Symposium Series; American Chemical Society: Washington, DC, 1985, Vol. 275; pp 63-72.

67. Sandstead, H. H.; Munoz, J. M.; Jacob, R. A.; Klevay, L. M.; Reck, S. J.; Logan, G. M.; Dintzis, F. R.; Inglett, G. E.; Shuey, W. C. *Am. J. Clin. Nutr.*, **1978**, *31*, S180-S184.
68. Lynch, S. R.; Dassenko, S. A.; Cook, J. D.; Juillerat, M. A.; Hurrell, R. F. *Am. J. Clin. Nutr.*, **1994**, *60*, 567-572.
69. Davies, N. T. In *Dietary Fiber in Health and Disease*; Vahouny, G., Kritchevsky, D., Eds.; Washington Symposium on Dietary Fiber; Plenum Press: New York, NY, 1982; pp 105-116.
70. Wise, A. *Int. J. Food Sci. Nutr.*, **1995**, *46*, 53-63.
71. Hunt, J. R.; Johnson, P. E.; Swan, P. B. *J. Nutr.*, **1987**, *117*, 1913-1923.
72. Hallberg, L.; Rossander-Hulten, L.; Brune, M.; Gleerup, A. *Euro. J. Clin. Nutr.*, **1992**, *46*, 317-327.
73. Hallberg, L.; Brune, M.; Erlandsson, M.; Sandberg, A. S.; Rossander-Hulten, L. *Am. J. Clin. Nutr.*, **1991**, *53*, 112-119.
74. Sandberg, A. S.; Larsen, T.; Sandstrom, B. *J. Nutr.*, **1993**, *123*, 559-566.
75. Lee, D. Y.; Schroeder, J. D.; Gordon, D. T. *J. Nutr.*, **1988**, *118*, 712-717.
76. Turnlund, J. R.; King, J. C.; Gong, B.; Keyes, W. R.; Michel, M. C. *Am. J. Clin. Nutr.*, **1985**, *42*, 18-23.
77. Walker, A. R. P.; Fox, F. W.; Irving, J. T. *Biochem. J.*, **1948**, *42*, 452-462.
78. Cullumbine, H.; Basnayake, V.; Lemottee, J.; Wickramanayake, T. W. *Br. J. Nutr.*, **1950**, *4*, 101-111.
79. McCance, R.; Widdowson, E. *J. Physiol. (Lond.)*, **1942**, *101*, 44-85.
80. McCance, R. A.; Widdowson, E. M. *J. Physiol. (Lond.)*, **1942**, *101*, 304-313.
81. Brune, M.; Rossander, L.; Hallberg, L. *Am. J. Clin. Nutr.*, **1989**, *49*, 542-545.
82. Camus, M. C.; Laporte, J. C. *Ann. Biol. Biochem. Biophys.*, **1976**, *16*, 719-729.
83. Singh, M.; Krikorian, A. *J. Agric. Food Chem.*, **1982**, *30*, 799-800.
84. Lathia, D.; Hoch, G.; Kievernagel, Y. *Plant Foods Hum. Nutr.*, **1987**, *37*, 229-235.
85. Knuckles, B. E.; Kuzmicky, D. D.; Betschart, A. A. *J. Food Sci.*, **1985**, *50*, 1080-1082.
86. Knuckles, B. E.; Kuzmicky, D. D.; Gumbman, M. R.; Betschart, A. A. *J. Food Sci.*, **1989**, *54*, 1348-1350.
87. Barre, R.; Nguyen-van Hout, N. *Bull. Soc. Chim. Biol.*, **1965**, *47*, 1399-1409.
88. Barre, R.; Nguyen-van Hout, N. *Bull. Soc. Chim. Biol.*, **1965**, *47*, 1419-1427.
89. Carnovale, E.; Lugaro, E.; Lombard-Bocea, G. *Cereal Chem.*, **1988**, *65*, 114-117.
90. Ritter, M. A.; Morr, C. V.; Thomas, R. L. *J. Food Sci.*, **1987**, *52*, 325-327.
91. Reddy, N. R.; Sathe, S. K.; Pierson, M. D. *J. Food Sci.*, **1988**, *53*, 107-110.

92. Serraino, M.; Thompson, L. U.; Savoie, L.; Parent, G. *J. Food Sci.*, **1985**, *50*, 1689-1692.
93. Gupta, M.; Khetarpaul, N.; Chauhan, B. M. *Plant Foods Hum. Nutr.*, **1992**, *42*, 109-116.
94. Chitra, U.; Vimala, V.; Singh, U.; Geervani, P. *Plant Foods Hum. Nutr.*, **1995**, *47*, 163-172.
95. McDonald, B.; Lieden, S.; Hambraeus, K. *Nutr. Rep. Int.*, **1978**, *17*, 49-56.
96. Thompson, L. U.; Serraino, M. *J. Agric. Food Chem.*, **1986**, *34*, 468-469.
97. Yoshida, T.; Shinoda, S.; Matsumoto, T.; Watarai, S. *J. Nutr. Sci. Vitaminol.*, **1982**, *28*, 401-410.
98. Satterlee, L.; Abdul-Kadir, R. *Lebensm. Wiss. Technol.*, **1983**, *16*, 8-?
99. Atwal, A.; Eskin, N.; McDonald, B.; Vaisey-Genser, M. *Nutr. Rep. Int.*, **1980**, *21*, 257-267.
100. Sandberg, A. S.; Andersson, H.; Kivisto, B.; Sandstrom, B. *Br. J. Nutr.*, **1986**, *55*, 245-254.
101. Sandstrom, B.; Andersson, H.; Kivisto, B.; Sandberg, A. S. *J. Nutr.*, **1986**, *116*, 2209-2218.
102. Wolever, T. M. S. *World Rev. Nutr. Diet.*, **1990**, *62*, 120-185.
103. Jenkins, D. J. A.; Jenkins, A. L.; Wolever, T. M. S.; Vuksan, V.; Rao, A. V.; Thompson, L. U.; Josse, R. G. *Eur. J. Clin. Nutr.*, **1995**, *49*, S68-S73.
104. Jenkins, D. J. A.; Wolever, T. M. S.; Taylor, R. H.; Barker, H. M.; Fielden, H. *Br. Med. J.*, **1980**, *2*, 578-580.
105. Jenkins, D. J. A.; Wolever, T. M. S.; Jenkins, A. L.; Thompson, L. U.; Rao, A. V.; Francis, T. In *Dietary Fiber: Basic and Clinical Aspects*; Vahouny, G. V., Kritchevsky, D., Eds.; Washington Symposium on Dietary Fiber; Plenum Press: New York, NY, 1986; pp 167-179.
106. Cawley, R. W.; Mitchell, T. A. *J. Sci. Food Agric.*, **1968**, *19*, 106-108.
107. Sharma, C. B.; Goel, M.; Irshad, M. *Phytochem.*, **1978**, *17*, 201-204.
108. Deshpande, S. S.; Cheryan, M. *J. Food Sci.*, **1984**, *49*, 516-519.
109. Thompson, L. U.; Yoon, J. H. *J. Food Sci.*, **1984**, *49*, 1228-1229.
110. Yoon, J. H.; Thompson, L. U.; Jenkins, D. J. *Am. J. Clin. Nutr.*, **1983**, *38*, 835-842.
111. Bjorck, I. M.; Nyman, M. E. *J. Food Sci.*, **1987**, *52*, 1588-1594.
112. Thompson, L. U.; Button, C. L.; Jenkins, D. J. *Am. J. Clin. Nutr.*, **1987**, *46*, 467-473.
113. Demjen, A.; Thompson, L. U. In *Proceedings of the 34th Canadian Federation of Biological Sciences*, 1991; p 53.
114. Cummings, J. H.; Macfarlane, G. T. *J. Appl. Bacteriol.*, **1991**, *70*, 443-459.
115. Newark, H. L.; Lupton, J. R. *Nutr. Cancer*, **1990**, *14*, 161-173.
116. Jenkins, D. J. A.; Wong, G. S.; Patten, R. P.; Bird, J.; Hall, M.; Buckley, G.; McGuire, V.; Reichert, R.; Little, J. A. *Am. J. CLin. Nutr.*, **1983**, *38*, 567-573.

117. Sharma, R. D. *Nutr. Rep. Int.*, **1986**, *33*, 669-677.
118. Klevay, L. M. *Nutr. Rep. Int.*, **1977**, *15*, 587-593.
119. Sharma, R. D. *Atherosclerosis*, **1980**, *37*, 463-468.
120. Jariwalla, R. J.; Sabin, R.; Lawson, S.; Herman, Z. S. *J. Appl. Nutr.*, **1990**, *42*, 18-28.
121. Sharma, R. D. *Nutr. Rep. Int.*, **1984**, *29*, 1315-1322.
122. Katayama, T. *Biosci. Biotech. Biochem.*, **1995**, *59*, 1159-1160.
123. Sakamoto, K.; Venkatraman, G.; Shamsuddin, A. M. *Carcinogenesis*, **1993**, *14*, 1815-1819.
124. Nielsen, B. K.; Thompson, L. U.; Bird, R. P. *Cancer Lett.*, **1987**, *37*, 317-325.
125. Shamsuddin, A. M.; Ullah, A. *Carcinogenesis*, **1989**, *10*, 625-626.
126. Shamsuddin, A. M.; Ullah, A.; Chakravarthy, A. K. *Carcinogenesis*, **1989**, *10*, 1461-1463.
127. Pretlow, T. P.; MA, O. R.; Somich, G. A.; Amini, S. B.; Pretlow, T. G. *Carcinogenesis*, **1992**, *13*, 1509-1512.
128. Shamsuddin, A. M.; Elsayed, A. M.; Ullah, A. *Carcinogenesis*, **1988**, *9*, 577-580.
129. Ullah, A.; Shamsuddin, A. M. *Carcinogenesis*, **1990**, *11*, 2219-2222.
130. Thompson, L. U.; Zhang, L. *Carcinogenesis*, **1991**, *12*, 2041-2045.
131. Hirose, M.; Hoshiya, T.; Akagi, K.; Futakuchi, M.; Ito, N. *Cancer Lett.*, **1994**, *83*, 149-156.
132. Vucenik, I.; Yang, G. Y.; Shamsuddin, A. M. *Carcinogenesis*, **1995**, *16*, 1055-1058.
133. Shamsuddin, A. M.; Baten, A.; Lalwani, N. D. *Cancer Lett.*, **1992**, *64*, 195-202.
134. Shamsuddin, A. M.; Yang, G. Y. *Carcinogenesis*, **1995**, *16*, 1975-1979.
135. Vucenik, I.; Tomazic, V. J.; Fabian, D.; Shamsuddin, A. M. *Cancer Lett.*, **1992**, *65*, 9-13.
136. Jariwalla, R. J.; Sabin, R.; Lawson, S.; Bloch, D. A.; Prender, M.; Andrews, V.; Herman, Z. S. *Nutr. Res.*, **1988**, *8*, 813-827.
137. Hirose, M.; Ozaki, K.; Takaba, K.; Fukushima, S.; Shirai, T.; Ito, N. *Carcinogenesis*, **1991**, *12*, 1917-1921.
138. Takaba, K.; Hirose, M.; Ogawa, K.; Hakoi, K.; Fukushima, S. *Food Chem. Toxicol.*, **1994**, *32*, 499-503.
139. Seaman, W. E.; Erickson, E.; Dobrwo, R.; Imeoden, J. B. *Proc. Natl. Acad. Sci. USA*, **1987**, *84*, 4239-4243.
140. Baten, A.; Ullah, A.; Tomazic, V. J.; Shamsuddin, A. M. *Carcinogenesis*, **1989**, *10*, 1595-1598.
141. Babich, H.; Borenfreund, E.; Stern, A. *Cancer Lett.*, **1993**, *73*, 127-133.
142. Thompson, L. U.; Nielsen, B. K. *FASEB J.*, **1988**, *3*, A1084.
143. Reddy, B. S. *Adv. Exper. Med. Biol.*, **1990**, *270*, 159-167.
144. Eriyamremu, G. E.; Osagie, V. E.; Alufa, O. I.; Osaghae, M. O.; Oyibu, F. A. *Ann. Nutr. Metab.*, **1995**, *39*, 42-51.
145. Thompson, L. U. *Crit. Rev. Food Sci. Nutr.*, **1994**, *34*, 473-497.

146. Nelson, R. L.; Yoo, S. J.; Tanure, J. C.; Andrianopoulos, G.; Misumi, A. *Anticancer Res.*, **1989**, *9*, 1477-1482.
147. McSherry, C. S.; Cohen, B. I.; Bokkenheuser, Y. D.; Mosbach, E. H.; Winter, J.; Matoba, N.; Scholes, J. *Cancer Res.*, **1989**, *49*, 6039-6043.
148. Behling, A. R.; Kaup, S.; Choquette, L. L.; Gregor, J. L. *Br. J. Nutr.*, **1990**, *64*, 505-513.
149. Berridge, M. J. *Mol. Cell. Endocrinol.*, **1994**, *98*, 119-124.
150. Cole, K.; Kohn, E. *Cancer Metastasis Rev.*, **1994**, *13*, 31-44.
151. Durham, A. C. H.; Walton, J. M. *Biosci. Rep.*, **1982**, *2*, 15-30.
152. Sorenson, A. W.; Slattery, M. L.; Ford, M. H. *Nutr. Cancer*, **1988**, *11*, 135-145.
153. Pence, B. C.; Buddingh, F. *Carcinogenesis*, **1988**, *91*, 187-190.
154. Whitfield, J. F. *Crit. Rev. Oncogen.*, **1992**, *3*, 55-90.
155. Lupton, J. R.; Steinbech, G.; Chang, W. C.; O'Brien, B. C.; Wiese, S.; Stoltzfus, C. L.; Glober, G. A.; Wargovich, M. J.; McPherson, R. S.; Winn, R. J. *J. Nutr.*, **1996**, *126*, 1421-1428.

Chapter 18

Anticarcinogenic Effects of Saponins and Phytosterols

A. V. Rao and R. Koratkar

Department of Nutritional Sciences, Faculty of Medicine, University of Toronto, 150 College Street, Toronto, Ontario M5S 3E2, Canada

There is growing evidence to indicate that some anti-nutritional plant constituents inhibit the process of carcinogenesis and reduce the risk of developing certain cancers. Plant foods contain, in addition to their common macro nutrients and dietary fiber, a wide variety of biologically active micro components. Saponins and phytosterols are examples of the known phytochemicals that may favorably alter the likelihood of carcinogenesis. Saponins are claimed to have hypocholesterolemic, immunostimulatory and anticarcinogenic properties. Several *in vitro* and *in vivo* studies have recently confirmed the anticarcinogenic properties of saponins of medicinal plants. The proposed mechanisms for the anticarcinogenic properties of saponins include antioxidant effect, direct and select cytotoxicity of cancer cells, immune-modulation, acid and neutral sterol metabolism and regulation of cell proliferation. Phytosterols on the other hand are structurally related to cholesterol and are present in plant oils, nuts and seeds, cereals and legumes. They are poorly absorbed and compete with cholesterol for its absorption. Recent studies have demonstrated the anticarcinogenic properties of phytosterols. Alterations in the absorption and metabolism of acid and neutral sterols is considered to be the main mechanisms by which phytosterols afford their beneficial properties against cancer. Saponins and phytosterols constitute important chemopreventative agents that are present in human diets.

Many recent epidemiological and experimental studies have implicated dietary factors in the causation and prevention of chronic disease, including cancer and cardiovascular disease. It is now estimated that diet could be responsible for 30-60% of cancers in the industrialized countries of the world *(1)*. Recent dietary guidelines recommend increased intake of fruits, vegetables, cereals and legumes in combating the incidence of cancer *(2-4)*. Several human and case-control studies have shown convincing association between the intake of fruits and vegetables and decreased incidence of gastrointestinal, lung, breast, cervical, ovarian, pancreatic and bladder cancers as well as mortality rates *(5-9)*. Research is now being directed to identify the phytochemicals that influence cancer risk and to understand the underlying mechanisms of their action.

Growing evidence indicates that some non nutritional plant constituents might, in fact, inhibit the process of carcinogenesis, or reduce the risk of developing certain cancers *(10)*. Plant foods contain in addition to the traditional macro nutrients and dietary fiber, a wide variety of biologically active micro components. Saponins, phytosterols, glucoisonolates, indols, flavones, lignans, tannins, terpenes, lycopenes, carotenoids, phenols, folates, protease inhibitor and some sulfur-containing compounds, are some examples of the known phytochemicals that may alter the likelihood of carcinogenesis, usually in a favorable direction. These bioactive substances in plant foods represent "extranutritional" constituents and may evoke physiological, behavioral and immunological effects, by influencing intestinal transit time; modifying nutrient absorption and excretion; modulating xenobiotic metabolizing and cholesterol synthetic enzymes; exhibiting immunostimulating, antihypertensive and antihyperlipidemic activities.

This paper discusses the anticancer effects of saponins and phytosterols, and to understand their mode of action, that may lead to new emphasis on cancer prevention.

Saponins

Saponins are glycosides occurring primarily in plants, and a particular species of marine starfish *(11)*. Structurally they are composed of a lipid-soluble aglycone consisting of either a sterol or more commonly a triterpenoid structure, attached to water-soluble sugar residues differing in their type and amount *(12)*. Thus, different types of saponins may be present in a single plant species.

Upon oral administration, saponins are poorly absorbed. They are either excreted unchanged or metabolized in the gut. Detailed information on the fate of saponins in the animal gut is lacking, but breakdown of saponins in the alimentary canal is most likely to be caused by enteric bacteria, intestinal enzymes and/or gastric juices *(13,14)*. Saponins being amphiphilic compounds act as natural surfactants and interact readily with cell membranes. Due to their strong hemolytic properties, saponins were traditionally considered as 'antinutrients' and toxic.

There is however renewed interest in saponins as recent evidence suggests that they also possess some therapeutic properties. On the bases of experimental investigations saponins are said to have hypocholesterolemic, immunostimulatory, antioxidant and antitumor properties. Other properties include antiviral *(15-17)*, antifungal and antibacterial *(18-20)*, piscicidal *(21)*, cardiovascular *(22,23)*, antidiabetic *(24,25)*, and may also alleviate symptoms of Alzheimer-type senile dementia *(26)*.

Anti-tumor activity. Several *in vitro* and *in vivo* studies have recently been reported in literature that indicate cytotoxic properties of saponins against cancer cells.

Monoglycosides of 14,18-cyclopoeuphane triterpenes exhibit potent selective cytotoxicity against MOLT-4 human leukemia cells, with ED50 values as low as

0.00625 mg/ml *(27)*. The yamogenin glycosides from the seeds of *Balanites aegyptiaca* (Balanitaceae) exhibited cytostatic activity against the P-388 lymphocytes leukemia cell line *(28)*, while solamargine and solasonine posses inhibitory effects against JTC-26 cells in vitro, and solamargine and khasianine exhibit cytotoxicity against human hepatoma PLC/PRF/5 cells *(29)*. The holothurinosides A, C and D from *Holothuria forskalii* are toxic to P-388, A-549, HeLa and B-16 cells in vitro *(30)*. Steroidal saponins from the tuber of *Brodiaea californica (31)* from the bulbs of *Lilium longiflorum (32)*, from the roots of *Hosta longipes (33)*, and from *Allium macleanii* and *A. senescences (34)* were found to inhibit the proliferation of various kinds of human malignant tumor cells and inhibit phorbol-13-acetate stimulated 32P incorporation into phospholipids of HeLa cells. The above mentioned, are some of the recent studies that demonstrate the cytotoxic actions of structurally different saponins.

Although not many studies have been designed to understand the structure-activity relationships, some that have addressed the issue have not given conclusive results. Experiments conducted by Ota and coworkers *(35)* to study the relationship between melanin production and cell morphology in B16 melanoma cells in the presence of ginsenosides Rh1 and Rh2, indicated Rh2 to inhibit the cell growth and to stimulate melanogenesis, and increase cell-to-cell adhesiveness, leading to organized non overlapping monolayers. In contrast Rh1, showed no effect on cell growth, but stimulated melanin production, without effects on cell-adhesiveness or other morphological changes in B16 cells. Both these ginsenosides differ only in the binding site of the glucose molecule (Rh1 at C-6 and Rh2 at C-3). Also, only Rh2 was found to be incorporated in the lipid fraction of B16 melanoma cells. The triterpenoid saponins from *Hedera delix* L were shown to have cytotoxic effects towards B16 and HeLa cells only in their monodesmoide forms *(36)*, whereas, only the bisdesmoidal saponins extracted from the leaves of *Aralia elata,* having 5 or more monosaccharide units were shown to be effective in the cytoprotective action towards carbon tetrachloride-treated hepatocytes *(37)*. Another report on ginseng-derived saponins indicated that the cytotoxic activity was inversely related to the number of sugars linked to the sapogenin; the diol-type sapogenins being more cytotoxic than triol-type ones *(38)*. In another study investigating the effects of various components of *Sho-saiko-to* on HuH-7 cells, it was shown that baicalein, baicalin and saikosaponin-a, dose dependently inhibited cell proliferation, independent of cell cycle *(39)*. In the same study, the other components, saikosaponin-c, ginsenoside Rb1 and ginsenoside Rg1 had no effect on cell proliferation. Studies performed with soya bean and gypsohila saponins on the effect of cell proliferation of HCT-15 human colon cancer cells, indicated that soya bean saponins were less cytotoxic to these cells in culture compared to gypsophila saponins, as indicated by DNA damage, also soya bean saponins had only marginal effects on RNA and protein synthesis. Gypsophila saponin, on the other hand, showed significant dose-dependent effects on DNA, RNA and protein synthesis *(40)*. It was also observed that gypsophila saponins exerted its effects on the cell membrane, while the effects induced by soya bean saponins appeared to be more cytoplasmic in nature *(41)*.

Ginseng, a widely used Oriental medicine for treatment of cancer, diabetes and hepatic and cardiovascular disease, is largely attributed to its triterpenoid saponins. These saponins were found to be cytotoxic to a variety of cancer cells in vitro and in vivo *(42- 44)*. Ginsenoside Rb2 was found to inhibit tumor angiogenesis and metastasis *(45)* and another study reported that this particular saponin suppressed the formation of sister-chromatid exchanges in human blood lymphocytes *(46)*.

The antimutagenic activity of saponins from both *Calendula arvensis* (Asteraceae) and *Hedera helix* (Araliaceae) against BaP, was comparable to the antimutagen chlorophyllin *(47)*. Also, saponins isolated from alfalfa roots and clover *Trifolium incarnatum* seeds (medicagenic acid, medicagenic acid 3-O-glucopyranoside and soyasaponin I), tested with *S.typhimurium* strains TA97, TA98, TA100, TA102 in the absence and presence of metabolic activation were found to be non-toxic and non-mutagenic *(48)*.

In vivo antitumor actions of tubeimoside I, a triterpenoid saponin from the bulb of *Maxim franquet* showed a significant dose-dependent inhibition of edema induced by arachidonic acid and tetradecanoylpharbol acetate (TPA), and also significantly decreased the number of tumor-bearing mice and the number of tumors in each mouse through either topical application or oral administration *(49)*. Saponin extracted from *Gleditsia japonica (50)* and soyasaponin I extracted from *Wisteria brachybotrys* (Leguminosae) *(51)* effectively inhibited the growth of *DMBA*-TPA induced mouse skin papilloma without any toxic effects. Saponin D and its parent aglycone, soyasapogenol E, also from *Wistaria brachybotrys* (Leguminosae) inhibited TPA-induced EBV early antigen activation *(52)*. Feeding soya bean saponins to mice, initiated with azoxymethane to develop colon tumors, resulted in a significant reduction both in the number of aberrant crypt formation per colon and the number of aberrant crypts per focus *(53)*.

At doses of 50 mg/kg b/w., aescine, obtained from the dried seeds of horse chestnut tree *Aesculus hippocastanum* L. (Hippocastanaceae), protected the gastric mucosa against lesions induced by absolute ethanol and also by pyloric ligature *(54)*. Rats given 200 mg of ginsenosides Ro orally per kg of b.w. 1h before i.p. administration of 400 mg galactosamine/kg were protected against hepatotoxic effects *(55)*.

The glycoside 4″, 6″-di-O-acetylsaikosaponin-d from *Bulpeurum kunmingense* (Umbelliferae) is cytotoxic to Ehrlich tumor cells (56), while 2″-O-, 3″-O- and 4″-O-acetylsaikosaponin a inhibit the growth of HeLa and L-1210 leukemia cells *(57)*. Antitumor formulations containing saikosaponins isolated from *Bupleurum falcatum, B. longeradiatum, B. nipponicum and B. ttriradiatum* (Umbelliferae) are the subject of a Japanese patent *(58)*. Other patents deal with the neoplastic inhibition of ginsenosides Rg1, Rg2, Rf, Rb1, Rb2, Ra and Ro, on the sarcoma growth *(59)* and the use of pharmaceuticals containing ginsenoside Ro as antithrombotic agents *(60)*.

Clinical application of Ginsenosides Rg1 and RB1 against stomach cancer, soyasaponins from *Glycine max* (Leguminosae) against a number of tumors *(61)* and gypensosides from *Gynostemma pentaphyllum* (Cucurbitaceae) for treatment of carcinoma of the skin, uterus and liver *(62)* have also been proposed.

In general, the anticarcinogenic effects of saponins may also be related to their immune-modulatory effects. The active component, formosanin-C, a saponin from *Yunan Bai Yao*, a Chinese herbal drug, injected intraperitoneally inhibited the growth of hepatoma cells implanted in C3H/HeN mice. Blood samples from these animals showed that the activity of natural killer cells and the production of interferon were significantly increased *(63)*. In another study beta-ecdysone was found to be a better immunostimulator than formosanin-C *(64)*. The ginsenoside Rg1 from the root of *Panax ginseng* was shown to increase both humoral and cell-mediated immune response *(65)*. Spleen cells recovered from ginsenoside-treated mice showed significant higher plaque-forming response and hemagglutinating antibody titer to sheep red cell antigen. Also, Rg1 increased the number of antigen-reactive T helper cells and T lymphocytes. There was also a significant increase in natural killer activity and lymph node indexes in Rg1 treated animals. Another study indicated that the wild ginseng had a better effect on immunological activity compared to cultured ginseng *(66)*. A recent study reported that saponin from *Quillaja saponaria* increased antibody production, tumor necrosis factor and interferon-gamma when injected into mice *(67)*.

Saponins can form ordered particulate structures of around 35 nm diameter (about the size of a virus particle) with the surface protein from enveloped virus, called as immunostimulating complexes. Such vaccines against influenza, measles, EBV, HIV-1, have been prepared. Crude preparations of quillaja saponins have been used to boost the response to BSA, keyhole limpet haemocyanin, SRBC and aluminum hydroxide-based vaccines *(68)*.

In addition, saponins also act as antioxidants. There is convincing evidence to indicate that free radicals are involved in the carcinogenic process *(69,70)*. Saponins from *Uncaria tomentosa* (Rubiaceae), a quinovic acid glycoside *(71)* and from *Crossopteryx febrifuga* (Rubiaceae), crossoptine A and B *(72)*, and 28-O-glycoside of 23-hydroxytormentic acid from *Quercus suber* (Fagaceae) *(73)*, exhibited antiinflammatory and antioedemic activities. Nishida's group *(74)* have shown soya bean saponins to have an inhibitory effect on radical-initiated lipid peroxidation in mouse liver microsomes. Other studies have also reported the anti-lipid peroxidative effects of ginsenoside Rb1 and Rg1 *(75)*, and antioxidative effects of chromosaponin I *(76)*. The protective effect of gypenosides (saponins of *Gynostemma pentaphyllum*) as an antioxidant was reported using various models of oxidative stress in phagocytes, liver microsomes and vascular endothelial cells *(77)*. Glycyrrhizin inhibits both cortisone degradation in the liver and generation of reactive oxygen species by neutrophils. Since there is no effect on reactive oxygen species generated in a cell-free system, the free radical scavenging action of glycyrrhizin is thought to be mediated by inhibiting neutrophil metabolism *(78)*. Liposomes that are surface modified with glycyrrhizin have been used to target the liver as glycyrrhizin demonstrates high accumulation in the liver, exhibits antioxidant properties and is even said to enhance antibody formation *(79)*.

Apart from a more direct action of saponins as antitumor agents, they may also act to delay the initiation and progression of the cancer process. Metabolic epidemiological and animal studies have shown a strong association between cancer incidence and a high concentration of cholesterol metabolites and bile acids in the feces *(80-82)*. In most cases, saponin intake substantially reduced plasma cholesterol concentration, and this was associated with an increase in fecal bile acids and neutral sterol output *(83-85)*. In vitro observations give convincing evidence that saponins bind and form large mixed micelles with bile acids and cholesterol *(86)*. Since saponins are not absorbed from the GI tract, these interactions in vivo with bile acids and cholesterol, may reduce the formation of secondary bile products, and also prevent cholesterol oxidation in the colon. Both, oxidised cholesterol products and secondary bile acid metabolites are known to act as promoters of colon cancer *(87,88)*.

Therefore, the anticarcinogenic actions of saponins could be the result of a direct interaction of the saponin on the carcinogen, or the tumor cell or mediated via its antioxidant, immuno-stimulatory and/or bile acid binding/hypocholestrolemic properties. Thus, the magnitude of the biological activity is diverse and depends on the source and type of saponin. Although saponins from various sources, are found to be cytotoxic to cancer cells, more studies are needed to elucidate the mechanism(s) involved and the physiological significance of these changes. So far the therapeutic properties of saponins have been studied primarily from medicinal plant. However, there is increased interest in the role of dietary sources of saponins in the management of cancer. Dietary saponins represent long term chronic intake patterns that may provide an important chemopreventive strategy in lowering risk of human cancers.

Toxicity. The toxicity of saponins is extremely important as a result of their widespread occurrence in foods. Mean daily intakes vary according to diet, ranging from 10-15 mg to about 100-110 mg on a western or vegetarian diet respectively. The intake may be as high as 215 mg in vegetarians in Asia *(89)*. Prolonged exposure to excessive amounts has been known to produce hypertension, flaccid quadriplegia and fulminate congestive heart failure, but this is very rare as saponins are feebly absorbed by the intestinal cells *(90)*. Solanidan and spirosolan, the steroidal glycosides of some species of Solanum have been found to cause congenital craniofacial malformations in hamsters administered high doses of these potato sprouts *(91)*, whereas acute toxicity experiments using extremely high doses of saponins from ginseng, reported the LD50 value to be more than 5000 mg/kg p.o for mice *(92)*.

Phytosterols. Many types of dietary phytosterols include campesterol, β-sitosterol and stigmasterol. They are structurally related to cholesterol, differing only in their side chain configuration- substitution of methyl (campesterol) or an ethyl (sitosterol) at position 24 or an additional bond at position 22 (stigmasterol). Their 5-alpha saturated derivatives, campestanol and sitostanol, occur in nature only in trace amounts. Phytosterols are not endogenously synthesized in the body, but are derived solely from the diet. Plant oils, nuts and seeds, cereals and legumes constitute the main sources of phytosterols in human diet *(93)*.

Clinical interest in phytosterols stems from studies done by Peterson who showed hypocholesterolemic effects of phytosterols in chicks *(94)*. Most of the experimental studies have shown that phytosterols reduce serum and/or plasma total cholesterol and LDL cholesterol levels *(95-97)*. Other therapeutic effects of phytosterols include- antiinflammatory, antibacterial and antifungal activities *(98)*, antiatherogenic and also antiulcerative properties *(99)*.

Less than 5% of dietary phytosterols are absorbed under normal conditions and only 0.3-1.7 mg/dl of phytostrols are found in normal human serum, under conditions of daily dietary intakes of 160-360 mg *(100)*. The uptake decreases with an increasing number of C-atoms at C24 of the sterol chain *(101)*. Phytosterol elimination takes place via the biliary route and appears to be more rapid than that of cholesterol. Due to the poor absorption in the intestine and faster excretion via bile the endogenous phytosterol pool size is low compared to cholesterol. Sitosterolemia, is a rare inherited disorder characterized by the increased absorption of plant sterols in plasma and tissue with the development of premature atherosclerosis *(102)*.

Anti-tumor property. Studies performed with β-sitosterol, isolated from *Solanum indicum* L. (Solanaceae), exhibited cytotoxicity on several cancer cell lines *(103)*. Awad and coworkers have indicated that the growth inhibition induced by β-sitosterol, was mediated through alterations in the lipid composition of cell membranes, since phospholipids are involved in signal transduction *(104)*.

Animal studies have also shown phytosterols to inhibit chemically induced tumors. Feeding dietary β-sitosterol to rats treated with methylnitrosourea, a direct-acting carcinogen, significantly decreased the rate of colonic epithelial cell proliferation and compressed the proliferative compartment, of the crypts and suppressed the incidence of tumors *(105,106)*. In another study a dose-dependent inhibition of elevated cell proliferation induced by dietary cholic acid was demonstrated where dietary phytosterol (60% β-sitosterol + 30% campesterol + 5% stigmasterol), was fed at levels of 0.3-2% *(107)*. Further, in the presence of a high fat diet phytosterol at level of 1% in the diet normalized cell proliferative indices but did not alter total fecal bile acid excretion. In another study, performed with carcinogen (azyxomethane) treated mice, (1%) phytosterol fed mice showed a reduction in cell proliferation and aberrant crypt formation, which is an early marker for colon cancer development *(108)*. Studies conducted by Yasukawa and his group *(109)*, indicated that suppression of TPA-induced inflammation and ODC activity caused by sitosterol, paralleled its inhibitory activities against tumor promotion.

Epidemiological studies have shown that populations consuming high levels of phytosterols are at decreased risk for colon cancer *(110,111)*. A recent study also indicated that phytosterol was effective in the treatment of benign prostatic hyperplasia *(112)*. Nair and coworkers *(113)* have shown that populations consuming increased levels of dietary phytosterol also excrete elevated levels of fecal cholesterol and less of its bacterial metabolites relative to populations at increased risk for colon cancer. Controlled human feeding studies also support the observation that increased intake of dietary phytosterol also resulted in increased fecal cholesterol excretion *(100,114)*. This may be one of the mechanisms through which phytosterols bring about hypocholesterolemic actions. The antitumor property of phytosterols may also be mediated via its hypocholesterolemic properties. Several studies suggest that sitosterol and sitostanol may be useful hypolipidimic agents *(115-117)*. Furthermore, ring-saturated phytosterols or corresponding esters appear to inhibit cholesterol levels more effectively than unsaturated phytosterols *(95,96)*. Other studies indicated that not only was the total concentration of cholesterol reduced but also LDL cholesterol levels reduced with dietary intake of phytosterol while HDL cholesterol levels remain unchanged, which inturn brought a reduction in the atherogenic index (LDL/HDL cholesterol) *(117)*.

At present, reduced cholesterol solubilisation in bile salt micelles is proposed as a major factor in inhibiting absorption of cholesterol by phytosterols *(117,118)*. Thus, phytosterols may act as cancer preventing agents by first, decreasing cholesterol absorption in the small intestine by inhibiting bile salt-cholesterol micellar formation and by competing with the brush border enzymes for cholesterol uptake. They may also shunt endogenous cholesterol excretion into the duodenum, and therefore into the colon. Since majority of the phytosterol also enters the colon, because it is unabsorbed, it may act to competitively inhibit cholesterol dehydrogenase, and other bacterial metabolizing enzymes in the colon, resulting in increased fecal levels of cholesterol and lower levels of its metabolites. Thus, phytosterol may also decrease the formation of secondary bile acids. With a decrease in the absorption of cholesterol in the small intestine, the liver will synthesis cholesterol, forming more of cholic acid, than chenodeoxycholic acid, which is derived primarily from exogenous cholesterol. This shift in the primary bile acid profile is significant, as studies measuring bile acids as risk markers for colon cancer, have shown that, total fecal bile acids are not as relevant as the ratio of the secondary bile acids, lithocholic-to-deoxycholic acid ratio, where chenodeoxycholic and cholic acids are the precursors, respectively *(88,120)*. Phytosterols may also decrease cell proliferation directly by inhibiting bile acids binding to the colonocytes, or may also be incorporated into the cell membrane, thus altering membrane lipid composition, fluidity, and/or receptor properties. However, further studies are required to understand the anticancer mechanisms of phytosterols.

Toxicity. Chronic administration of β-sitosterol subcutaneously to rats for 60 days was well tolerated with no microscopic lesions in the liver and kidney. Except for a marked reduction of serum cholesterol levels in both sexes (suggesting an intrinsic hypocholesterolemic effect of phytosterol), all the other blood/serum parameters were in the normal range *(115)*.

Conclusion

In conclusion, it may be said, that consumption of different plant foods, means high intake of a wide variety of phytochemicals that keep the body 'tuned' to handle the carcinogen intake; block activation of some carcinogens; act as substrates for endogenous production of anticarcinogens, that reduce the likelihood of DNA damage, inhibit the proliferation of the transformed cell in preventing/delaying the process of carcinogenesis.

Literature Cited

1. Doll, R. *Cancer Res.*, **1992**, *52*, 2024S-2029S.
2. Health and Welfare Canada. *Nutritional recommendations: the report of the scientific review committee.* Department of National Health and Welfare, Ottawa, ON, **1990**.
3. Kritchevsky, D. Cancer, **1993**, *72S*, 1011-1014.
4. Nutrition Science Policy. *Nutr. Rev.*, **1993**, *51*, 90-93.
5. Tomatis, L. Scientific Publ No. 100 Lyon, IARC, **1990**.
6. Block, G.; Patterson, B.; Subar, A. *Nutr. Cancer*, **1992**, *18*, 1-29.
7. Steinmetz, K.A.; Potter, J.D.; Folsom, A.R. *Cancer Res.*, **1993**, *53*, 536-543.
8. Miller, A.B.; Berrino, F.; Hill, M.; Pietinen, P.; Riboli, E.; Wahrendorf, J. *Eur. J. Cancer*, **1994**, *30A*, 207-220.
9. Dragsted, L.O.; Strube, M.; Larsen, J.C. *Pharmacol. Toxicol.*, **1993**, *72S*, 116-135.

10. Wattenberg, L.W. *Cancer Res.*, **1992**, *52*, 2085S-2091S.
11. Hostettmann, K.; Marston, A. In *Chemistry and Pharmacolgy of Natural Products*. Phillipson, J.D., Ayres, D.C., and Baxter, H., Eds. Cambridge University Press, New York, **1995**.
12. Price, K.R.; Johnson, I.T.; Fenwick, G.R. *CRC Crit. Rev. Food. Sci. Nutr.*, **1987**, *26*, 27-135.
13. Karikura, M.; Miyase, T.; Tanizawa, H.; Takino, Y.; Taniyama, T.; Hayashi, T. *Chem. Pharm. Bull.*, **1990**, *38*, 2859-2861.
14. Karikura, M.; Miyase, T.; Tanizawa, H.; Taniyama, T.; Takino, Y. *Chem. Pharm. Bull.*, **1991**, *39*, 400-404.
15. Ushio, Y.; Abe, H. *Planta. Medica*, **1992**, *58*, 171-173.
16. De Tommasi, N.; Conti, C.; Stein, M.L.; Pizza, C. *Planta. Medica.*, **1991**, *57*, 250-253.
17. Simoes, C.M.O.; Amoros, M.; Schenkel, E.P.; Shin-Kim, J.S.; Rucker, G.; Girre, L. *Planta. Medica.*, **1990**, *56*, 652-653.
18. Takechi, M.; Shimada, S.; Tanaka, Y. *Phytochemistry*, **1990**, *30*, 3943-3944.
19. Shimoyamada, M.; Susuki, M.; Sonta, H.; Maruyama, M.; Okubo, K. *Agric. Biol. Chem.*, **1990**, *54*, 2553-2557.
20. Okunji, C.O.; Okeke, C.N.; Gugnani, H.C.; Iwu, M.M. *Inter. J. Crude Drug Res.*, **1990**, *28*, 193-199.
21. Marston, A.; Hoststtmann, K. In *Methods in Plant Biochemistry*, Vol 6, Hostettmann K., Academic Press, London, **1991**, pp. 153-178.
22. Schopke, T.H.; Hiller, K. *Pharmazie.*, **1990**, *45*, 313-342.
23. Zang, W.; Wojta, J.; Binder, B.R. *Arteriosclero. Thrombo.*, **1994**, *14*, 1040-1046.
24. Espada, A.; Rodriguez J.; Villaverde, M.C.; Riguera, R. *Can. J. Chem.,* **1990**, *68*, 2039-2044.
25. Kamel, M.S.; Ohitani, K.; Kurokawa, T.; Assaf, M.H., El-Shanawany, M.A.; Ali, A.A.; Kasai, R.; Ishibashi, S.; Tanaka, O. *Chem. Pharm. Bull.*, **1991**, *39*, 1229-1233.
26. Pang, P.K.T.; Wang, L.C.H.; Benishin, C.G.; Liu, H.J. *Chem. Abstr.*, **1991**, *114*, 17584.
27. Kashiwada, Y.; Fujioka, T.; Chang, J.J.; Chen, I.; Mihashi, K.; Lee, K.H. *J. Org. Chem.*, **1992**, *57*, 6946-6953.
28. Pettit, G.R.; Doubek, D.L.; Herald, D.L.; Numata, A.; Takahasi, C.; Fujiki, R.; Miyamoto, T. *J. Nat. Prod.*, **1991**, *54*, 1491-1502.
29. Lin, C.N.; Lu, C.M.; Cheng, M.K.; Gan, K.H.; Won, S.J. *J. Nat. Prod.*, **1990**, *53*, 513-516.
30. Rodriguez, J.; Castro, R.; Riguera, R. *Tetrahedron*, **1991**, *47*, 4753-4762.
31. Mimaki, Y.; Sashida, Y.; Kuroda, M.; Nishino, A.; Satomi, Y.; Nishino, H. *Biol. Pharm. Bull.*, **1995**, *18*, 467-469.
32. Mimaki, Y.; Nakamura, O.; Sashida, Y.; Satomi, Y.; Nishino, A.; Nishino, H. *Phytochemistry*, **1994**, *37*, 227-232.
33. Mimaki, Y.; Kanmoto, T.; Kuroda, M.; Sashida, Y.; Nishino, A.; Satomi, Y.; Nishino, H. *Chem. Pharm. Bull.*, **1995**, *43*, 1190-1196.
34. Inoue, T.; Mimaki, Y.; Sashida, Y.; Nishino, A.; Satomi, Y.; Nishino, H. *Phytochemistry*, **1995**, *40*, 521-525.

35. Ota, T.; Maeda, M.; Odashima, S. *J. Pharm. Sci.*, *1991*, *80*, 1141-1146.
36. Quetin-Leclercq, J.; Elias, R.; Balansard, G.; Bassleer, R.; Angenot, L. *Planta. Medica*, **1992**, *58*, 279-281.
37. Saito, S.; Ebashi, J.; Sumita, S.; Furumoto, T.; Nagamura, Y.; Nishida, K.; Isiguro, I. *Chem. Pharma. Bull.*, **1993**, *41*, 1395-1401.
38. Baek, N.I.; Kim, D.S.; Lee, Y.H.; Park, J.D.; Lee, C.B.; Kim, S.I. *Arch. Pharm. Res.*, **1995**, *18*, 164-168.
39. Okita, K.; Li, Q.; Murakamio, T.; Takahashi, M. *Eur. J. Cancer Prevent.*, **1993**, *2*, 169-175.
40. Koratkar, R.; Sung, M.K.; Rao, A.V. *Cancer Letters*. In press.
41. Rao, A.V.; Sung, M.K. *J. Nutr.*, **1995**, *125*, 717S-745S.
42. Yokozawa, T.; Iwano, M.; Dohi, K.; Hattori, M.; Oura, H. *Jpn. J. Nephrol.*, **1994**, *36*, 13-18.
43. Tode, T.; Kikuchi, Y.; Kita, T.; Hirata, J.; Imaizumi, E.; Nagata, I. *J. Cancer Res. Clin. Oncol.*, **1993**, *120*, 24-26.
44. Kikuchi, Y.; Sasa, H.; Kita, T.; Hirata, J.; Tode, T.; Nagata, I. *Anti-Cancer Drugs*, **1991**, *2*, 63-67.
45. Mochizuki, M.; Yoo, Y.C.; Matsuzawa, K.; Sato, K.; Saiki, I.; Tono Oka, S.; Samakawa, K.I.; Azuma, I. *Biol. Pharm. Bull.*, **1995**, *18*, 1197-1202.
46. Zhu, J.H.; Takeshita, T.; Kitagawa, I.; Morimoto, K. *Cancer Res.*, **1995**, *55*, 1221-1223.
47. Elias, R.; De Meo, M.; Vidal-Ollivier, E.; Laget, M.; Balansard, G.; Dumenil, G. *Mutagenesis*, **1990**, *5*, 327-331.
48. Czeczot, H.; Rahden-Staron, I.; Oleszek, W.; Jurzysta, M. *Acta. Polonoae. Pharm.* **1994**, *51*, 133-136.
49. Yu, L.; Ma, R.; Wang, Y.; Nishino, H. *Planta. Medica.*, **1994**, *60*, 204-208.
50. Tokuda, H.; Konoshima, T.; Kozuka, M.; Kimura, T. *Oncology*, **1991**, *48*, 77-80.
51. Konoshima, T.; Kozuka, M.; Haruna, M.; Ito, K. *J. Nat. Prod.*, **1991**, *54*, 830-836.
52. Konoshima, T.; Kokumai, M.; Kozuka, M.; Tokuda, H.; Nishino, H.; Iwashima, A. *J. Nat. Prod.*, **1992**, *55*, 1776-1778.
53. Koratkar, R.; Rao, A.V. *Nutr. & Cancer*, **1997**, in press.
54. Marhuenda, E.; Martin, M.J.; Alarco de la Lastra, C. *Phytother. Res.*, **1993**, *7*, 13-16.
55. Kubo, M.; Matsuda, M. *Jap.Pat.* **1992**, 92 05,235.
56. Otake, N.; Ra, S.; Zeni, F.; Kin, K.; Kyo, B.; Kawai, H.; Kobayashi, E. *Jap.Pat.* **1988**, *62*, 240,696.
57. Otake, N.; Seto, H.; Ra, S.; Sen, F.; Jo, H.; Han, T.; Kyo, B. *Jap. Pat.* **1987**, *61*, 282,395.
58. Kono, H.; Odajima, T. *Jap. Pat.* **1988**, *62*, 187,408.
59. Hsuan, C.-P.; Chang, H.-H. *Jap. Pat.* **1981**, *81*, 46,817.
60. Nikan, K.; Jinsen, K.K.; Kubo, M. *Jap. Pat.* **1983**, *82*, 163,315.
61. Osaka, Y.; Kenkyusho, K.K. *Jap. Pat.* **1983**, *58*, 72,523.
62. Takemoto, T.; Odashima, T. *Jap. Pat.* **1983**, *58*, 59,921

63. Wu, R.T.; Chiang, H.C.; Fu, W.C.; Chien, K.Y.; Chung, Y.M.; Horng, L.Y. *Int. J. Immunopharmacol*, **1990**, *12*, 777-786.
64. Chiang, H.C.; Wang, J.J.; Wu, R.T. *Anticancer Res*, **1992**, *12*, 1475-1478.
65. Kenarova, B.; Neychev, H.; Hadjiivanova, C.; Petkov, D. *Jpn. J. Pharmacol.*, **1990**, *54*, 447-454.
66. Mizuno, M.; Yamada, J.; Terai, H.; Kozukue, N.; Lee, Y.S.; Tsuchida, H. *Biochem. Biophy. Res. Commun.*, **1994**, *200*, 1672-1678.
67. Gebara, V.C.; Petricevich, V.L.; Raw, I.; da Silva, W.D. *Biotech. Applied Biochem.*, **1995**, *22*, 31-37.
68. Kensil, C.R.; Patel, U.; Lennick, M.; Marciani, D. *J. Immunol.*, **1991**, *146*, 431-437.
69. Ceutti, P.A. *Lancet*, **1994**, *344*, 862-863.
70. Collins, A.; Duthie, S.; Ross, M. *Proc. Nutr. Soc.*, **1994**, *53*, 67-75.
71. Aquino, R.; De Feo, V.; De Simone, F.; Pizza, C.; Cirino, G. *J. Nat. Prod.*, **1991**, *54*, 453-459.
72. Gariboldi, P.; Verotta, L.; Gabetta, B. *Phytochemistry*, **1990**, *29*, 2629-2635.
73. Romussi, G.; Bignardi, G.; Pizza, C.; De Tommasi, N. *Arch. Pharm.*, **1991**, *324*, 519-524.
74. Nishida, K.; Ohta, Y.; Araki, Y.; Ito, M.; Nagamura, Y.; Ishiguro, I. *J. Clin. Biochem. Nutr.*, **1993**, *15*, 175-184.
75. Deng, H.I.; Zang, J.T. *Chinese Med. J.*, **1991**, *104*, 395-398.
76. Tsujino, Y.; Tsurumi, S.; Yoshida, Y.; Niki, E. *Biosci. Biotech. Biochem.*, **1994**, *58*, 1731-1732.
77. Li, L.; Jiao, L.; Lau, B.H. *Cancer Biotherapy*, **1993**, *8*, 263-272.
78. Akamatsu, H.; Komura, J.; Asada, Y.; Niwa, Y. *Planta. Medica*, **1990**, *57*, 119-121.
79. Tsuji, H.; Osaka, S.; Kiwida, H. *Chem. Pharm. Bull.*, **1991**, *39*, 1004-1008.
80. Owen, R.W.; Day, D.W.; Thompson, M.H. *Eur. J. Cancer Prevent*, **1991**, *1*, 105-112.
81. Biasco, G.; Panganelli, G.M.; Owen, R.W.; Hill, M.J. *Eur. J. Cancer Prevent*, **1991**, *1S*, 63-68.
82. Lapre J.A.; Van der Meer, R. *Carcinogenesis*, **1992**, *13*, 41-44.
83. Sauvaire, Y.; Ribes, G.; Baccou, J.C.; Loubatieres-Mariani, M.M. *Lipids*, **1991**, *26*, 191-197.
84. Harwood, H.J., Jr.; Chandler, C.E.; Pellarin, L.D.; Bangerter, F.W.; Wilkins, R.W.; Long, C.A.; Cosgrove, P.G.; Malinow, M.R.; Marzetta, C.A.; Pettini, J.L.; Savoy, Y.E.; Mayer, J.T. *J. Lipid Res.*, **1993**, *34*, 377-395.
85. Amarowicz, R.; Shimoyamada, M.; Okubo, K. *Roczniki Panstwowego Zakladu Higieny*, **1994**, *45*, 125-130.
86. Oakenfull, D.; Sidhu, G.S. *Eur. J. Clin. Nutr.*, **1990**, *44*, 79-88.
87. Kendall, C.W.; Koo, M.; Sokoloff, E.; Rao, A.V. *Cancer Lett.*, **1992**, *66*, 241-248.
88. Reddy, B.S.; Engle, A.; Simi, B.; Goldman, M. *Gastroenterol.*, **1992**, *102*, 1475-1482.
89. Ridout, C.L.; Wharf, S.G.; Price, K.R.; Johnson, I.T.; Fenwick, G.R. *Food Sci. Nutr.*, **1988**, *42F*, 111-116.

90. Spinks, E.A.; Fenwick, G.R. *Food Additivies and Contaminants*, **1990**, *7*, 769-778.
91. Keeler, R.F.; Baker, D.C.; Gaffield, W. In *Handbook of Natural Toxins*, Vol 6, Keeler, R.F.; Tu, A.T., Eds. Marcel Dekker, New York, **1991**, pp. 83-99.
92. Sonnenborn, U.; Hansel, R. In *Adverse Effects of Herbal Drugs*, Vol. 1, de Smet, P.A.G.M.; Keller, K.; Hansel, R.; Chandler, R.F. Eds., Springer-Verlag, Berlin, **1992**, pp. 179-192..
93. Weilhrauch, J.; Gardner, J. *J. Am. Diet. Assoc.*, **1978**, *73*, 39-47.
94. Peterson, D.; Nichols, C.W., Jr.; Schneour, D. *J. Nutr.*, **1952**, *47*, 51-65.
95. Miettinen, T.A.; Vanhanen, H. *Atheroscler.*, **1994**, *105*, 217-226.
96. Becker, M.; Staab, D.; von Bergmann, K. *Pediatr.*, **1993**, *122*, 292-296.
97. Fraser, G.E. *Am. J. Clin. Nutr.*, **1994**, *59*, 1117S-1123S.
98. Padmaja, V.; Thankamany, V.; Hisham, A. *J. Ethnopharmacol.*, **1993**, *40*, 181-186.
99. Zhao, J.; Zhang, C.Y.; Xu, D.M.; Huang, G.Q.; Xu, Y.L.; Wang, Z.Y.; Fang, S.D.; Chen, Y.; Gu,Y.L. *Thrombosis Res.* **1990**, *57*, 957-966.
100. Miettinen, T.; Tilvis, R.; Kesaniemi, Y. *Metabolism*, **1989**, *38*, 136-140.
101. Heinemann, T.; Axtmann, G.; von Bergmann, K. *Eur. J. Clin. Invest.*, **1993**, *23*, 827-831.
102. Salen, G.; Shefer, S.; Nguyen, L.; Ness, G.C.; Tint, G.S.; Shore, V. *J. Lipid Res.*, **1992**, *33*, 945-955.
103. Chiang, H.C.; Tseng, T.H.; Wang, C.J.; Chen, C.F.; Kan, W.S. *Anticancer Res*, **1991**, *11*, 1911-1917.
104. Awad, A.B.; Chen, Y.C.; Fink, C.S.; Hennessey, T. In *FASEB J., Exper. Biol.*, **1996**, Washington D.C., April 14-17.
105. Deschner, E.; Cohen, B.; Raicht, R. *J. Cancer Res. Clin. Oncol.*, **1982**, *103*, 49-54.
106. Raicht, R.; Cohen, B.; Fazzini, E.; Sarwal, A.; Takahashi, M. *Cancer Res.*, **1980**, *40*, 403-405.
107. Janezic, S.A.; Rao, A.V. *Fd. Chem. Toxicol.*, **1992**, *30*, 611-616.
108. Rao, A.V.; Janezic, S.A. *Nutr. Cancer*, **1992**, *18*, 43-52.
109. Yasukawa, K.; Takido, M.; Matsumoto, T.; Takeuchi, M.; Nakagawa, S. *Oncol.*, **1991**, *48*, 72-76.
110. Nair, P. *Am. J. Clin. Nutr.*, **1984**, *40*, 880-886.
111. Hirai, K.; Shimazu, C.; Takezoe, R.; Ozek, Y. *J. Nutr. Sci. Vitaminol.*, **1986**, *32*, 363-372.
112. Berges, R.R.; Windeler, J.; Trampisch, H.J.; Senge, T. *Lancet*, **1995**, *345*, 1529-1532.
113. Nair, P.; Turjman, N.; Kessie, G.; Calkins, B.; Goodman, G.; Davidovitz, H.; Nimmagadda, G. *Am. J. Clin. Nutr.*, **1984**, *40*, 927-930.
114. Kudchodhkar, B.; Horlick, L.; Sodhi, H. *Atherosclerosis*, **1976**, *23*, 239-248.
115. Malini, T.; Vanithakumari, G. *J. Ethnopharmac.*, **1990**, *28*, 221-234.

Author Index

Ali, Rashda, 223
Amarowicz, Ryszard 127
Barkholt, V., 44
Daun, James K., 152
DeClercq, Douglas R., 152
Friedman, Mendel, 61
Frøkiær, H., 44
Hagerman, Ann E., 209
Hatano, Tsutomu, 245
Igarashi, Kiharu, 260
Jadhav, S. J., 94
Johnson, Sarah, 209
Jørgensen, T. M. R., 44
Koratkar, R., 313
Liener, Irvin E., 31
Lutz, S. E., 94
Mazza, G., 94
Miyamoto, Ken-ichi, 245
Murayama, Tsugiya, 245
Naczk, Marian, 127,186
Okubo, Kazuyoshi, 260
Okuda, Takuo, 245
Plhak, Leslie C., 115
Rao, A. V., 313
Rickard, Sharon E., 273,294
Rosendal, A., 44
Salunkhe, D. K., 94
Sayeed, Syed Asad, 223
Shahidi, Fereidoon, 1,127,152,171,186
Sporns, Peter, 115
Thompson, Lilian U., 273,294
Tonsgaard, M. C., 44
Wanasundara, P. K. J. P. D., 171
Whitaker, John R., 10
Yoshida, Takashi, 245
Yoshiki, Yumiko, 260
Yotsuhashi, Kazuhiko, 260
Zhao, Yan, 209

Affiliation Index

Agricultural Research Service, 61
Agriculture and Agri-Food Canada, 94
Canadian Grain Commission, 152
Chugai Pharmaceutical Company, Limited, 260
Food Processing Development Centre (Canada), 94
Hokuriku University, 245
Kanazawa Medical University, 245
Louisiana State University, 115
Memorial University of Newfoundland, 1,127,152,171,186
Miami University, 209
Okayama University, 245
Polish Academy of Sciences, 127
St. Francis Xavier University, 127,186
Technical University of Denmark, 44
Tohoku University, 260
University of Alberta, 115
University of California—Davis, 10
University of Karachi, 223
University of Minnesota, 31
University of Toronto, 273,294,313
U.S. Department of Agriculture, 61
Utah State University, 94
Yamagata University, 260

Subject Index

A

Acid butanol method, analysis of condensed tannins, 216
Active oxygen species, role in chemiluminescence
catechins and gallic acid, 260–266
soybean saponins, 266–271

Aescine, antitumor activity, 316
Agrimonia pilosa, anticarcinogenic activities of polyphenols, 245–257
Agrimoniin
anticarcinogenic activities, 245–257
in vitro cytotoxicity, 253,255
induction of cytotoxic cells, 254*f*,255
stimulation of interleukin-1 induction, 255–257
Aldehyde, role in chemiluminescence of catechins and gallic acid, 260–266
Allergy
definition, 44
description, 54
Amino acid linkage, mode for lawsone, 233
α-Amylase inhibitors
discovery, 10
higher plants
amino acid sequences, 17–21,23
mechanism of action, 23,25
quaternary structure, 22*f*,23
secondary structure, 22*f*,23
tertiary structure, 22*f*,23
Antiatherogenic mechanisms, beneficial effects of phytoestrogens and lignans in chronic disease, 285–286
Anticarcinogenic activity
agrimoniin, 249,253
oenothein B, 249,253
polyphenols in foods and herbs, 247–257
potato polyphenols, 85
tannins on mouse sarcoma 180, 247–253
Antimicrobial activity
henna extract, 240–242*t*
lawsone, 240–242*t*
Antinutritional, definition, 44
Antinutritional proteins
lectins, 45–50
properties, 44
proteinase inhibitors, 49,51–53
Antioxidant(s), potential effect in prevention and control of free radical formation, 3,4*f*
Arcelin, defense mechanism against predators, 41

B

Blackspot bruise resistance, potato, 75
Bound cyanide, description, 179
Bowman–Birk inhibitors, 49,51*t*
Brassica oilseeds, glucosinolates, 152–169

C

Caffeic acid, antioxidant activity, 2–3
Calcium, role of phytic acid in bioavailability, 295–298
Cancer protective effect, role of phytoestrogens and lignans, 283–284
Canola, description, 157,159
Canola condensed tannins
chemical structure, 187–188
content in canola hulls, 192–193
effect on nutrition
carbohydrates, 203,204*f*
methods of quantification of tannin–protein interaction, 195,197
minerals, 193–195
molecular interactions with proteins and polyphenols, 195
pH, 196–199,201
protein-precipitating capacity of canola tannins, 201–203
proteins, 195
tannin concentration, 197,200–201
processing, 203,205
quantification, 190*f*,191–192
recovery vs. solvent system, 189,191
Carbohydrates
interaction with canola condensed tannins, 203,204*f*
role in flatulence of legumes, 131,133
Catabolism, cyanogenic glycosides of flaxseeds, 179–181*f*
Catechins, chemiluminescence, 260–266
Cattle, exposure to estrogenic compounds in plants, 278
Cereal grains, content of α-galactosides, 138,139*t*

Chemiluminescence
catechins, 260–266
gallic acid, 260–266
soybean saponins, 266–271
Chlorogenic acid
antioxidant activity, 2–3
content in potato polyphenols, 63–71
Cholinesterase inhibition, potato glycoalkaloids, 122
Colon, anticancer effect of phytic acid, 302–303
Concanavalin A, 32*f*,33
Condensed tannins
analysis using functional group methods, 216–217
determination methods, 209–219
See also Canola condensed tannins
Copper, role of phytic acid in bioavailability, 295–298
p-Coumaric acid, antioxidant activity, 2–3
Crossoptines A and B, antitumor activity, 317
Cyanogenesis, 171
Cyanogenic compounds of plants, 171
Cyanogenic glycosides
content of released HCN, 172,174*t*
function, 171–172
of flaxseeds
analyses, 179,183
biosynthesis, 177*f*,178
catabolism, 179–181*f*
chemistry, 172,175–176,178
removal, 182*f*,183–184
role in selenium toxicity, 184
structures, 172,173*f*
toxicity, 179
sources, 172,174*t*
Cytotoxic immunocytes, role of polyphenols, 245–257

D

Diet
development of chronic diseases, 273,313–314
flatulence of legumes, 129
Diet—*Continued*
lectins, 37–39,40*f*,47
potato polyphenols, 62–87
Dietary protein, role in mineral bioavailability, 298

E

Enzyme inhibition, role in mechanism of action of phytic acid, 304
Enzyme-linked immunosorbent assay
analysis of protease inhibitors, 53
lectin analysis, 47,49,50*f*
potato glycoalkaloids, 103,105
Erythroleukemia cells, anticancer effect of phytic acid, 303
Estrogenic compounds in plants, exposure of domestic animals, 277–278
Estrogenic effects in postmenopausal women, role of phytoestrogens and lignans, 282–283

F

Flatulogenic factors in legumes
flatulence, 128–129
role of carbohydrates, 131,133
role of diet, 129
role of microorganisms, 129–131,132*f*
Flavonoids, antioxidant activity, 2–3
Flaxseeds
cancer risk, 283–284
cyanogenic glycosides, 172–184
role of phytoestrogens and lignans on reproduction and chronic disease, 273–286
Folin–Denis method, tannin analysis, 214
Food(s)
anticarcinogenic activities of polyphenols, 245–257
composition of α-galactosides, 133–139
α-galactosides of sucrose, 127–147
health effects of antinutrients and phytochemicals, 1–7

Food allergens
characterization, 54–55
evaluation of allergenicity of novel food, 55
induction of tolerance, 56–57
mechanisms of food intolerance, 54
prevalence of food allergy, 54
types, 53–54
Food components, interaction with phytic acid, 295,296*f*
Food processing, role in mineral bioavailability, 295,297
Formosanin C, antitumor activity, 316–317
Free cyanide, description, 179
Free radicals, contribution in diseases, 3,4*f*
Functional lectin immunoassay, lectin analysis, 49,50*f*

G

Galactinol, function, 127–128
α-Galactoside composition in foods of plant origin
chemical structure, 132*f*,133
content
cereal grains, 138,139*t*
legumes, 134–137
oilseeds, 138,139*t*
vegetables, 138,139*t*
quantification, 133–134
α-Galactosides of sucrose in foods
composition, 133–139
flatulogenic factors in legumes, 128–133
processing
cooking, 142,144
dehulling, 138,140,141*t*
enzymatic treatments, 144–146
extrusion, 146
irradiation, 146–147
soaking, 140–144
Gallic acid, chemiluminescence, 260–266
Gas–liquid chromatography
potato glycoalkaloids, 103
potato polyphenols, 63
Ginsenosides Rg1 and RB1, antitumor activity, 316–317
Ginsenosides Rh1 and Rh2, antitumor activity, 315
Glucose, role of kidney bean lectins in absorption, 35,37,38*f*
Glucosinolates
antinutritional factors, 152
chemical structure, 152–153
in *Brassica* oilseeds
antinutrients
detoxification methods, 159,161
methanol–ammonia–water–hexane extraction process, 161,163–165,167
canola variety vs. glucosinolate content and nature of toxicity, 159,162*t*
canola vs. rapeseed, 157
conventional processing of seeds, 157,158*f*,160*t*
total content, 159,160*t*
toxicity of degradation products, 155–157
Glycoalkaloids, potato, *See* Potato glycoalkaloids
Glycoconjugates, use of lectins for detection and identification, 41
Glycoside
4' ',6' '-di-*O*-acetylsaikosaponin d, antitumor activity, 316
23-hydroxytormentic acid, antitumor activity, 317
See also Cyanogenic glycosides of flaxseeds

H

Hemagglutinin(s), 7,31
Hemagglutination assays, lectin analysis, 47,49,50*f*
Henna
applications, 223
extract, 240–242*t*
factors affecting quantitative variation, 225,226*t*

Henna—*Continued*
growth, 224
origin, 224
use
cosmetic industry, 227
edible products, 225,228*t*
nonfood products
pharmaceutical industry, 227
textile industry, 225,227
Herbs, anticarcinogenic activities of polyphenols, 245–257
High-performance liquid chromatography
potato glycoalkaloids, 99,101,103
potato polyphenols, 63
High-performance–thin-layer chromatography, potato glycoalkaloids, 99,101
Hydrolyzable tannins
analysis using functional group methods, 217–218
chemistry, 212–213
determination methods, 209–219
2-Hydroxy-1,4-naphthaquinone, *See* Lawsone

I

Ileostomates, definition, 300
In vitro cytotoxicity
agrimoniin, 253,255
oenothein B, 253,255
myo-Inositol-1,2,3,4,5,6-hexakisdihydrogen phosphate, *See* Phytic acid
Interleukin-1 induction, stimulation using polyphenols, 255–257
Iodate assay, analysis of hydrolyzable tannins, 217
Iron, role of phytic acid in bioavailability, 295–298
Irradiation, role in α-galactoside composition, 146–147
Isoflavonoids, antioxidant activity, 2–3
Isolectins
description, 46
structural features, 33,34*f*

J

Juglone, 231,232*f*

K

Kampo medicine, description, 245
Khasianine, antitumor activity, 315
Kidney bean(s), role of lectin in toxicity, 39,40*f*
Kidney bean lectin
bacterial colonization, 37
effect on absorption of nutrients, 35,37,38*f*
interaction with intestines, 35,36*f*
physiological effects in animals, 35–37,38*f*
structural features, 33,34*f*
Kunitz soybean inhibitor, 49,51*t*

L

Lawsone
antimicrobial activity, 240–241,242*t*
identification by NMR and MS, 229
interaction with protein molecules, 233
isolation, 228–231
purification, 228–229
toxicity, 240
use as staining agent for proteins, 235*t*,236
Lawsone–protein adduct
description, 237
in vitro digestibility, 238*t*,239–240
pH, 239
specificity of protein interaction, 239
Lawsonia inermis, *See* Henna
Lectins
analysis, 47,49,50*f*
biochemical properties, 31–33
biological activities, 45
definition, 45
description, 7
detection and identification of diverse glycoconjugates, 41

Lectins—*Continued*
effect on immune system, 46–48*f*
isolectins with different specificity, 46
legume lectins, 45
physiological effects
animals, 33–38
plants
defense mechanism for plants against predators, 41
interaction of legumes with N-fixing bacteria, 39,41
prevention of graft rejection in bone marrow transplantation, 41–42
responses in small intestine, 45–46
retardation of growth of transplantable tumor, 42
role in cells of immune system, 46–48*f*
role in intoxication, 39
significance in human diet, 37–39,40*f*
structural features, 32*f*,33,34*f*
type 2 ribosome-inactivating lectins, 45
use for disease control, 1–2
Legume(s)
content
α-galactosides, 134–137
oligosaccharides, 133,135–137,139
interaction with N-fixing bacteria, 39,41
Lignans
adverse effects, 273
beneficial effects in chronic disease, 284–286
description, 6
exposure of experimental animals, 279–280
exposure of human females
alterations in menstrual cycling, 282
estrogenic effects in postmenopausal women, 282–283
gastrointestinal effects, 281
immunosuppression, 281
mode of action, 275–276
reduction in risk of chronic disease, 273
renal impairment, 281
sources, 275,277*t*
tumor growth, 280–281
types of compounds, 274*f*,275
Lipids, role of phytic acid in reduction, 301–302

M

Macrocyclic ellagitannin dimer, anticarcinogenic activities, 245–257
Mammary gland, anticancer effect of phytic acid, 303
Menstrual cycling, role of phytoestrogens and lignans, 282
Methanol–ammonia–water–hexane extraction process, glucosinolates in *Brassica* oilseeds, 161,163–165,167
Mineral(s)
absorption enhancers, 297–298
interaction with canola condensed tannins, 193–195
role of phytic acid in bioavailability, 295–298
Mineral binding, role in mechanism of action of phytic acid, 304–305
Monoglycosides of 14,18-cyclopoeuphane triterpenes, antitumor activity, 314–315
Mouse sarcoma 180, anticancer activity of tannins, 247–253
Multidomain–multisite inhibitors
amino acid sequence, 24*f*,25–27
examples, 24*f*,25
structures, 25,26*f*

N

N-fixing bacteria, interaction with legumes, 39,41
Naphthaquinones, 231–233
Nitrous acid assay, analysis of hydrolyzable tannins, 217
Nonglycosidic cyanide, 179
Nontannin phenolics, chemistry, 213
Nontoxic reactions, types, 53–54

O

Oenothein B
anticancer activity, 245,249,253
anticarcinogenic activities, 245
in vitro cytotoxicity, 253,255

Oilseed(s)
α-galactosides, 138,139*t*
glucosinolates, 152–169
phenolic acids, 187,189*t*
Oligomeric hydrolyzable tannins, anticarcinogenic activities, 245–257
Oligosaccharides, content in legumes, 133,135–137,139
Oral immunization, induction of food allergens, 56–57
Oxygen species, *See* Active oxygen species

P

Pakarli, 236–237
Phenolic acids
antioxidant activity, 2–3
content in oilseed products, 187,189*t*
Phenolic compounds
antioxidant activity in foods and biological systems, 2–3
content in potato polyphenols, 63–71
function, 2
Phlorotannins, chemistry, 213
Phytates, 3,5
Phythiocol, 231–233
Phytic acid
anticancer effects
colon, 302–303
erythroleukemia cells, 303
mammary gland, 303
mechanisms of action, 304–305
prostate cells, 303
beneficial effects, 3,5
digestion and absorption
protein, 298–300
starch, 300–301
function, 3
interaction with food components, 295,296*f*
mineral bioavailability
added vs. endogenous phytic acid, 297
food processing, 295,297
level and type of dietary protein, 298
metabolic adaptation, 298
presence of other minerals, 298
Phytic acid—*Continued*
occurrence in foods, 294–295
reduction in serum lipids, 301–302
structure, 294,296*f*
use for disease control, 1–2
Phytoestrogens
adverse effects, 273
beneficial effects in chronic disease, 284–286
description, 6
exposure of experimental animals, 279–280
exposure of human females
alterations in menstrual cycling, 282
estrogenic effects in postmenopausal women, 282–283
mode of action, 275–276
reduction in risk of chronic disease, 273
sources, 275,276*t*
tumor growth, 280–281
types of compounds, 274*f*,275
Phytosterols
antitumor activity, 319–320
mechanism of action, 318–319
structure, 318
Plant(s)
exposure of domestic animals to estrogenic compounds, 277–278
lectins, 39,41
potato polyphenols, 62–82
protease and α-amylase inhibitors, 10–27
Plant lectins, *See* Lectins
Plumbagin, 231,232*f*
Polyacrylamide gel electrophoresis, henna, 223
Polyphenols
anticarcinogenic activities in foods and herbs, 245–257
potato, *See* Potato polyphenols
Potato
cultivation for use as food, 115
toxicity, 117
Potato glycoalkaloids
analytical methods, 101–103,105
biological activities, 121–123
biosynthesis, 104–108,117

Potato glycoalkaloids—*Continued*
chemistry, 95,97–98
colorimetry, 101–102
distribution, 117–119
extraction, 99,100*f*
hydrolysis products, 95,97*t*
metabolism, 108
role in toxicity of potato, 117
separation, 99
structures, 95–96,116–118
toxicity, 94
Potato polyphenols
analysis, 63
composition, 63,65–71
enzyme-catalyzed browning reactions, 62*f*,63
function, 63
postharvest events
after-cooking darkening, 76–77
browning, 75
browning prevention, 71*t*,77–78
γ-radiation effect, 80–81,82*t*
light effect, 79–80
preharvest events
biosynthesis, 69,72–75
host–plant resistance, 72
properties, 63
role in diet
anticarcinogenic effects, 85
antiglycemic properties, 87
antimutagenic effects, 85
antioxidant activity, 82–86
flavor, 81,83
taste, 81,83
structures, 63,64*f*
Proanthocyanidins, 210–211
Prostate cells, anticarcinogenic effect of phytic acid, 303
Protease inhibitors
analysis, 53
description, 5
function, 44
in higher plants
amino acid diversity at active site, 13,15*f*
classification, 11–13,14*f*
enzyme–inhibitor complex, 16*f*,17,18*f*
mechanism of action, 13,17
Protein
interaction with lawsone, 233
role of phytic acid in digestion and absorption, 298–300
use of rapeseed as source, 186–187
Proteinase inhibitors
analysis, 53
long-term dietary effects, 53
mode of action, 49
nutritional effects, 51
occurrence, 49,51*t*
physicochemical properties, 52–53
Protocatechuic acid, antioxidant activity, 2–3
Prussian blue method, tannin analysis, 214

Q

Quinovic acid glycoside, antitumor activity, 317

R

Raffinose, 127
Rapeseed, 186–187
Rapeseed protein product, content of phenolic compounds, 187,189*t*
Rhodanine assay, hydrolyzable tannin analysis, 217
Ricin, description, 31

S

Saikosaponins, antitumor activity, 316
Saponins
antitumor activity, 314–318
chemiluminescence, 266–271
description, 6
structure, 314
therapeutic properties, 314
toxicity, 318
use for disease control, 1–2
Sarcoma 180, anticancer activity of tannins, 247–253
Selenium toxicity, role of cyanogenic glycosides of flaxseeds, 184

Serum lipids, role of phytic acid in reduction, 301–302
β-Sitosterol, antitumor activity, 319
Solamargine, antitumor activity, 315
Solanidane, structure, 95
Solanum species, *See* Potato
Solasonine, antitumor activity, 315
Soy, cancer risk, 283
Soyasaponin I, antitumor activity, 316
Soybean, role of phytoestrogens and lignans on reproduction and chronic disease, 273–286
Soybean lectin, physiological effects in animals, 33–38
Soybean saponins, chemiluminescence, 266–271
Spirosolane, structure, 95
Stachyose, distribution, 127
Starch, role of phytic acid in digestion and absorption, 300–301
Starch malabsorption, role in mechanism of action of phytic acid, 304
Sucrose, α-galactosides, 127–147
Syringic acid, antioxidant activity, 2–3
Systemic effects, kidney bean lectin, 37

T

Tannins
- analytical methods
  - functional group methods, 215–218
  - mixture evaluation, 218–219
  - precipitation methods, 215
  - total phenolic methods, 214–215
- anticarcinogenic activity, 245,247–253
- biological activities, 245
- chemistry
  - hydrolyzable tannins, 212–213
  - nontannin phenolics, 214
  - phlorotannins, 213
  - proanthocyanidins, 210–211
- definition, 210
- distribution, 209–210

Tannins—*Continued*
- functions, 210
- *See also* Canola condensed tannins

Thin-layer chromatography
- potato glycoalkaloids, 99,102–103
- potato polyphenols, 63

Titrimetry, potato glycoalkaloids, 102
Tocopherols, antioxidant activity, 2
Total phenolics methods, analysis of tannins, 214–215
Toxic, definition, 44
Toxicity
- cyanogenic glycosides of flaxseeds, 179
- lawsone, 240
- potato glycoalkaloids, 94,121–122
- saponins, 318

Triterpenoid saponins, antitumor activity, 315–316
Trypsin inhibitors
- analysis, 53
- long-term dietary effects, 53
- nutritional effects, 51
- occurrence, 49,51*t*
- physicochemical properties, 52–53

Tubeimoside I, antitumor activity, 316
Type 2 ribosome-inactivating lectins, description, 45

U

UV spectrophotometry, potato polyphenols, 63

V

Vanillic acid, antioxidant activity, 2–3
Vanillin assay, analysis of condensed tannins, 216
Verbascose, distribution, 127

W

Walker carcinoma, growth retardation by lectins, 42

Y

Yamogenin glycosides, antitumor activity, 315

Z

Zinc, role of phytic acid in bioavailability, 295–298